Synthesis and Properties of Silicones and Silicone-Modified Materials

ACS SYMPOSIUM SERIES **838**

Synthesis and Properties of Silicones and Silicone-Modified Materials

Stephen J. Clarson, Editor
University of Cincinnati

John J. Fitzgerald, Editor
General Electric Silicones

Michael J. Owen, Editor
Michigan Molecular Institute

Steven D. Smith, Editor
Procter and Gamble Company

Mark E. Van Dyke, Editor
Southwest Research Institute

American Chemical Society, Washington, DC

Library of Congress Cataloging-in-Publication Data

Synthesis and properties of silicones and silicone-modified materials / Stephen J. Clarson, editor ...[et al.].

p. cm.—(ACS symposium series ; 838)

Developed from a symposium sponsored by the Division of Polymer Chemistry, Inc., at the 221st National Meeting of the American Chemical Society, San Diego, California, April 1–5, 2001.

Includes bibliographical references and index.

ISBN 0–8412–3804–9

1. Silicones—Congresses.

I. Clarson, Stephen J. II. American Chemical Society. Division of Polymer Chemistry, Inc. III. American Chemical Society. Meeting (221st : 2001 : San Diego, Calif.) IV. Series.

QD383.S54 S96   2003
668.4′227—dc21   2002038305

The paper used in this publication meets the minimum requirements of American National Standard for Information Sciences—Permanence of Paper for Printed Library Materials, ANSI Z39.48–1984.

PRINTED IN THE UNITED STATES OF AMERICA

Foreword

The ACS Symposium Series was first published in 1974 to provide a mechanism for publishing symposia quickly in book form. The purpose of the series is to publish timely, comprehensive books developed from ACS sponsored symposia based on current scientific research. Occasionally, books are developed from symposia sponsored by other organizations when the topic is of keen interest to the chemistry audience.

Before agreeing to publish a book, the proposed table of contents is reviewed for appropriate and comprehensive coverage and for interest to the audience. Some papers may be excluded to better focus the book; others may be added to provide comprehensiveness. When appropriate, overview or introductory chapters are added. Drafts of chapters are peer-reviewed prior to final acceptance or rejection, and manuscripts are prepared in camera-ready format.

As a rule, only original research papers and original review papers are included in the volumes. Verbatim reproductions of previously published papers are not accepted.

ACS Books Department

J. Anthony Semlyen (1936-2001)

A Tribute*

David M. Goodall

Department of Chemistry, University of York, Heslington, York, YO1 5DD, United Kingdom

My friendship with Tony goes back over 30 years, and dates from the late 1960s when we both joined the University as lecturers in physical chemistry.

As a scientist, Tony had a broad sweep of scientific interests, ranging from theory to practical applications. He was the author of four books [1-4] in the field of cyclic polymers and large ring molecules, opening up a whole new field which had previously not been explored. Cyclics are closely related to the more familiar linear polymers, which are extensively used today. They have the same structural units, but the constraints of ring form create interesting differences in properties. Tony was interested in the history of science as well as in his state-of-the-art polymers, and often discussed the relationship between science and religion. He saw beauty in rings, and regarded them as exemplars of perfection. Unlike the linears which have end groups as well as the polymer repeat units, rings have just the repeat units, joined in a circle with neither end nor beginning. He loved to point out that Nature uses cyclic DNA to hold the genetic code for many organisms, and cyclic sulfur is present in large amounts in one of the moons of Jupiter.

Secondly, some reflections on Tony as a person. Our offices were close together in D Block, one of the four wings of the Chemistry Department, so we saw a lot of each other. I remember particularly the 1980s, the time when Tony's research was blossoming and there was a super atmosphere in the coffee club, with everyone arguing passionately about the three key issues of life - politics, religion and sport. Tony was a football fanatic, and shared the highs and lows of his beloved Spurs. He would have been cock a hoop this weekend to hear they've got through to the Cup semi finals!

Petar Dvornic, Stephen Clarson, Tony Semlyen, and Michael Owen at the *Silicones and Silicone-Modified Materials* Symposium, 215th American Chemical Society National Meeting in Dallas, Texas, in 1998. (The photograph is courtesy of Stephen Clarson.)

There are many other things about Tony which were special: his mischievous sense of humour; his encyclopedic recall of past events; his tidy desk, with its neatly arranged rows of pens; his genuine interest and concern for all those around him.

Tony's philosophy was based on friendship and co-operation. He had 150 co-workers drawn from over five continents. Tony wanted everyone to be happy and to be friends. His vision of science as a team effort, the excitement and enjoyment of collaborating with other groups rather than competing against them, serves as a model for all of us.

Tony always said he felt blessed to have such a wonderful family, his faith, and his science. I was privileged to visit Tony the week before he died, and very moved when he shared with me his experiences over Christmas. The fact that he had been able to get to church when he hadn't expected to, and his joy at being there with friends around him and sunshine streaming through the windows.

He always saw beauty and perfection in the natural world and in God's creation, and one of his favourite quotations was from another scientist who had an unshakeable belief in God, Sir Isaac Newton: "I know not what I

Stephen Clarson and Tony Semlyen at the *Silicones and Silicone-Modified Materials* Symposium, 215th American Chemical Society National Meeting in Dallas, Texas, in 1998. (The photograph is courtesy of Stephen Clarson.)

may appear to the world, but to myself I seem to have been only like a boy playing on the sea-shore, and diverting myself in now and then finding a smoother pebble or prettier shell than ordinary, whilst the great ocean of truth lay all undiscovered before me".

We shall miss him terribly.

*Given at the funeral service for Tony Semlyen held on Tuesday 13 March, 2001 in the Holy Trinity Church, Elvington, United Kindgom.

1. Semlyen, J. A. (editor), *Cyclic Polymers*, Elsevier Applied Science, London, **1986**
2. Clarson, S. J.; Semlyen, J. A. (editors), *Siloxane Polymers*, Prentice Hall, Englewood Cliffs, NJ, **1993**
3. Semlyen, J. A. (editor), *Large Ring Molecules*, John Wiley & Sons, Chichester, **1996**
4. Semlyen, J. A. (editor), *Cyclic Polymers (2nd Edition)*, Kluwer Academic Publishers, Dordrecht, **2000**

Contents

Synthesis

Characterization

Elastomers and Reinforcement

Surfaces and Interfaces

Copolymers

Reinforcing Fillers

Preface

The second *Silicones and Silicone-Modified Materials* symposium was held April 2–5, 2001 at the American Chemical Society (ACS) National Meeting in San Diego, California. The symposium was a major success just like our previous meeting that was held in Dallas, Texas in the spring of 1998 that led to the book entitled *Silicones and Silicone-Modified Materials:* edited by S. J. Clarson, J. J. Fitzgerald, M. J. Owen, and S. D. Smith; ACS Symposium Series 729; American Chemical Society: Washington, D.C.; ISBN 0–8412–3613–5.

The four full-day symposium consisted of 55 oral presentations and an evening poster session of 37 presentations. We are delighted that we are now able to share 31 of these contributions with you in this book and note that the summaries of all the contributions to the symposium can be found in *Polymer Preprints,* Volume 42, Number 1, 2001, pages 85–257.

Our sincere thanks go to the ACS Division of Polymer Chemistry, Inc. and especially to our patient Program Chair Carrington Smith and to our gracious Treasurer Kathleen Havelka. I thank Siddharth Patwardhan and Hyeon Woo Ahn for help with some of the manuscripts that arrived on my "i-desk" in electronic form. At the ACS Books Department Kelly Dennis and Stacy Vanderwall in acquisitions and Margaret Brown in editing/production have been simply delightful to work with on this project and all were kind, helpful, and patient beyond words during the preparation of this book. We also thank Dow Corning, Procter and Gamble, and the ACS for kindly providing financial support and thus making it possible for many of our presenters to be able to attend this symposium.

We were all deeply saddened that one of our dear colleagues and an invited speaker Dr. Tony Semlyen left us in this world shortly before our symposium in San Diego, California. We thereby humbly dedicate this book to Tony and we are sure that his pioneering contributions in the areas of ring-chain equilibria of siloxane systems and cyclic poly(siloxanes) will be appreciated by future generations of scientists. We were pleased that Tony Dagger bravely agreed to give Tony Semlyen's oral presentation in San Diego and that he also ensured that we were able to include this presentation as a chapter in this book.

We trust that that the material herein will be of widespread interest to the global silicones community and we hope to see you at the third Silicones and Silicone-Modified Materials symposium at the ACS National Meeting in Anaheim, California from March 28–April 2, 2004.

Stephen J. Clarson
Department of Chemical and
  Materials Engineering
  and the Polymer Research Center
College of Engineering
University of Cincinnati
Cincinnati, OH 45221–0012
513–556–5430 (telephone)
513–556–2569 (fax)
sclarson@uceng.uc.edu (email)

John J. Fitzgerald
GE Silicones
Mail Drop 12/11
260 Hudson River Road
Waterford, NY 12188
518–233–3884 (telephone)
518–233–3866 (fax)
john.fitzgerald@gepex.ge.com (email)

Michael J. Owen
1505 West Saint Andrews Road
Midland, MI 48640
980–631–7339 (telephone)
michaelowen@chartermi.net (email)

Steven D. Smith
Procter and Gamble Company
Miami Valley Laboratories
11810 East Miami River Road
Cincinnati, OH 45061
513–627–2102 (telephone)
513–627–1259 (fax)
smith.sd@pg.com (email)

Mark E. Van Dyke
Southwest Research Institute
6220 Culebra Road
San Antonio, TX 78238–5166
210–522–3068 (telephone)
210–522–5122 (fax)
mvandyke@swri.org (email)

Chapter 1

Silicones and Silicone-Modified Materials:
A Concise Overview

Stephen J. Clarson

Department of Chemical and Materials Engineering and the Polymer
Research Center, University of Cincinnati, Cincinnati, OH 45221–0012

Silicones have become established in a wide variety
of applications ranging from electronics to personal
care products. It has been estimated recently that
there are approximately 3000 commercial silicone
products. These are made up from silicone fluids,
elastomers and resins. Many hybrid materials also
exist whereby organic and/or inorganic materials
are incorporated into silicones or vice versa. The
literature on silicones continues to grow with large
numbers of patents and publications appearing each
year. Some aspects of the science and technology of
silicones and silicone-modified materials are
described in this concise overview.

The world-wide sales of polysiloxanes or silicones at the beginning of this new millennium is around ten billion dollars per year and is still growing. Commercial products range from those entirely composed of silicone to products where the silicone is a low level but key component. Below is a short overview, the purpose of which is to serve as an introduction to the recent academic and technological developments behind silicones and silicone-modified materials that are described in the various contributions in this book.

Nomenclature

The term silicone was first introduced to describe what was believed to be the silicon analog of the carbon based ketones, which are represented thus $R_1R_2C=O$. This assignment proved to be incorrect for the silicon containing systems but, despite this error, the name has persisted to this day. The correct structure for the linear silicones is $-[R_1R_2SiO]_y-$ where R_1 and R_2 are organic groups and silicon and oxygen atoms form the backbone of the system.

The nomenclature of silicon-based polymers has been reviewed recently by Teague in the *Polymer Preprints* that accompanied this symposium and was published therein as Macromolecular Nomenclature Note Number 20 (1). It was noted that the M, D, T and Q notation is not used for naming or registering siloxanes or silicones in the CAS databases. As this terminology is widely used, it will be described here. For the silicones these letters correspond to the monofunctional (M), difunctional (D), trifunctional (T) and quadrifunctional (Q) units that make up these materials. Thus the linear siloxane

$$(CH_3)_3Si\text{-}[O\text{-}Si(CH_3)_2]_5\text{-}O\text{-}Si(CH_3)_3$$

may be denoted MD_5M and the cyclic siloxane octamethylcyclotetrasiloxane may be denoted D_4.

Chemical and Physical Properties of Silicones

From the commercialization of silicones in the early 1940's up to the present date the most widely recognized silicone is the system where $R_1=R_2=CH_3=Me$ namely poly(dimethylsiloxane) or PDMS (2-11). This material has become the workhorse of an industry based on silicon containing polymers and the terms silicone, siloxane polymer, poly(diorganosiloxane), dimethicone and PDMS are equivalent when describing the commercial products derived from this system.

The linear silicones (PDMS) are fluids at room temperature (12) and the molar mass of the system governs a number of key physical properties. For example, low molar mass materials have low viscosities, whereas high molar mass materials are gum-like and exhibit the viscoelastic behavior that is characteristic of long chain molecules in the melt state. Such materials will dissolve in common organic solvents such as toluene.

Silicone elastomers (or networks or rubbers) are formed by introducing chemical crosslinks to the system that was described above and there are many ways that have been long established to do this. Such materials show high extensibility with complete recovery – namely rubberlike elasticity. Furthermore silicone elastomers swell when immersed in common organic solvents such as toluene but do not dissolve owing to the chemical crosslinks having formed a three-dimensional network.

Silicones which are highly crosslinked exhibit the properties of resins in that they show low extensibility and very low degrees of swelling when placed in common organic solvents such as toluene.

The silicone fluids, elastomers and resins exhibit high stability – being thermally stable in air up to 250°C and in N_2/Ar/vacuum even up to 350°C. They each also find many applications due to being hydrophobic, a property which makes them very useful for water repellant uses (13-15). Their low surface energy also means that they can compete for Teflon replacement applications due to their non-stick characteristics. Thus many coating applications result for silicones: one example being mold release agents for plastic parts. Silicones also exhibit excellent dielectric properties which leads to them have applications as electronic encapsulants, electrical wire insulation and transformer fluids.

Commercial Applications of Silicone Products

Listed in Table 1. are some examples of large businesses for silicones and in each category some typical applications are given for illustrative purposes.

Table 1. Commercial Silicone Products

Silicone Technology	Application
1. Elastomers	Baby bottle teats
2. Defoamers	Food preparations
	Beer manufacture
3. Coatings	Paper (post-it notes)
	Cooking
	Mold release agents
4. Fluids and Lubricants	Greasing hypodermic needles
5. Personal Care Products	Deodorants
6. Medical products	Catheters
	Finger joints
	Breast implants
	Pacemaker leads

It has been stated in the inaugural issue of the journal *Silicon Chemistry* that there are ca. 3000 silicone products. These applications form the basis of the large world-wide sales of silicones.

Silicone-Modified Organic Systems

As the thermodynamic interactions are not favourable for making miscible blends of silicones with typical organic polymers, other strategies must be used for making silicone-modified organic systems.

Successful approaches to date have included chemically grafting siloxanes/silanes on to existing organic macromolecules, the synthesis of organic-siloxane copolymers and organic-siloxane interpenetrating networks (IPN's).

An extremely useful review of organic-siloxane copolymers is the one by Yilgor and McGrath (16). The addition of small amounts of organic-siloxane block copolymers to the parent organic homopolymer can lead to improved fire resistance, dielectric properties, toughness, crack resistance and surface properties (16-28). Experimental techniques such as contact angle measurements, angular-resolved X-ray Photoelectron Spectroscopy (XPS) and time-of-flight secondary ion mass spectrometry (TOFSIMS) have been used to investigate the surface segregation of the siloxane blocks / siloxane repeat units in various organic-siloxane copolymer /organic homopolymer systems. In the case where the siloxane is PDMS, saturation of the surface with siloxane occurs above a critical concentration of PDMS in the bulk.

Silicone-Modified Inorganic Systems

A wide variety of hybrid systems can be prepared by incorporating inorganic materials into a silicone matrix and vice versa. In many of their commercial applications siloxane polymers are filled with silica (SiO_2) in order to improve their mechanical properties (30-32). This is commonly achieved by simply mechanically mixing the inorganic filler into the silicone. A series of investigations have shown the feasibility of precipitating silica into PDMS elastomers by the catalyzed hydrolysis of an alkoxysilane or silicate (33-38). In the case where the elastomers are swollen by using tetraethoxysilane (TEOS), the reaction is simply

$$Si(OC_2H_5)_4 + 2H_2O \longrightarrow SiO_2 + 4C_2H_5OH$$

The *in-situ* precipitated silica particles typically have diameters in the range 15-25 nm (39-41). Furthermore, such *in-situ* generated particles have been shown to be unagglomerated, whereas the conventional techniques of blending pre-reacted silica (30) or the 'wet process' silica (42-44) methods employed commercially, may produce large agglomerates of the filler. These methods have also been used to successfully generate reinforcement by the *in-situ* precipitation of SiO_2

and TiO$_2$ into other silicones such as poly(methylphenylsiloxane) (41,45,46).

The technique of hydrolyzing an alkoxy silane can be generalized to make the silica the continuous phase, with domains of PDMS dispersed in it. This requires relatively high concentrations of the silane precursor. By varying the silicone/inorganic composition, composite materials can be obtained ranging from relatively soft elastomers, to tough hybrid materials, to brittle ceramics (47). Additional properties that are dependent on composition include impact resistance, ultimate strength, and maximum extensibility.

Silicon Biochemistry

In the list of elements that are known to be essential to life, silicon (Si) is one of the required trace elements. It has been estimated that humans ingest 30mg Si per day, with 60% coming from cereals and 20% from water and drinks. While the exact role(s) of silicon in biological systems remains to be determined, the strong interactions between aluminum and silicic acid Si(OH)$_4$ at circumneutral pH have been cited as being of vital importance (48,49). It was shown that silicic acid may mitigate the toxic effects of aluminum in experiments involving young Atlantic Salmon (50). Experiments have also shown that silicic acid restricts the absorption of aluminum in the human gut (51).

Ornate biosilica structures occur in diatoms, sponges and grasses. Investigations of such systems are beginning to reveal some of the key aspects of biosilicification. Biosilica entraps certain biomacromolecules that are thought to be responsible for catalyzing/ templating/ scaffolding the biosilica formation. Selective dissolution of the biosilica has led to the isolation of biomacromolecules from diatoms (52,53), sponges (54) and grasses (55). It has been demonstrated that proteins such as silaffins (from diatoms) and silicateins (from sponges) can precipitate silica *in vitro* and thus it is proposed that they can mediate biosilicification. A transporter molecule for carrying silicic acid across cell membranes has also been identified (56).

A 19-amino acid R5 polypeptide from the silaffin-1A protein has been used to microfabricate a novel organic/inorganic hybrid optical device (57). The concept that biomacromolecules can act as catalysts/

templates/ scaffolds for generating biosilica (*58*) has also been utilized in a bioinspired synthetic macromolecule system which can generate ordered silica structures at neutral pH and under ambient conditions (*59-62*).

References

1.   Teague, S. J. *Polymer Preprints*, **2001**, 42(1), pp xviii-xxiii.

2.   Rochow, E. G. *An Introduction to the Chemistry of the Silicones*; John Wiley & Sons: New York, 1946.

3.   Noll, W. *Chemistry and Technology of Silicones*; Academic Press: New York, 1968.

4.   Liebhafsky, H. A. *Silicones Under the Monogram*; Wiley-Interscience: New York, 1978.

5.   Rochow, E. G. *Silicon and Silicones*; Springer-Verlag: Berlin, 1987.

6.   Warrick, E. L. *Forty Years of Firsts - the Recollections of a Dow Corning Pioneer*; McGraw Hill: New York, 1990.

7.   *The Analytical Chemistry of Silicones*; Smith, A. L., Ed.; Wiley-Interscience: New York, 1991.

8.   *Siloxane Polymers*; Clarson, S. J.; Semlyen, J. A., Eds.; Prentice Hall: Englewood Cliffs, NJ, USA, 1993.

9.   *Silicones and Silicone-Modified Materials*; Clarson, S. J.; Fitzgerald, J. J.; Owen, M. J.; Smith, S. D., Eds.; ACS Symposium Series Vol. 729; Oxford University Press: New York, 2000.

10.   *Silicone Surfactants*, Hill, R., Ed.; Marcel Dekker: New York, 2000.

11.   Brook, M. A. *Silicon in Organic, Organometallic and Polymer Chemistry*; Wiley-Interscience: New York, 2000.

8

12. Clarson, S. J.; Dodgson, K.; Semlyen, J. A. *Polymer* **1985**, *26*, 930.

13. Owen, M. J. *Chemtech* **1981**, *11*, 288.

14. Owen, M. J. *Comm. Inorg. Chem.* **1988**, *7*, 195.

15. Owen, M. J. In *Siloxane Polymers*; Clarson, S. J.; Semlyen, J. A., Eds.; Prentice Hall: Englewood Cliffs, NJ, USA, 1993.

16. Yilgor, I.; McGrath, J. E. *Adv. Polym. Sci.* **1988**, *86*, 1.

17. Saam, J. C.; Gordon, D. J.; Lindsey, S. *Macromolecules* **1970**, *3*, 1.

18. LeGrand, D. G.; Gaines, G. L. *Polymer Preprints* **1970**, *11*, 442.

19. Gaines, G. L.; Bender, G. W. *Macromolecules* **1972**, *5*, 82.

20. Jones, F. R. *European Polym. J.* **1974**, *10*, 249.

21. Clark, D. T.; Dilks, A. *J. Polym. Sci.: Polym. Chem. Ed.* **1976**, *14*, 533.

22. Gaines, G. L. *Macromolecules* **1979**, *12*, 1011.

23. Krause, S.; Iskander, M.; Iqbal, M. *Macromolecules* **1982**, *15*, 105.

24. Wang, B.; Krause, S. *Macromolecules* **1987**, *20*, 2201.

25. Dwight, D. W.; McGrath, J. E.; Riffle, J. S.; Smith, S. D.; York, G. A. *J. Electron Spectroscopy* **1990**, *52*, 457.

26. Chen, X.; Gardella, J. A.; Kumler, P. L. *Macromolecules* **1992**, *25*, 6621.

27. Selby, C. E.; Stuart, J. O.; Clarson, S. J.; Smith, S. D.; Sabata, A.; van Ooij, W. J.; Cave, N. G. *J. Inorganic and Organometallic Polymers* **1994**, *4(1)*, 85.

28. Clarson, S. J.; Selby, C. E.; Stuart, J. O.; Sabata, A.; Smith, S. D.; Ashraf, A.; *Macromolecules* **1995**, *28*, 674.

29. Clarson, S. J.; Mark, J. E. In *The Polymeric Materials Encyclopedia*; Salamone, J. C., Ed.; CRC Press: Boca Raton, Florida, **1996**, *10*, 7663.

30. Warrick, E. L.; Pierce, O. R.; Polmanteer, K. E.; Saam, J. C. *Rubber Chem. Tech.* **1979,** *52,* 437.

31. Polmanteer, K. E. *Rubber Chem. Tech.* **1981,** *54,* 1051.

32. Polmanteer, K. E. *Rubber Chem. Tech.* **1988,** *61,* 470.

33. Mark, J. E.; Pan, S. -J. *Makromol. Chemie: Rapid Comm.* **1982,** *3,* 681.

34. Jiang, C. –Y.; Mark, J. E. *Colloid Polym. Sci.* **1984,** *262,* 758.

35. Mark, J. E.; Jiang, C. -Y.; Tang, M. -Y. *Macromolecules* **1984,** *17,* 2613.

36. Sur, G. S.; Mark J. E. *Makromol. Chem.* **1986,** *187,* 2861.

37. Clarson, S. J.; Mark J. E.; Dodgson, K. *Polymer Communications* **1988,** *29,* 208.

38. Mark, J. E. *Chemtech* **1989,** *19,* 230.

39. Ning, Y. -P.; Tang, M. -Y.; Jiang, C. -Y.; Mark, J. E.; Roth, W. C. *J. Appl. Polym. Sci.* **1985,** *29,* 3209.

40. Mark, J. E.; Ning, Y. -P.; Jiang, C. -Y.; Tang, M. -Y.; Roth, W. C. *Polymer* **1985,** *26,* 2069.

41. McCarthy, D. W.; Mark, J. E.; Clarson, S. J.; Schaefer, D. W. *J. Polym. Sci.: Part B: Polym. Phys. Ed.* **1998,** *36,* 1191.

42. Lutz, M. A.; Polmanteer, K. E.; Chapman, H. L. *Rubber Chem. Tech.* **1985,** *58,* 939.

43. Chapman, H. L.; Lutz, M. A.; Polmanteer, K. E. *Rubber Chem. Tech.* **1985,** *58,* 953.

44. Polmanteer, K. E.; Chapman, H. L.; Lutz, M. A. *Rubber Chem. Tech.* **1985,** *58,* 965.

45. Clarson, S. J.; Mark, J. E. *Polymer Communications* **1987,** *28,* 249.

46. Clarson, S. J.; Mark, J. E. *Polymer Communications* **1989,** *30,* 275.

47.   Mark, J. E.; Sun, C. -C. *Polym. Bulletin* **1987**, *18*, 259.

48.   Birchall, J. D. *Chemical Society Reviews* **1995**, *24*, 351.

49.   Perry, C. C.; Keeling-Tucker, T. *J. Inorg. Biochem.* **1998**, *69*, 181.

50.   Birchall, J. D.; Exley, C.; Chappell, J. S.; Phillips, M. J. *Nature* **1989**, *338*, 146.

51.   Edwardson, J. A.; Moore, D. B.; Ferrier, I. N.; Lilley, J. S.; Newton, G. W. A.; Barker, J.; Templar, J.; Day, J. P. *Lancet* **1994**, *342*, 211.

52.   Kroger, N.; Deutzmann, R.; Sumper, M. *Science* **1999**, *286*, 1129.

53.   Kroger, N.; Deutzmann, R.; Bergsdorf, C.; Sumper, M. *PNAS* **2000**, *97*, 14133.

54.   Shimizu, K.; Cha, J. N.; Stucky, G. D.; Morse, D. E.; *PNAS* **1998**, *95*, 6234.

55.   Harrison (formerly Perry), C. C. *Phytochemistry* **1996**, *41(1)*, 37.

56.   Hildebrand, M.; Volcani, B. E.; Gassmann, W.; Schroeder, J. I. *Nature* **1997**, *385*, 688.

57.   Brott, L. L.; Pikas, D. J.; Naik, R. R.; Kirkpatrick, S. M.; Tomlin, D. W.; Whitlock, P. W.; Clarson, S. J.; Stone, M. O. *Nature* **2001**, *413*, 291.

58.   Tacke, R. *Angew. Chem. Int. Ed.* **1999**, *38(2)*, 3015.

59.   Patwardhan, S. V.; Mukherjee, N.; Clarson, S. J. *J. Inorg. and Organometallic Polymers*, **2001**, *11(2)*, 117.

60.   Patwardhan, S. V.; Mukherjee, N.; Clarson, S. J. *J. Inorg. and Organometallic Polymers*, **2001**, *11(3)*, 193.

61.   Patwardhan, S. V.; Mukherjee, N.; Clarson, S. J. *Silicon Chemistry*, **2002**, *1(1)*, 47.

62.   Patwardhan, S. V.; S. J. Clarson, S. J. *Polymer. Bull.* **2002**, *48*, 367.

Synthesis

Chapter 2

Controlled Synthesis of All Siloxane-Functionalized Architectures by Ring-Opening Polymerization

J. Chojnowski, M. Cypryk, W. Fortuniak, K. Kaźmierski, K. Rózga-Wijas, and M. Ścibiorek

Center of Molecular and Macromolecular Studies, Polish Academy of Sciences, Sienkiewicza 112, 90–363 Łódź, Poland

Ring-opening copolymerization of cyclotrisiloxanes was used for the synthesis of various well-defined all-siloxane block and gradient copolymers. These copolymers contain functional groups in organic radicals and are specifically functionalized at the chain end. They were also further exploited for synthesis of the star-branched and dendritic-branched structures. They were also attached to the surface of silica to obtain silica-siloxane hybrid materials.

There has been an increasing interest in using polysiloxanes as fragments of various macromolecular architectures, due to unusual combination of properties of these polymers. The polysiloxane chain has exceptionally high dynamic and static flexibility, which is related to very low barriers of rotation around the Si-O bond and of linearization of the Si-O-Si angle (*1*). Thus, the polysiloxane chain has a very high conformational freedom easily adopting various shapes. It adapts itself readily to its surrounding and the functional groups attached to polysiloxanes are available for the interaction with neighboring molecules. Taking into account that substituents appear only at every second atom in the chain, the Si-O bond is relatively long (1.63 Å) and the SiOSi angle unusually large (145°), the polysiloxane flexibility is not much restricted by substituents unless they are very bulky. The nature of the polysiloxane backbone is inorganic, which gives this polymer a high thermal stability and also, in some sense, an amphiphilic character. Its inorganic skeleton is formed of strongly polar Si-O bonds, but it bears nonpolar organic groups. Due to this feature, the polysiloxane tends to go to the interface, adopting a conformation in which its polar skeleton sticks to more hydrophilic surface, while organic groups are directed towards more hydrophobic surface. In this way it decreases the interfacial surface tension.

Hybrids of polysiloxanes with many organic polymers are very attractive materials. Particular attention has been paid to siloxane-organic block and graft copolymers (2-4). Polysiloxanes in combination with organic polymers are also exploited for the construction of more complex branched structures, such as star shape (5) and dendritic (cascade shape) copolymers (6,7) as well as for the formation of cross-linked materials (8). In contrast to very broad literature on siloxane-organic hybrids, relatively little attention has so far been devoted to the generation of all-siloxane macromolecular architectures, see for ex. refs (6,9,10). Complex structures based on all-siloxane inorganic skeleton may be attractive as new materials of a high thermal stability, good solubility, expected interesting morphology and surface properties. Polysiloxanes may be readily modified within organic groups, which can dramatically change their behavior. For example, introduction of hydrophilic groups may make the polymer water soluble (11-13). Combination of such a polymer in one hybrid with hydrophobic polydimethylsiloxane (PDMS) may lead to very interesting amphiphiles (11). On the other hand, combination of PDMS as soft segment with a stiff crystalline polysiloxane may give materials of interesting morphology leading to a high mechanical strength and improved thermal stability (10). The purpose of our study is to elaborate methods of the controlled synthesis of all-siloxane copolymers of various topologies, such as diblock, triblock, multiblock, star-branched and dendritic-branched structures. Our general approach is to synthesize various structures being combinations of the two types of segments, the one composed exclusively of hydrophobic, inert dimethylsiloxane units and the other containing siloxane units functionalized in organic groups. The functionalized units could appear as the sole units in the segment, but it is often preferable to "dilute" them with dimethylsiloxane units. Thus, the segment itself may be the siloxane-siloxane copolymer. The functions in organic groups are to give the polymer special physical or biological properties, such as hydrophilicity, biocidal properties, spectral properties and others or provide the polymer with specific chemical reactivity, such as complex formation ability or catalytic properties.

Generation of Functionalized Polysiloxane Segments

Anionic ring opening polymerization and copolymerization of cyclotrisiloxanes is the method giving the best possibility of the control of the structure of the polysiloxane product. It is well known that a significant reduction of back biting and chain transfer, leading to a narrow molecular weight distribution, may be achieved in these processes (14). High precision of functionalization of chain end is also possible, which is crucial for the controlled generation of macromolecules of more complex topologies. The controlled functionalization of side groups may also be achieved by the polymerization and copolymerization of cyclotrisiloxanes containing functions in organic substituents. Two types of functionalized cyclotrisiloxanes were used, these having functional groups in all siloxane units and those having two dimethylsiloxane groupings and one functionalized siloxane unit. The representative of the second type is 2-imidazolopropylpentamethylcyclotrisiloxane (equation 1). Its polymerization leads to polysiloxane with imidazole functions (15a).

$$\text{BuLi} + n \quad \longrightarrow \quad \text{BuMe}_2\text{Si}\left[\text{OSi O Si O Si}\right]_n \text{O}^- \text{Li}^+ \tag{1}$$

The polymerization of a monomer with the target functional group is often inconvenient or impossible. Acidic or electrophilic groups do not tolerate anionic propagation centers. Some functional monomers are difficult to purify. For example, the removal of traces of a protic contamination, such as water, alcohol, acid from the imidazole-substituted cyclotrisiloxane is troublesome (15a). Interaction of propagation center with the functional group may promote chain transfer or back biting (16). For these reasons, the preferred synthetic strategy is to use monomers bearing precursor groups which are transformed into target functions in polymer. The most common precursors are: $\equiv$Si-CH=CH$_2$, $\equiv$Si-H and $\equiv$Si(CH$_2$)$_3$Cl. The anionic ROP of (ViMeSiO)$_3$, **1** (11), ViMeSiO(Me$_2$SiO)$_2$, **2** (17) and Vi$_2$SiO(Me$_2$SiO)$_2$, **3** (18) on lithium silanolate centers was studied. The polymerization proceeds with full preservation of vinyl groups and leads to a narrow molecular weight distribution. Recently, Paulasaari and Weber (19) succeeded in synthesis of HMeSiO(Me$_2$SiO)$_2$, **4**, and showed that this monomer may be chemoselectively and regioselectively polymerized on the lithium silanolate center in THF at -70°C. Although the chloropropyl bonded to silicon was not expected to tolerate the silanolate propagation center (*20*), both monomers [Cl(CH$_2$)$_3$MeSiO]$_3$, **5**, and Cl(CH$_2$)$_3$MeSiO(Me$_2$SiO)$_2$, **6**, polymerized smoothly on the lithium silanolate centers in THF at ambient temperature with full control of the polymer product (*21*).

Polymerization of the monomers with mixed units, such as **2**, **3**, **4** and **6**, leads to the copolymer of functional siloxane with dimethylsiloxane. Since the propagation is not accompanied by any chain cleavage reaction and the monomer enters the chain undivided, the distribution of functional units in macromolecule is uniform. Each monomer unit contains one functional group. The order of units may be regular if the propagation occurs regioselectively, i.e., the ring is opened exclusively at one site. This is the case of the polymerization of **4** at low temperature (*22*). Usually, the regioselectivity is limited as various structural factors may be important in choosing of the place of ring opening. There are three nonequivalent sites of the ring opening noted by *a*, *b* and *c* in equation 2.

Basing on ^{29}Si NMR data, the sequential analysis was made at the pentad level using 1st order Markov chain statistics for the polymerization of [Ph$_2$SiO(Me$_2$SiO)$_2$], **7**, (*23*) and **6** (*22*). Since the method does not differentiate between the opening at *a* and *b*, additional experiment of initiation was performed allowing to determine the order in the first monomer unit. These combined methods gave relative rates of ring opening at *a*, *b* and *c* (Table I).

The site of the ring opening depends on both the substituents and the active propagation center. Substituent X is electron-withdrawing as compared with methyl, thus, substituted silicon is the most electrophilic center to which the attack of the silanolate anion is likely to be directed, leading to opening at *a*. This is the case of the polymerization of vinyl substituted monomer **2** on the lithium silanolate center in THF, where the opening at *a* prevails, although the regioselectivity is low. However, it significantly increases when the reaction is performed at lower temperature (Table I). Another course of the polymerization in the same system ($\equiv$SiOLi/THF) was observed for monomer 7. The ring is preferentially opened at c, which may be explained either by a smaller steric effect or by formation of an intermediate complex between monomer and counter-ion (23). On the other hand, potassium silanolate complexed by crown ether opens the monomer 7 mostly at b which is interpreted by preferential formation of more stable silanolate anion, the one with negative charge delocalized to phenyl groups.

Table I. Regioselectivity of monomer opening in various polymerization systems

Monomer	Initiator	Temp	% of the monomer ring opening						Ref.
			Markov 1st order statistics (%)			Initiation experiment			
			a	b	c	a	b	c	
7	Me₃SiOLi/THF	50°	17	35	48	33	18	49	23
			35	17	48				
7	Me₃SiOK/ 18-crown-6	50°	72	13	15	10	73	17	23
			13	72	15				
2	BuLi/THF	-30°	89	8	3				22
			8	89	3				
2	BuLi/THF	25°	67	21	12	78	11	11	22
			21	67	12				

The copolymerization of a functionalized cyclotrisiloxane with hexamethylcyclotrisiloxane (D_3) leads to a copolymer of dimethylsiloxane and siloxane containing functional group which, similarly to the copolymer obtained by polymerization of cyclotrisiloxane with mixed units, contains macromolecules of uniform structures with regard to the size, monomer unit composition and sequencing. However, the distribution of units along the chain is different. The more reactive comonomer enters the chain preferentially, thus the density of the units derived from this monomer in polymer is high at the beginning of the chain formation. It generally decreases during the chain growth as the contribution of the more reactive monomer in the feed decreases. This leads to gradient distribution of the functional groups along the copolymer chain.

Synthesis of Functionalized All-Siloxane Block Copolymers

Block coupling is the method often used for synthesis of siloxane block copolymers (*2,3*). However, the higher precision of the synthesis may be achieved by building blocks either using macroinitiator or exploiting the sequential copolymerization. The latter method, i.e., sequential copolymerization of cyclotrisiloxanes according to general equation 3, is particularly suitable for the precision synthesis of diblock and triblock all-siloxane functionalized copolymers. This reaction was used as early as in sixties by Bostick in attempts to generate silicone plastic elastomers (*9*). For this purpose he obtained in a controlled way block copolymers composed of flexible polydimethylsiloxane segments and rigid crystalline polydiphenylsiloxane segments (*9*). This concept was continued later by Meier et al. (*10*). Elastomers with good mechanical properties were obtained, but too high melting point of the crystalline domains formed by poly(Ph$_2$SiO) segments was the obstacle to their practical use.

$$(3)$$

The sequential copolymerization of cyclotrisiloxanes permitted us to obtain block copolymers of dimethylsiloxane and a functionalized siloxane of high topological purity, defined size of blocks, low polydispersion and precise functionalization at chain termini (11,12). Using a monofunctional initiator leads to the diblock copolymer, while triblock copolymers are formed with bifunctional initiators. It is advantageous to polymerize the less reactive monomer, which is usually D_3, in the first step. Thus, polydimethylsiloxane block is usually formed first. The comonomer should be introduced not later than at 90-95% of conversion of the first monomer to avoid the processes leading to chain randomization. Monomers having functional groups, such as vinyl or 3-chloropropyl are more reactive towards the propagation center, as compared with D_3, therefore, the cross propagation to residual D_3 hardly occurs in the second stage of the copolymerization. Conversely, if more reactive monomer is polymerized first, the second block contains units derived from this monomer (22).

The selectivity of the first step of the polymerization was checked by MALDI and SEC analyses of the polymer obtained by the quenched polymerization D_3 initiated with in situ formed lithium butyldimethylsilanolate. The reaction was quenched at about 90% of monomer conversion by introduction of a small excess of vinyldimethylchlorosilane. The polymer having $\overline{M}_n = 1.0 \cdot 10^4$ and $M_w/M_n = 1.1$ showed in MALDI TOF spectrogram exclusively those signals which belonged to polydimethylsiloxanes having one fragment of initiator and one fragment of terminator, i.e., $BuMe_2Si(OSiMe_2)_nOSiViMe_2$. Thus, no chain cleavage leading to mixing of the chain ends was observed. The chains are functionalized with the silanolate group converted quantitatively to the vinylsilane group, being exclusively at one chain terminus. This leads to precise topology of diblock copolymers in experiments of sequential copolymerization. Indeed, SEC and NMR analyses confirmed a high precision of the synthesis of block copolymer. Examples of synthesis of some diblock siloxane-siloxane copolymers are presented in Table II.

Vinyl groups pendant to polysiloxane chain may be transformed by various addition reactions such as hydrosilylation (24), thiol-ene addition (11,17,25) and hydrophosphination (26). Thus, various functional groups may be introduced to the block copolymer containing the segments with the vinyl groups. For example, thiol-ene addition was successfully used to introduce carboxyl groups pendant to the chain, according to equation 4 (11). Amphiphilic diblock copolymers of interesting solution properties and morphologies were obtained.

3-Chloropropyl groups bonded to silicon may be transformed by nucleophilic substitution of chlorine. This reaction was used to introduce to one segment of diblock and triblock siloxane copolymers quarternary ammonium salt groups to obtain amphiphilic all-siloxane copolymers of bacteriocidal properties (27).

$$\text{BuMe}_2\text{Si}(\text{OSiMe}_2)_n(\overset{\underset{\displaystyle \text{CH}}{|}}{\underset{\underset{\displaystyle \text{CH}_2}{||}}{\text{OSiMe}}})_m\text{OSiMe}_3 \quad + \quad \text{HSCH}_2\text{COOH} \quad \xrightarrow[\text{toluene } 70\,^\circ\text{C}]{\text{AIBN}} \tag{4}$$

$$\longrightarrow \quad \text{BuMe}_2\text{Si}(\text{OSiMe}_2)_n(\underset{\underset{\displaystyle \text{SCH}_2\text{COOH}}{|}}{\overset{\overset{\displaystyle (\text{CH}_2)_2}{|}}{\text{OSiMe}}})_m\text{OSiMe}_3$$

A great potential of synthesis of new block siloxane-siloxane copolymers bearing various functional groups has anionic sequential copolymerization of monomer 4 (*19,28*). It would generate a segment with the Si-H group making possible the introduction of variety of functional groups by hydrosilylation.

Table II. Examples of functionalized diblock all-siloxane copolymers obtained by sequential anionic copolymerization of cyclotrisiloxanes. Block A is polydimethylsiloxane.

Block B	$\overline{M}_n$	*PDI*	$\overline{M}_n(A)/\overline{M}_n(B)$	*Ref.*
-[MeViSiO]$_m$-	18 700	1.1	1.1	*11*
-MeViSiO(Me$_2$SiO)$_2$]$_m$-	8 600	1.1	0.95	*11*
-[MeSiO]$_m$- ^{a} (CH$_2$)$_2$SCH$_2$COOH	25 500	1.1	0.54	*11*
-[MeSiO(Me$_2$SiO)$_2$]$_m$- ^{a} (CH$_2$)$_2$SCH$_2$COOH	10 200	1.1	0.70	*11*
-[MeSiO(Me$_2$SiO)$_2$]$_m$- ^{b} (CH$_2$)$_3$-Im	17 000	1.4	0.80	*15b*
-[MeSiO]$_m$- (CH$_2$)$_3$Cl	10 900	1.08	0.91	*21*
-[MeSiO(Me$_2$SiO)$_2$]$_m$- (CH$_2$)$_3$Cl	9 500	1.37	1.71	*21*

^{a} Functional groups were introduced by addition of HSCH$_2$COOH to silylvinyl groups in the copolymer.
^{b} Im = imidazole.

Synthesis of All-Siloxane Functionalized Branched Structures

Anionic copolymerization of cyclotrisiloxane leads to well defined various siloxane copolymers functionalized in side and end groups, which may further serve as bricks for building more complex branched structures. This concept was used by us to generate all-siloxane functionalized star-branched and dendritic-branched copolymers. General approach used for synthesis of dendritic-branched copolymers, based on graft and star technique is presented in equation 5.

(5)

The first step is the formation of living siloxane gradient or block copolymer $M_n = 0.5 \div 1 \cdot 10^4$, $M_w/M_n = 1.1 \sim 1.2$ having vinylmethylsiloxane and dimethylsiloxane units with lithium silanolate center at one end. This living copolymer obtained in the THF solution is quenched by 2,2,5,5-tetrachloro-2,5-disilahexane, **8**, as a tetrafunctional quencher, which leads to tetrabranched star polysiloxane. Compound **8** was shown earlier by Wilczek and Kennedy (*29*) to be more suitable for this purpose than tetrachlorosilane. The substitution of the fourth branch in SiCl$_4$, used in the stoichiometric amount, proceeds very slowly. Even after 30 h the reaction in room temperature is not completed, while the full substitution in **8** occurs in about 3 h under the same conditions. The star copolymer had almost theoretical molecular weight, PDI was 1.2÷1.3 and the content of vinyl groups was close to theoretical, since the living block copolymer was performed by sequential copolymerization using vinyl substituted monomer in the first step and D$_3$ in the second step, the vinyl groups were in the outside part of the branched copolymer. Similarly, in the simultaneous copolymerization of monomers **2** and D$_3$, the larger reactivity of the vinyl substituted monomer ensured a larger density of the vinyl groups in the part of the gradient copolymer chain opposite to the silanolate center. This led to desired placement of the vinyl functions mostly in the outer part of the branched copolymer.

Vinyl groups in the star-branched copolymer were transformed into silyl chloride functions by hydrosilylation with Me$_2$SiHCl. The Si-Cl functionalized star branched copolymers were subjected to reactions with the living copolymer obtained in analogous way to that in the first stage of the synthesis. The resulting dendritic branched copolymers had a content of vinyl groups close to stoichiometric, 5-8 per branch, however, molecular weights were higher than theoretical, $\overline{M}_n \approx 2 \div 3 \cdot 10^5$. In spite of using vacuum line technique, it was difficult to avoid coupling reactions of branched macromolecules. The copolymers were fully soluble and did not contain microgels. The presence of vinyl functions allows further modification of the outer part of the macromolecule.

Synthesis of Branched Functionalized Siloxane Copolymers on Silica Surface

The concept of generation of functionalized branched siloxane copolymers on the surface of silica particles was related to the idea of the preparation of reactive polysiloxanes which could be readily separated. The way of the generation of the functionalized branched polysiloxane-silica hybrid is shown schematically in equation 6.

Two types of commercial Merck porous silica gel were used as the polymer support: silica 1; grains d = 0.2÷0.5 mm 35/70 mesh, porosity d'=100Å and silica 2; grains d = 0.05÷0.2 mm 70/325 mesh, porosity d'=60Å. The silica gel grains were silylated with vinyltriethoxysilane. Thus, the vinyl groups bound to silicon were introduced to the silica surface. The vinyl groups were hydrosilylated with dimethylchlorosilane which permitted us to introduce reactive silyl chloride groups which were bonded through the ethylsilyl bridge to the silica surface. For example, 0.34 mmol of Cl per 1 g of SiO_2 was introduced to silica 1. The SiCl containing silica in a THF suspension was reacted with a living gradient vinylmethylsiloxane-dimethylsiloxane copolymer ended with lithium silanolate, $\overline{M}_n \approx 7 \cdot 10^3$, having 6 vinyl groups on average. The copolymer synthesis was described in one of the previous sections. The copolymer (0.024 mmol) was added to 1 g of SiO_2. The repeated hydrosilylation followed by reaction with the living copolymer permitted introduction of over 300 mg of the vinyl-functionalized branched siloxane copolymer per 1 g of SiO_2 (0.24 mmol Vi/g SiO_2).

(6)

These functionalized polysiloxane-silica hybrids were successfully used by Michalska and Rogalski as a support for platinum and rhodium complexes used further as solid-supported homogeneous catalysts for hydrosilylation reactions (30). Direct complexation, ligand displacement and bridge splitting reactions were used to attach the metal complexes to the vinyl groups of the polysiloxane-silica hybrids. The obtained catalysts showed very high activity.

Cationic Ring Opening Polymerization as the Method of Formation of Complexed Functionalized All-Siloxane Macromolecules

Equilibrium cationic ring opening polymerization has often been used in synthesis of end- and side-functionalized polysiloxanes, which may be used further as fragments of more complex architectures. Since, as a rule, the initiator should not influence the products of the equilibrium polymerization, the cationic and anionic polymerizations lead to the same results (*14*). End functionalization is accomplished by using of functionalized blockers according to equation 7.

$$X(CH_2)_n \overset{|}{\underset{|}{Si}}O\overset{|}{\underset{|}{Si}}(CH_2)_nX \ + \ m\,(Me_2SiO)_4 \longrightarrow \tag{7}$$

$$\longrightarrow \ X(CH_2)_n\overset{|}{\underset{|}{Si}}(OSiMe_2)_zO\overset{|}{\underset{|}{Si}}(CH_2)_nX \ + \ \sum_{p\geq 4} r\,(Me_2SiO)_p$$

Polysiloxane functionalized at both ends is obtained. Polymerization leads to the formation of considerable amounts of cyclic oligomers, which decrease the yield of the polymer and make the control of its molecular weight difficult. The separation of cyclics is troublesome and not always possible. Molecular weight distribution of 2 is broader than that of the polymer obtained by the quenched anionic polymerization of cyclotrisiloxane. Instead, the substrate for the equilibrium polymerization may be cyclic siloxanes of optional ring size. Mixture of cyclics of various sizes and even linear polysiloxanes can be used as substrates. The copolymerization of siloxanes with various substituents leads to copolymers having statistical distribution of units in the polymer chain (*31*), although some deviations from this distribution may be observed for bulky or possibly polar substituents (*32*).

General conclusion is that the cationic equilibrium polymerization and copolymerization, similarly as the corresponding anionic equilibrium processes, may be used as the method of the synthesis of functionalized siloxane copolymers for the generation of more complex all-siloxane structures, but the control in synthesis of these copolymers is limited. The question arises whether quenched cationic polymerization and copolymerization of cyclotrisiloxanes may be applied for these purposes similarly to the corresponding anionic process. It was shown that the polymerization of some cyclotrisiloxanes initiated with protic acids occurs without processes leading to the cleavage of the polysiloxane chain (*33*). However, this polymerization does not allow to avoid equilibration of chain ends and formation of cyclic oligomers by end to end ring closure or by cyclic oxonium ion transformation (34,35). It is not possible to avoid permanent initiation by protic acid being reproduced as the polymerization proceeds, either (33,36). Recently, we made an attempt of sequential copolymerization of monomer 6 with D_3. 3-Chloropropyl substituted monomer was polymerized first using CF_3SO_3H as initiator, then D_3 was introduced. The products were analyzed after the first and second steps by SEC using the RI detector. The analyses were performed in two solvents, toluene and methylene chloride. The refractive index of the latter is between this of PDMS and that of chloropropyl-substituted siloxane blocks. The results showed that the sequential copolymerization leads to considerable amounts of homopolymers. The polymerization of D_3 is mostly initiated by CF_3SO_3H and not by the activated end groups in the polymer obtained in the first step (37). Thus, perspectives of using cationic sequential polymerization of cyclotrisiloxanes for the precision polymerization of block copolymers do not seem to be attractive.

Acknowledgments: The research is supported by the State Committee for Scientific Research (KBN), Grant no. PZB-KBN 15/T09/99/01c.

References

1. Grigoras, M. *Computational Modeling of Polymers*, Bicerano J., Ed. Marcel Dekker: New York, 1993, p. 161.
2. Belorgey, G.; Sauvet, G. In *Silicon Containing Polymers*; Jones, R. G.; Ando, W.; Chojnowski, J., Eds.; Kluwer Academic Publishers: Dodrecht, 2000, p. 43.
3. Yilgör, I.; McGrath, J. E. *Adv. Polym. Sci.* **1988**, *86*, 1.
4. Wagener, K.B.; Zulunga, F.; Wanigatunga, S. *Trends Polymer Sci.* **1996**, *4*, 157.
5. Bellas, S.; Iatrou, H.; Hadjichristidis, N. *Macromolecules* **2000**, *33*, 6993.
6. Majoral, J. P.; Caminade, A. H. *Chem. Rev.* **1999**, *99*, 845.
7. Ouali, N.; Mery, S.; Skoulios, A.; Noires, L. *Macromolecules* **2000**, *33*, 6185.
8. Mazurek, M. In *Silicon Containing Polymers*; Jones, R. G.; Ando, W.; Chojnowski, J., Eds.; Kluwer Academic Publishers: Dodrecht, 2000, p. 113.
9. Bostick, E.E. *ACS Polymer Preprints* **1965**, *10*, 877.
10. Ibemesi, J.; Gvozdic, N. V.; Keumin, M.; Lynch, M. J.; Meier, D. J. *ACS Polymer Preprints* **1985**, *26*, 18.
11. Ścibiorek, M.; Gladkova, N.K.; Chojnowski, J. *Polym. Bull.* **2000**, *44*, 377.
12. Zeldin, M.; Rubinsztajn, S.; Fife, W. K. *J. Inorg. Organomet. Polym.* **1992**, *2*, 319.

13. Lin, J. K.; Wnek, G. E. *Macromolecules* **1994**, *27*, 4080.
14. Chojnowski, J.; Cypryk, M. In *Polymeric Materials Encyclopedia*, Salamone, J., Ed., CRC Press, 1996; Vol. 2, p. 1682.
15. (a) Fortuniak, W.; Chojnowski, J. *Polymer Bull.* **1997**, *38*, 371. (b) Fortuniak, W.; Chojnowski, J. in preparation.
16. Chojnowski, J.; Rózga, K.; Fortuniak, W.; Kowalewska, A. *Makromol. Chem. Macromol. Symp.* **1993**, *73*, 183.
17. Rózga-Wijas, K.; Chojnowski, J. Zundel, T.; Boileau, S. *Macromolecules* **1996**, *29*, 2711.
18. Weber, W. P.; Cai, G. *Macromolecules* **2001**, *34*, 4355.
19. Paulasaari, J. K.; Weber, W. P. *Macromolecules* **1999**, *32*, 6574.
20. Eaborn, C.; Bott, R.W. In *Organometallic Compounds of the Group IV Elements. The Bond to Carbon*, Vol. 1; MacDiarmid, Ed.; Marcel Dekker Inc.: New York, 1968, Part 1, p. 389.
21. Fortuniak, W.; Chojnowski, J.; Sauvet, G. *Macromol. Chem. Phys.* **2001**, *202*, 2306.
22. Chojnowski, J.; Cypryk, M.; Fortuniak, W.; Rózga-Wijas, K.; Ścibiorek, M. *Polymer* **2002**, *43*, 1993.
23. Cypryk, M.; Kaźmierski, K.; Fortuniak, W.; Chojnowski, J. *Macromolecules* **2000**, *33*, 1536.
24. Hempenius, M.A.; Lammerting, M.G.H.; Vancso, G.L. *Macromolecules* **1997**, *30*, 2306.
25. Rózga-Wijas, K.; Chojnowski, J.; Boileau, S. *J. Polym. Sci. Polym. Chem.* **1997**, *35*, 879.
26. Chojnowski, J.; Rózga, K. *J. Inorg. Organomet. Polym.* **1992**, *2*, 297.
27. (a) Sauvet, G.; Dupond, S.; Kaźmierski, K.; Chojnowski, J. *J. Appl. Polym. Sci.* **2000**, *75*, 1005. (b) Sauvet, G.; Fortuniak, W.; Kaźmierski, K.; Chojnowski, J. in preparation.
28. Paulasaari, J. K.; Weber, W. P. *Macromolecules* **1999**, *32*, 5217.
29. Wilczek, L.; Kennedy, J. P. *Polymer J.* **1987**, *19*, 531.
30. Michalska, Z.; Rogalski, Ł. in preparation.
31. Ziemelis, M. J.; Saam, J. C.; *Macromolecules* **1989**, *22*, 2111.
32. Kaźmierski, K.; Cypryk, M.; Chojnowski, J. *Macromol. Symp.* **1998**, *132*, 405.
33. Chojnowski, J.; Kaźmierski, K.; Cypryk, M.; Fortuniak, W. In *Silicones and Silicone-Modified Materials*; Clarson, S. J.; Fitzgerald, J. J.; Owen, M. J.; Smith. S. D., Eds.; American Chemical Society: Washington, DC, 2000; p. 20.
34. Nicole, P.; Masure, M.; Sigwalt, P. *Macromol. Chem. Phys.* **1994**, *195*, 2327.
35. Chojnowski, J.; Kurjata, J. *Macromolecules* **1994**, *24*, 2302.
36. Sigwalt, P.; Gobin, C.; Nicole, P.; Moreau, M.; Masure, M. *Makromol. Chem. Macromol. Symp.* **1991**, *42/43*, 229.
37. Kaźmierski, K. to be published.

Chapter 3

Novel Methods of Catalyzing Polysiloxane Syntheses

C. J. Embery[1], J. G. Matisons[1,2], and S. R. Clarke[1,2]

[1]Polymer Science Group, Ian Wark Research Institute, University
of South Australia, Mawson Lakes, SA 5095, Australia
[2]Current address: School of Chemistry, Physic and Earth Sciences,
The Flinders University of South Australia, G.P. Office 2001,
Adelaide, South Australia 5001, Australia

Abstract

Lewis Acid catalysis has also been presented, including a review of acid and
base catalyzed equilibration polymerization of siloxane monomers has been
presented, including the current Lewis acid perspective on equilibration
polymerization by phosphoronitrile chloride catalysts. An overview of our
research demonstrates that phosphoronitrile chloride equilibration catalysts
occurs by an acid catalytic role, in preference to the commonly accepted Lewis
acid pathway.

Introduction

Siloxane polymers, which are commonly referred to as 'silicones', are
commercially prepared by an equilibration polymerization reaction [1-4] ; where
hydrolyzate monomers are catalyzed with either strong bases (such as ammonia
or alkali metal hydroxides) or strong acids (such as mineral acids or Lewis
acids). Equilibration polymerization offers control over the resulting
conformation and molecular weight of the polymer. Disiloxanes, such as
hexamethyldisiloxane, are usually used as 'end blockers' in this equilibration
polymerization, resulting in inert trimethylsilyl end groups for each siloxane
polymer chain. When 1,3-difunctionaldisiloxanes are used as end blockers, then
α,ω end-chain functionality can be incorporated into the polymer chain.

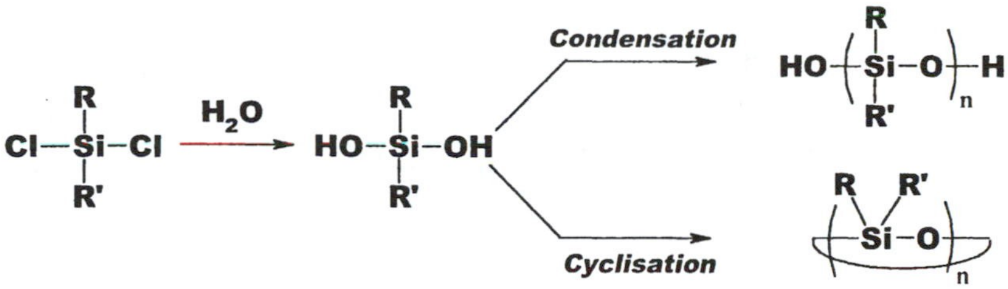

Figure 1. Reaction for the acid catalyzed condensation of organodichlorosilanes to prepare hydrolyzate monomer [1,2]

Commercial hydrolyzate monomer is a mixture of low molecular weight silanol end terminated linear siloxane oligomers (siloxanediols) and cyclic siloxane species, which is predominantly D_4 (with some D_3 and D_5) ; the hydrolyzate being obtained by pouring the respective organohalosilane into ice water [1,2]. (see figure 1).

Chain extension of hydrolyzate monomer in the equilibration polymerization reaction has often been attributed to a relatively rapid catalyzed ring-opening polymerization (ROP) reaction, which is then followed by a slower re-equilibration reaction. It is for this reason that the catalyzed ring-opening polymerization reaction of cyclic siloxanes, and in particular, octamethylcyclotetrasiloxane (D_4) has been extensively studied.

Base (Anionic) Catalyzed Ring-Opening Polymerization

Strong bases, such as alkali metal hydroxides, alkali metal silanolates and amines will catalyze the ring-opening polymerization of cyclic siloxanes, such as D_4. Commercially, this is the most commonly used reaction to prepare poly(dimethylsiloxane) (PDMS), which is without a doubt, the major poly(siloxane) in global production.

For the purposes of this article, a very simplified base catalyzed mechanism, [1-6] involving anionic attack on the siloxane bond to open the cyclic ring has been shown (see figure 2). However, Chojnowski and others [5,6] have studied the mechanism for this reaction, and have shown it to be far more complicated, involving the complexation of silanolate end-groups to form clusters, which influence the overall reaction.

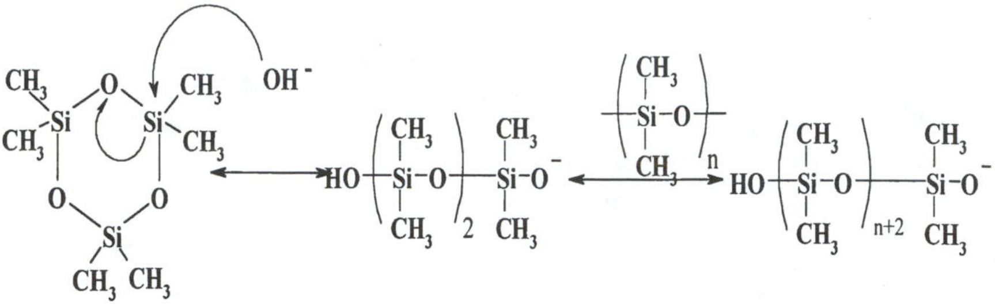

Figure 2. Proposed Mechanism of Base (Anionic) catalyzed ring-opening polymerization of Siloxane Polymerizations [1,2, 5,6]

The reaction undergoes a rapid, anionic initiated ring-opening reaction, which gives chain extension, followed by a slower, back-biting, re-equilibration reaction, resulting in increased polydispersity.

From the literature, [2-5,7-9], one particular group of basic catalysts has been particularly well documented. From the patent literature [10-15], such catalysts include oligomers of neutral phosphoryl phosphazenes of the general formula, such as $Cl[P(Cl)_2N]_nP(O)Cl_2$ and $HO[P(Cl_2)N]_nP(O)Cl_2$, (where 'n' is an integer). Hager and Weis [16] have recently proved such catalysts to be extremely effective in promoting the polycondensation reaction. Research has also been carried out using oligophosphazenium salts of the general formula $[Cl_3P(NPCl_2)_nCl]^+X^-$, where X can be a chlorine atom or a complex ion such as PCl_6^-, $SbCl_6^-$ or $AlCl_4^-$; examples of such compromising hexachloro-1-λ-diphosphaza-1-enium hexachloroantimonate salt $[Cl_3PNPCl_3]^+[SbCl_6]^-$ and p-trichloro-N-dichlorophosphoryl phosphazene $(Cl_3PNP(O)Cl_2)$ [7-8].

It has been shown [17-19] that the activity of this ring-opening of cyclic siloxane monomers increases strongly down the series $Cs^+ > Rb^+ > K^+ > Na^+ > Li^+$, which has been attributed to ion-pairing interactions becoming weaker as the cation size increases. As it stands, current commercial polymerizations of cyclic siloxanes by alkali metal hydroxides are disadvantageous from the viewpoint that the polymerization process is of a low polymerization rate at low reagent concentrations or temperature, and that the polymers thus obtained have poor thermal stability. This has been attributed in the literature [5,7-8] to association of the active catalyst (which is the alkali metal silanolate) with itself ; this 'self association' needing to be broken before ring-opening polymerization can occur [17-19]. Such bases are similar in structure, but undergo a different mechanism to the phosphoronitrile halides covered later.

In a recent development [17], phosphazene bases of the form $((NMe_2)_3PN)_3PNt\text{-}Bu$ were found to be strongly basic, with PK_b values of up to 10^{18} times stronger than most other commercially available bases, such as diazabicycloundecene. It has also been proposed that with the new phosphazene materials, the catalytic activity being much greater because of the larger, and hence softer cation exhibiting little or no self-association.

Research conducted with such bases has shown that upon contact with even trace amounts of water, the catalyst is activated and the highly active species $[((NMe_2)_3PN)_3PNt-Bu]^+[OH]^-$ is formed ; such water being contained within all silicone hydrolyzate feed stocks. It has been shown that such phosphazene and phosphoronitrile chloride catalysts can be used at sufficiently low concentrations, over a broad temperature window and yet, yield high molecular weight polymers – the molecular weight of which may be controlled by *end-capping* or *end-blocking* with monofunctional halosilanes.

Acid (Cationic) Catalyzed Ring-Opening Polymerization

Strong acids [20-21], such as mineral acids and acid modified clays [22] have been used to catalyze the ring-opening polymerization of cyclic siloxanes. Commercially, this process is used to a lesser extent than anionic polymerization, but is important when hydrido cyclic siloxanes need to be polymerized by a ring-opening reaction. This results in hydrido functionality along the backbone.

Base catalyzed polymerization of hydrido cyclic siloxanes results in unwanted hydrolysis of the silicon-hydrogen bond, causing silanol formation and ultimately, cross-linking across these silanol groups. The acid catalyzed ring-opening procedure does not result in this ; however, the reaction is much less understood because it also involves not only a polymerization step and re-equilibration (the reaction involving a polymerization reaction as shown in figure 3), but also, a slower, condensation step (involving the polymer chain ends) of the terminal silanol groups.

Chojnowski and others [1,2,5,6] have also studied this reaction in some detail in recent years.

However, after further research, it was proposed [2] that the reaction also includes two very different and distinct pathways ; those of equilibration (figure 4) and condensation (figure 5). [1,2, 5,6]

Lewis Acid Catalyzed Ring-Opening Polymerization

Lewis acids [23-35], particularly transition metal complexes, organometallics and metal halides have also been used to catalyze the ring-opening polymerization of cyclic siloxanes. The actual mechanism of catalysis was studied and proposed [2] as far back as the late 1960's. Lewis acid catalysis differs from other, conventional methods of siloxane polymerization in that no silanol species are generated – instead, donation of a lone pair from the oxygen atom to the metal occurs ; in the mechanism below, iron being an electron deficient atom, and chloride providing the leaving group.

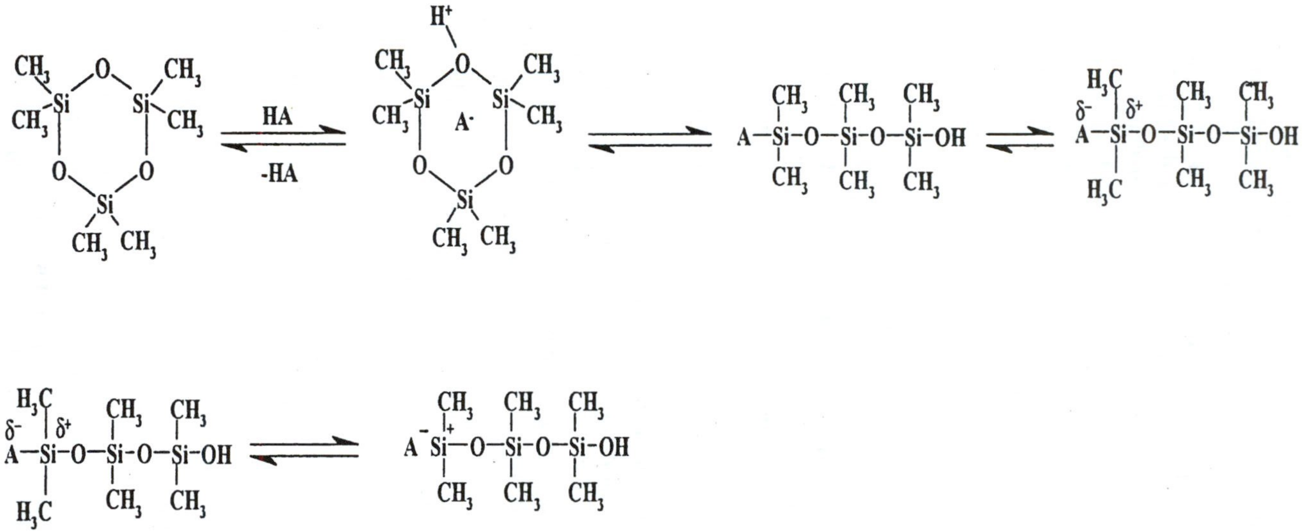

Figure 3. *Proposed Mechanism of Acid (Cationic) catalyzed ring-opening polymerization of Siloxane Polymerizations* [1,2, 5,6]

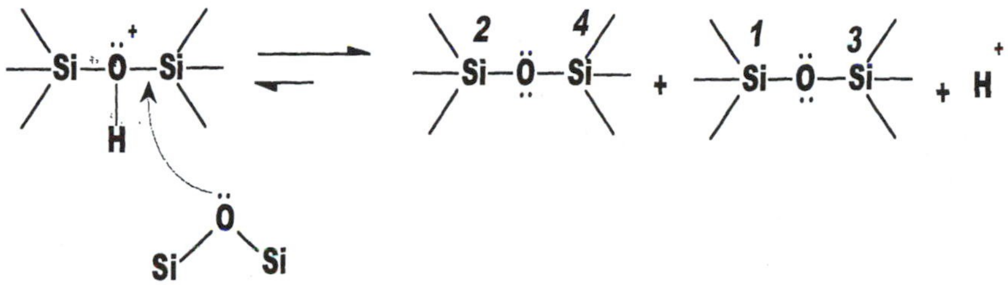

Figure 4. Equilibration Mechanism Pathway of
Acid / Base Catalyzed Reaction

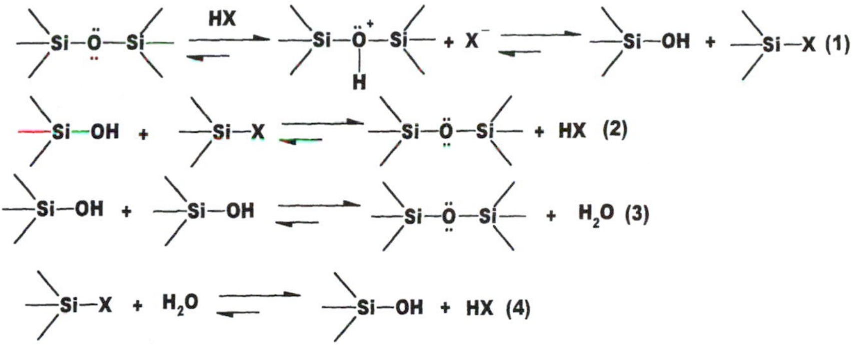

Figure 5. Condensation Mechanism Pathway of
Acid / Base Catalyzed Reaction

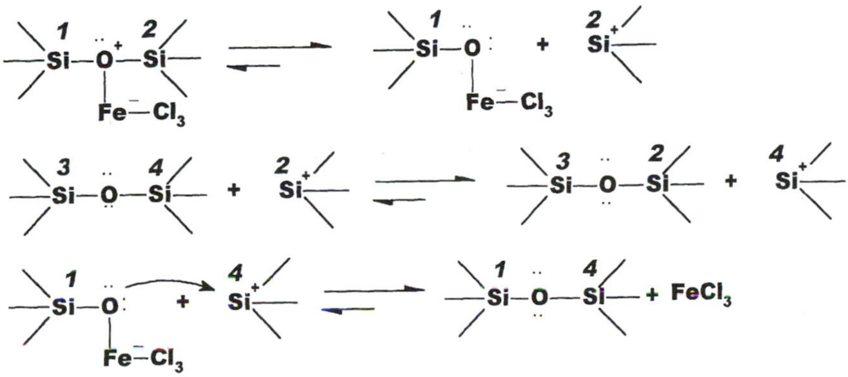

Figure 6. Proposed Mechanism of Lewis Acid catalyzed
ring-opening polymerization of Siloxane Polymerizations [2].

Purpose of Our Research - to Investigate whether Lewis Acid Catalysis or Protonic Catalysis by Phosphoronitrile Halides?

This use of the trimeric form of phosphoronitrile chloride in the ring-opening polymerization of cyclic siloxanes by Wacker-Chemie [10-11] has resulted in two patents being lodged, with respect to claims that the catalyst offers a number of advantages over conventional acid or base catalysis. These patents not only claim that the polycondensation product does *not* contain cyclic oligosiloxanes (due to the catalytic redistribution of linear polysiloxanes ; i.e. formation of a significant amount of cyclics), but also take account that the polymer had improved wettability characteristics and more uniform viscosities.

The aim of our research [3,4,9] was to investigate the nature of the industrial polymerization of commercially available silicone hydrolyzate by phosphoronitrile chloride type catalysts. It has been proposed that the catalysis of polymerisation by phosphoronitrile chloride oligomers occurs by a Lewis acid mechanism (figure 7), that is analogous to the mechanism detailed previously in figure 6 ; with phosphorous being an electron deficient atom, and chloride providing the leaving group.

The overall intention of our studies was to investigate the validity of the proposed Lewis acid catalysis mechanism.

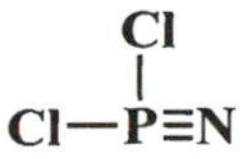

| Monomer | Linear Trimer | Cyclic Trimer |

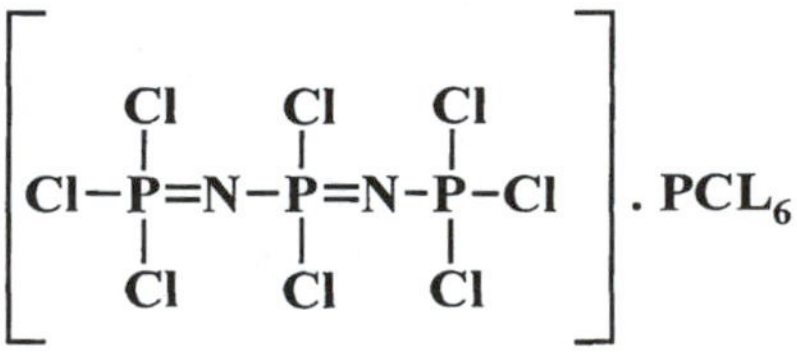

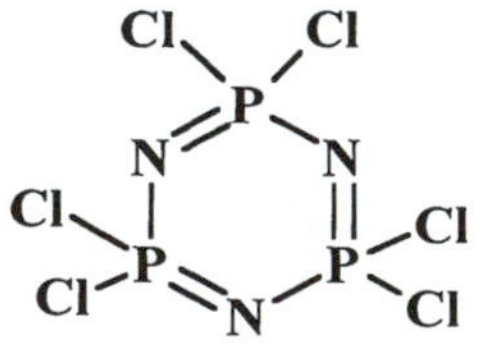

Figure 7. Phosphoronitrilic Chloride catalyst forms.

The polymeric form selected for this research was the linear trimer $Cl_3PNPCl_2NPCl_3.PCl_6$, because it has the highest catalytic activity of all forms, even when compared with the monomer. Presumably, the ionic nature of the trimer promotes the activity of this Lewis acid catalyst [15].

Synthesis of Phosphoronitrile Chloride Type (Lewis Acid) Catalysts.

The synthesis (and subsequent use) of Phosphoronitrile chloride type catalysts has been well documented throughout the literature in various patents and journals [36-44]. Detailed below is a typical synthetic procedure.

The synthetic procedure from the patent literature [10-11] was used to prepare the phosphoronitrile chloride [3-4, 9] ; Polymerizations of commercially available hydrolyzate were carried out, samples were removed from the reaction mix and quenched with zinc oxide at pre-set intervals, and the resultant polymers characterized by Gel Permeation Chromatography (GPC). Gel permeation chromatography was used to determine molecular weights and molecular weight distributions, $\overline{M}_w / \overline{M}_n$, of polymer samples with respect to polystyrene standards, as well as the relative masses of cyclic oligomeric and linear polymeric species.

Polymerization via Phosphoronitrile Chloride Catalysts.

Hydrolyzate was weighed into a steel container, then heated to 75°C, with mechanical stirring. Phosphoronitrile chloride catalyst was added to the reaction vessel (see Table i) as defined in Table i.

Zinc oxide (2:1 ratio with respect to the phosphoronitrile chloride catalyst) was used to neutralize the catalyst and subsequently terminates the reaction ; samples being taken at required time intervals (see figure 8).

From figure 8, *In-situ* molecular weight measurements made during the reaction [9] shows that from the start, the amount of D_4 falls because it is being incorporated into the polymer, due to the onset of the equilibration reaction. This is also reflected in figure 9, at **Stage 2** (discussed later). As the reaction

**Table i. Masses of Starting Materials Employed
in Polymerization Reactions.**

Polymerization	One	Two	Three	Four
Catalyst (ppm)	432	800	1503	3000
Hydrolyzate mass (grams)	1946	1000	1252.5	2000
Catalyst mass (grams)	0.841	0.770	1.883	6.00
Zinc Oxide mass (grams)	1.219	1.553	2.683	8.430
% D_4	44.1	40.5	31.6	40.7

proceeds, the amount of D_4 in the system decreases exponentially as it is reacting, with a corresponding increase in the percentage of linear polymer in the system.

From the molecular weight analysis shown in Figure 9, it is seen that **Stage 1** is representative of the initial condensation reaction, which consumed all of the linear oligomers at the start of the reaction, leading to a high molecular weight polymer.

Stage 2 shows a large number of D_4 molecules being incorporated into the polymer, due to the onset of the equilibration reaction, but this number (as well as the molecular weight of the polymer chains) decreased upon reaching **Stage 3**.

The equilibration process resulted in the most stable system at **Stage 4**, however a minimum molecular weight for the polymer was observed. The subsequent increase in molecular weight is due to the continued, but slow inclusion of D_4 into the polymer ; current thinking [9] assumes that a final equilibrium can be seen, accounting for the plateau at **Stage 5**.

Noll [2] proposed catalysis of the ring-opening polymerization reaction of cyclic siloxanes via Lewis acids such as phosphoronitrile halides was not possible. Conversely, the patent literature [10-15], independently suggests that this is necessarily be the case.

It was found that during the polymerization, a definite decrease in the D_4 content of the system occurred (as in figure 8), which points towards it being ring-opened, and included in the final product (see also figure 8). This is in agreement with the patent literature [10-15].

We believe that the polymerization of low molecular weight siloxanes by phosphoronitrile chloride acts via two different pathways – equilibration and condensation. During the course of the reaction, Hydrogen Chloride gas was evolved, and we are of the opinion that the catalyst is hydrolyzed by water and that H^+ is actually catalyzing the reaction.

Our Proposed Mechanism

The alternate, and less predominant pathway of catalysis is that of direct Lewis acid catalysis, which is a reaction sequence analogous to that mentioned previously from figure 6. It should be clearly noted at this stage, that the reaction pathways provided show two very distinct ends of a spectrum of proposed mechanisms. There is some experimental evidence to suggest that the hydrolyzed form of the catalyst (figure 10) was involved in the reaction, but this role is currently open to speculation.

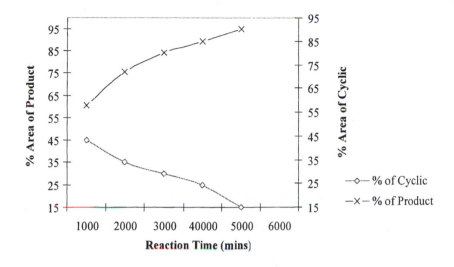

Figure 8. Schematic Representation of changes in percentages
of cyclic siloxane / polymeric siloxane product with time

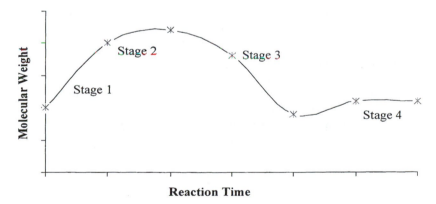

Figure 9. Diagram of Molecular Weight versus Reaction Time from
GPC analysis of quenched polymer samples.

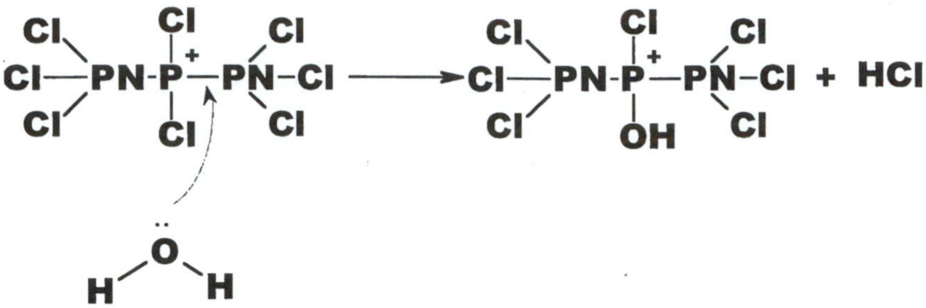

Figure 10. Schematic Diagram of Hydrolysis of a Phosphoronitrile Chloride Catalyst Molecule [3,4,9]

Results and Discussion

It seems that all of the species concerned in the catalysis reactions were involved in competing reactions, and one pathway was dominant in catalyzing the entire reaction ; alternatively, the catalyst may involve catalyzing just one of the reactions (equilibration or condensation). A third possibility is that one of the species involved may catalyze a reverse reaction, thus reducing the overall reaction rate.

In this situation, hydrolysis of a catalyst molecule (as detailed above in figure 8) leads to the release of a proton, and thus protonic acid catalysis is much more likely to be the dominant pathway. This may be more clearly understood when we consider that the rate of reaction with respect to protonic acid catalysis is several times higher than that with Lewis acid catalysis [10].

Recent literature [15] suggests that a 'back-biting' reaction may be occurring, thus cyclic molecules are produced from the polymer during the course of the equilibration, but further P^{31} NMR spectroscopy could provide further information on this. This is actually in good agreement with work done by Chojnowski et al., [5,7-8] who showed that in the first stage of the polymerization process, linear oligomers were formed as almost the sole product of the reaction (condensation), then consecutively, a redistribution reaction (equilibration) took place.

It was also noted that the formation of cyclic oligomers was considerably delayed with respect to linear species. Thus, Chojnowski's kinetic analysis was based on the logical assumption that the redistribution step leads exclusively to the equilibrium between linear species, and hence neglects the formation of cyclic species.

Current research by Dow Corning [12-14, 17] has steered away from the more common practices of using compounds such as barium hydroxide, acid clays and potassium and ammonium catalysts to drive their reactions ; however, such catalysts have been found to be non-specific with respect to the reaction pathway of condensation. Such phosphoronitrile halide catalysts have been extensively studied, particularly by Dow Corning, [12-14, 17] and it has been acknowledged from such studies that the active species for silanol condensation is species '3' in the reaction scheme proposed below. In this mechanism, the species is being formed by hydrolysis of the precursor complex '1' through to the oxo derivative '2', followed by further hydrolysis to the amidic acid complex '3'.

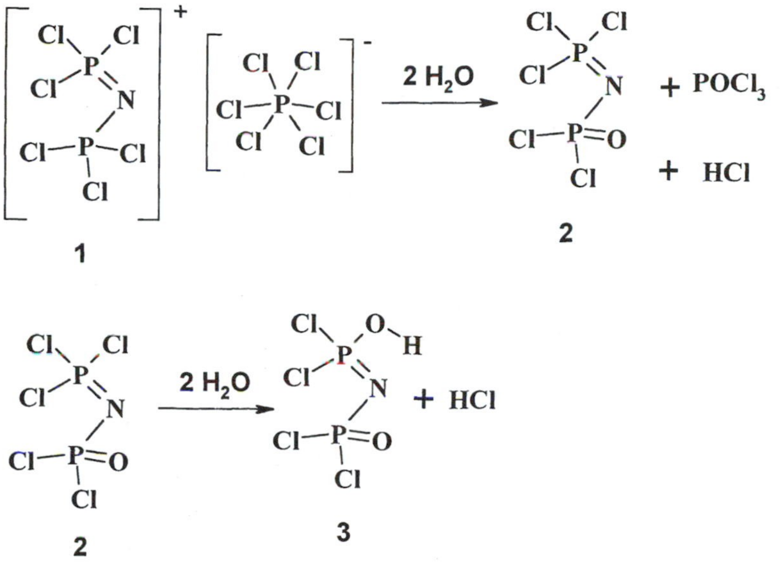

Figure 11. Proposed Dow Corning mechanism of Phosphoronitrile halide catalysis [12-14, 17]

Conclusions

It has been shown that the mode of catalysis achieved by phosphoronitrile chloride is a combination of condensation and equilibration. Therefore, it may be concluded that the active catalyst exists as the hydrolyzed form; and as a result, the actual catalytic species is the hydrogen ion. This may be properly understood when it is taken into account that the rate of reaction with respect to protonic acid catalysis is several times higher than that with Lewis acid catalysis.

The effect of the concentration of this catalyst has also been ascertained, and it appears to have little effect on the condensation process, but a definite effect on the equilibration process. This dependence was not linear, but a definite correlation has been shown to exist.

The difference in results between hydrolyzate batches shows that the system is complicated and there are many factors affecting the reaction that are yet to be discovered, or at least quantitated. Both H^1 and P^{31} NMR spectroscopy may provide valuable information about the active form of phosphoronitrile chloride in this system.

Acknowledgements.

The authors of this paper would like to thank Dow Corning Proprietary Limited, our industrial sponsors, for their much appreciated help and financial support in this investigation.

References

1. Brydson J.A, *Plastics Materials* ; Butterworth-Hall : New York, 1996.
2. Noll W., *The Chemistry and Technology of Silicones* ; Academic Press : 1968.
3. Fielden, M., Embery, C.J., Matisons, J.G., RACI Polymer Division, *24th RACI Australasian Polymer Symposium*, Beechworth, Victoria, Australia.
4. Fielden, M., Embery, C.J., Britcher, L.G., Clarke, S.R., Matisons, J.G., ACS Spring Meeting, San Diego, California. *Polym. Prepr.* (*Am. Chem. Soc., Div. Polym. Chem.*) **2001**, *42* (1), 165.
5. J. Chojnowski, *Siloxane Polymers* ; Prentice-Hall, Englewood Cliffs, New Jersey, 1993 ; p 1.
6. S. Boileau, *Ring-opening Polymerization* ; American Chemical Society : Washington, DC, 1985 ; p 23.
7. Chojnowski, J., Cypryk M., Fortuniak, W., Kazmierski, K., Taylor, R.G., *J. Organomet. Chem.*, **1996**, *526* (2), 351.
8. Chojnowski, J., Fortuniak, W., Habimana J., Taylor, R.G., *J. Organomet. Chem.*, **1997**, *534*, 105.
9. Fielden, M. : Applied Chemistry Project, University of South Australia.
10. Burkhardt J., US Patent No. 4,053,494.
11. Burkhardt J., US Patent No. 4,203,913.
12. Bischoff R., Currie J., Herron W., Taylor, R.G., Dow Corning Ltd., European Patent No. EP 0 860,461 A2.
13. Harkness B., Taylor, R.G., Dow Corning Ltd., European Patent Number EP 0 860,459 A2.
14. Bischoff R., Taylor, R.G., Dow Corning Ltd., European Patent Number EP 0 860,460 A2.
15. Nitzche S., US Patent No. 2830967.

16. Hager R., Weis, J., *Z. Naturforsch.*, **1994**, *49*, 1774.
17. Taylor, R.G., **2001**, 43 (ISPO Presentation 2001 ; Conference on silicon containing polymers at Canterbury, Kent, England).
18. Melikyan N.O., Tergazarova D.A., *Arm. Khim. Zh.*, **1969**, , *22*, 82.
19. Kuznetsova, A.G., Ivanov V.I., Golubtsov S.A., *Zh. Obsch. Khim.*, **1970**, *40*, 706.
20. Adrianov, K.A, Shkol'nik M.I., Kopylov V.M., Baravina N.N., *Vysokomol. Soedin Ser. B.*, **1974**,.*16*, 893.
21. Voronkov, M.G., Sviridova N.G., *Zh. Obshch. Khim.*, **1976**, *46*, 126.
22. Bennett D.R., Matisons J.G., Netting A.K.O., Smart R.St.C., Swincer A.G., *Polym. Int.* **1992**, *27*, 147
23. Mingotaud A.F., Cansell F., Gilbert N., Soum A., **1999**, *Polym. J. (Tokyo)*, *31 (5)*, 406.
24. Yoshioka H., German Patent No. DE 3932231.
25. Ohba T., European Patent No. EP 0393954.
26. Andrianov K., Izmailov B., *Zh. Obshch. Khim.* (*Journal of General Chemistry*), **1965**, *5*, *35 (2)*, 333.
27. Vdovin V., Nametkin N., Finkelshtein E., Oppengeim V., *Izv. Akad. Nauk SSSR, Ser. Khim.* (*Bulletin of The Academy of Sciences of The USSR, Chemical Series*), **1964**, *3*, 458.
28. Omietanski G., British Patent No. GB 1066574.
29. Scott D.W, *J. Am. Chem. Soc.,* **1946**, *68*, 2294.
30. Klebansky A.L., Fikhtengolts V.S., Karlin A.V., **1957**, *J. Gen. Chem. USSR*, *27*, 3321.
31. Prut E.V., Trofimova G.M., Yenikolopyan N.S., **1964**, *Vysokomolekul. Soedin. 6*, 2102
32. Prut E.V., Trofimova G.M., Yenikolopyan N.S., **1964**, *Polymer Sci. USSR*, *6*, 2331.
33. Borisov S.N., Sviridova N.G., *J. Organometal. Chem.,* **1968**, *11*, 27.
34. Tabuse A., Minowra Y., *Kogyo Kagaku Zasshi*, **1969**, *72*, 2133.
35. Pace, S.C., Riess J.G., *J. Organometal. Chem.*, **1976**, *121*, 307
36. Van Dyke M.E., Clarson S.J., *Polym. Prepr.* (Am. Chem. Soc. Div. Polym. Chem.), **1996**, *37* (2)*, 668.
37. Van Dyke M.E., Clarson S.J., *J. Inorg. and Organomet. Polymers*, **1996**, *8* (2), 111.
38. Schwesinger R. et al., *Liebigs Ann.*, **1996**, *7*, 1055.
39. Schwesinger R. et al., *Chem. Ber.*, **1994**, *127*, 2435.
40. Schwesinger R. et al., *Angew. Chem. Int. Ed. Engl.*, **1993**, *32*, 1361
41. Schwesinger R., *Tech. Lab.*,**1990** *38*,1214.
42. Schwesinger R. et al., *Angew. Chem.*, **1987**, *99* (11), 1210.
43. Esswein B., Molenberg A., Möller M., *Macromol. Symp.* **1996**, *107* (International Symposium on Ionic Polymerization)
44. Molenberg A., Möller M., *Macromol. Rapid Commun.* **1995**, *16*, 449.

Chapter 4

Anionic Ring-Opening Polymerization of *cis*-2,4,6-Trimethyl-2′,4′,6′-triphenyl Cyclotrisiloxane (*cis*-P₃) in Cyclohexane

H. W. Ahn and S. J. Clarson*

Department of Materials Science and Engineering, The University of Cincinnati, Cincinnati, OH 45221–0012

The anionic ring-opening polymerization of *cis*-2,4,6-trimethyl-2',4',6'-triphenyl cyclotrisiloxane (*cis*-P₃) in cyclohexane has been investigated. By employing a non-polar solvent with a small amount of tetrahydrofuran (THF) as a promoter, both an intermolecular reaction (propagation) and an intramolecular reaction (back-biting) occur. It is proposed that the reduced reactivity of the nucleophile at the chain end and the solvation of any phenyl-phenyl interactions may facilitate the intramolecular reaction. A mechanism for the intramolecular back-biting reaction in this system is described that is consistent with the stereochemistry of the linear PMPS structure formed.

INTRODUCTION

Linear siloxane polymers are typically prepared by either the polycondensation of difunctional silanols or by the ring-opening polymerization of cyclic siloxane monomers [1]. In ring-opening polymerization, the siloxane polymer can be synthesized by cationic or anionic catalysis or by radiation initiation [2-9]. Living anionic ring-opening polymerization is often the most preferred polymerization method since it is possible to obtain a narrow molecular weight distribution with good control of molar mass.

In anionic ring-opening polymerization, the polymerization center is the silanolate ion at the growing chain end(s). The counterion in anionic polymerization is also very important in controlling the rate of polymerization, and following the order of $Li^+ < Na^+ < K^+ < Rb^+ < Cs^+$ of rate has been shown [5, 10, 11]. The larger size of the counterion, the longer the distance from the nucleophile, resulting in weaker ion aggregation and ion pair interactions. These ion aggregates become looser when a polar solvent is used and thus, the rate of polymerization increases with solvent polarity [12].

Poly(methylphenylsiloxane) (PMPS) is an asymmetrically disubstituted siloxane of the structure $[MePhSiO]_x$ and has the possibility of stereochemical variability [13]. L. S. Bresler *et. al.* discussed the stereo-isomerism of PMPS for the first time [14]. They polymerized PMPS with *cis*-P_3 and *trans*-P_3 with sodium hydroxide at 100°C and found that there was a difference in the resulting peaks in ^{1}H-NMR spectroscopy. Curtis *et al.* also reported that they prepared stereoregular PMPS [15]. Mompher *et al.* first showed the existence of crystallinity in stereoregular PMPS by wide angle x-ray scattering (WAXS) and the PMPS was studied by nuclear magnetic resonance spectroscopy (NMR) and differential scanning calorimetry (DSC) [16]. Recently, Oishi *et.al.* prepared syndiotacticity rich PMPS by Rh-catalyzed stereoselective cross-dehydrocoupling polymerization of optically active 1,3-dimethyl-1,3-diphenyldisiloxane derivatives. However, the resulting polymer was very low in syndiotactic content [17].

For the poly[methyl(3,3,3-trifluoropropyl)siloxane] (PMTFPS) system, Saam and Kuo pioneered the synthesis of stereoregular PMTFPS using *cis*-1,3,5-trimethyl-1,3,5-tris(3',3',3',-trifluoropropyl)cyclotrisiloxane and the products were analyzed by ^{19}F-NMR [18]. It was found that stereoregular PMTFPS elatomers were able to undergo strain-induced crystallization and this was shown by WAXS [19]. A detailed review of the synthesis of polysiloxanes via ring-opening polymerization and possible stereochemical control in such systems has been published by Saam [20].

Here we investigate the polymerization of *cis*-P_3 in cyclohexane using *sec*-BuLi as the initiator. Both an intermolecular and an intramolecular reaction were observed under these conditions.

EXPERIMENTAL

Chemical Reagents

Pure *cis*-P$_3$ was prepared by procedures that we will report elsewhere. It was kept in a vacuum oven at 40°C for more than 5 hours before each reaction. The solvent, HPLC grade cyclohexane, was purchased from Fisher and distilled over phosphorous pentoxide at 85°C prior to use. The promoter, HPLC grade of tetrahydrofuran (THF), was distilled with sodium metal in paraffin (Aldrich) using benzophenone as the indicator. The *sec*-butyllithium (*sec*-BuLi), 1.3M solution in cyclohexane and the terminating agent, chlorotrimethylsilane,were purchased from Aldrich and were used as received. Methanol (non-solvent) was purchased from Fischer and used without further purification.

Synthetic Procedure

All glassware, needles and syringes were dried in an oven at 130°C for 24 hours and the glassware was flame-dried before each experiment. *Cis*-P$_3$ was added into the reactor and a magnetic stirrer bar placed inside. The reactor was purged with dry N$_2$ gas and then sealed. Distilled solvent was transferred directly into the reactor to avoid possible air contact. The reactor was then immersed into an oil bath or cold bath set at the desired temperature. A calculated amount of initiator, *sec*-BuLi, was injected at one time and the promoter, THF, was added after 10 minutes. Each time, 100 µl of *sec*-BuLi was used to produce PMPS with a molecular weight of ~8,000 g/mol. The polymerization was left for a desired time period and chlorotrimethylsilane was then added into the reactor to terminate the polymerization. The solvent and the terminating agent were removed using a rotary evaporator and a vacuum oven.

RESULTS & DISCUSSION

For *atactic* PMPS, there are three different peaks of the α-methyl proton region in in the ^{1}H-NMR spectrum [16,21,22]. By analogy of the α,α'-disubstituted vinylic polymer in the upfield to downfield scale, these three peaks were assigned to *racemic-racemic*, *meso-racemic*, and *meso-meso* triads, respectively. However, Baratova *et al.* reversed the assignment of the *meso-meso* and *racemic-racemic* triad peaks [23, 24].

Figure 1 shows the ^{1}H-NMR spectra for the α-methyl region with different reaction times in cyclohexane at 55°C. It is clearly seen that there are more than three peaks in ^{1}H-NMR spectra, indicating that an intramolecular reaction (backbiting) has occured. In Table 1, all the peaks seen in Figure 1 are summarized. A schematic view of the different isomers of P_3 and P_4 is presented in Figure 2 where the superscripts a, b, and c represent hydrogens with different environments.

As summarized in Table 1, the areas of the *meso-meso* and *meso-racemic* triads give a ratio of (a) and (b) in Figure 4 of 30:70, resulting in 76.6% isotacticity for the PMPS. The mechanism of the reaction is presented in Figure 3. In a non-polar solvent, the dissociation of the nucleophile and counter ion is difficult and this results in a much less reactive nucleophile. The less reactive nucleophile may prefer to attack its own growing chain. It is also proposed that non-polar cyclohexane may disrupt any phenyl-phenyl interactions.

The *trans*-P_4[IV] isomer is hardly seen in comparison with the other isomers of P_4. This is in accordance with the observation that the racemic-racemic triad is hardly found in the polymer. From the peaks at 0.035 and 0.085 ppm, which are assigned to the *meso-meso* and *meso-racemic* triads in the polymer, the growing polymer chains would be in the form (a) or (b), as illustrated in Figure 4. Thus when the nucleophile prefers to attack either the third or the fourth silicon atom in the backbone, there are fewer chances of forming the *trans*-P_4[IV] isomer.

Further experimental investigations and theoretical calculations on stereoregular PMPS are in progress.

REFERENCES

1. Noll, W. *'Chemistry and Technology of Silicones'*, Academic Press: New York, 1968.
2. Chojnowski, J. In: *Siloxane Polymers,* Clarson, S. J.; Semlyen. J. A. Eds; Prentice Hall, Englewood Cliffs, NJ, 1993.
3. Hyde, J. F. *U. S. Pat.*, **2490357** (1949).
4. Warrick, E.L. *U. S. Pat.,* **2634252** (1949).
5. Grubb, W. T.; Osthoff, R. C. *J. Amer. Chem. Soc.* **1955**, *77*, 1405.
6. Chojnowski, J.; Mazurek, M. *J. Makromol. Chem.* **1975**, *176*, 2999.
7. McGregor, R. R.; Warrick, E. *U. S. Pat.*, **2437204** (1948).
8. Lawton, E. J.; Grubb, W. T.; Balwit, J. S. *J. Polym. Sci.* **1956**, *19*, 455.
9. Lebrun, J. J.; Sauvet, G.; Sigwalt, P. *Makromol. Chem., Rapid Comm.* **1982**, *3*, 757.
10. Hurd, D. T.; Osthoff, R. C.; Corrin, M. C. *J. Amer. Chem. Soc.* **1954**, *76*, 249.

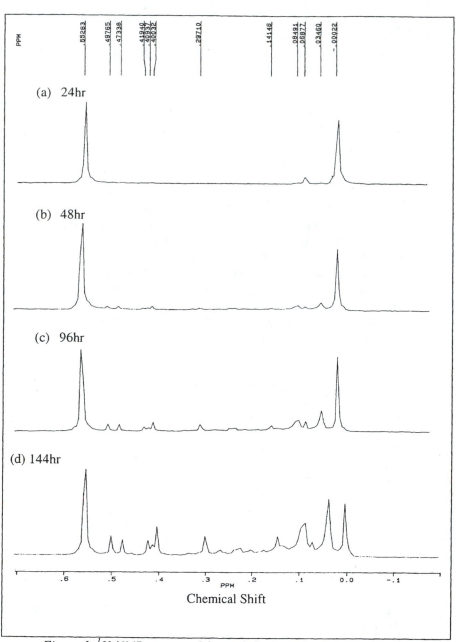

Figure 1. ¹H-NMR spectra of the methyl region for the ring-opening polymerization of cis-P₃ in cyclohexane at 55 °C as a function of reaction time

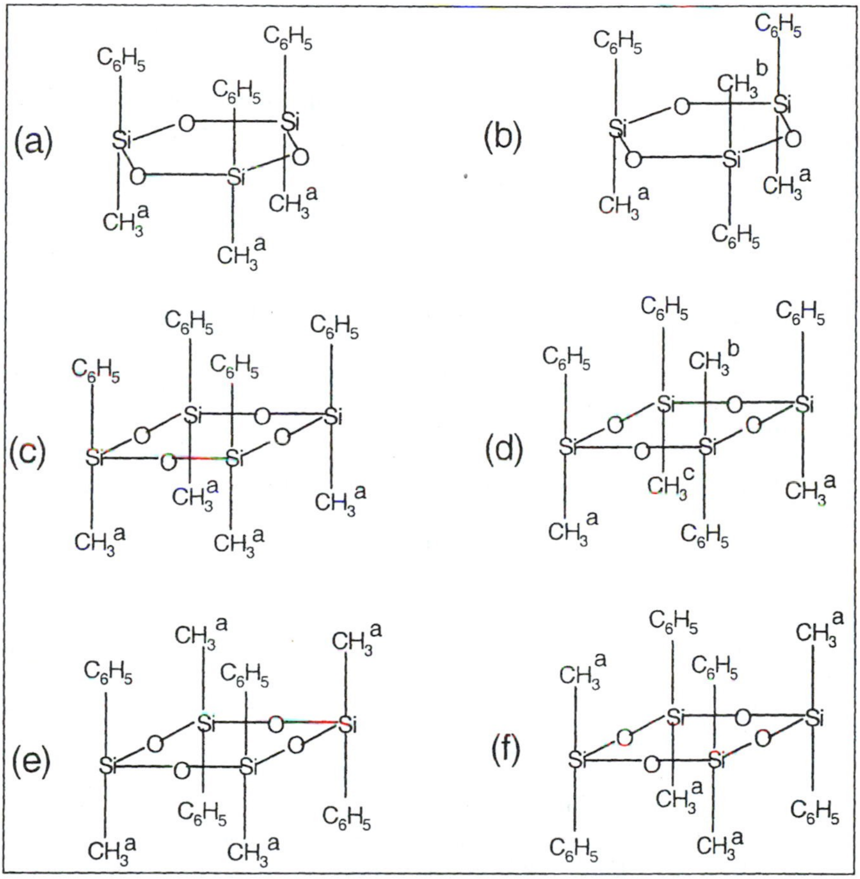

(a) 2,4,6-Trimethyl-2',4',6'-triphenylcyclotrisiloxane (*cis*-P$_3$)
(b) 2,4,6'-Trimethyl-2',4',6-triphenylcyclotrisiloxane (*trans*-P$_3$)
(c) 2,4,6,8-Tetramethyl-2',4',6',8'-tetraphenylcyclotetrasiloxane (*cis*-P$_4$[I])
(d) 2,4,6,8'-Tetramethyl-2',4',6',8-tetraphenylcyclotetrasiloxane (*trans*-P$_4$[II])
(e) 2,4,6',8'-Tetramethyl-2',4',6,8-tetraphenylcyclotetrasiloxane (*trans*-P$_4$[III])
(f) 2,4',6,8'-Tetramethyl-2',4,6',8-tetraphenylcyclotetrasiloxane (*trans*-P$_4$[IV])

Figure 2. The two diastereomers of P$_3$ and four diastereomers of P$_4$

Table 1. Proton NMR peak assignments

Chemical Shift (ppm)	Isomer Type	Hydrogen Type [e]	Peak Area [ζ]
0.035	Meso-meso triad[a]		5.36
0.085	Meso-racemic triad[a]		4.72
0.142	Trans-P_4[II]	b (3)	0.67 [0.22]
0.297	Trans-P_4[III]	a (12)	1.36 [0.11]
0.401	Trans-P_4[II]	a (6)	1.33^β [0.22]
0.410	Trans-P_3	b (3)	0.36^γ [0.12]
0.420	Trans-P_4[II]	c (3)	0.67^β [0.22]
0.474	Trans-P_3	a (6)	0.72 [0.12]
0.498	Cis-P_4[I]	a (12)	0.96 [0.08]
0.553	Cis-P_3	a (9)	5.95^δ [0.66]

[a] Peaks from the polymer

[β] Calculated after subtracting the area of the hydrogen b in trans-P_3

[γ] The hydrogen b in trans-P_3 was calculated from the fact that the ratio of the hydrogen a to b in trans-P_3 is 2 to 1.

[δ] This area also includes unreacted cis-P_3

[e] The number in the bracket corresponds to the number of hydrogens in the methyl groups in Fig 2.

[ζ] The peak area was obtained from Figure 1(d). with the number in the bracket corresponding to the area of one hydrogen.

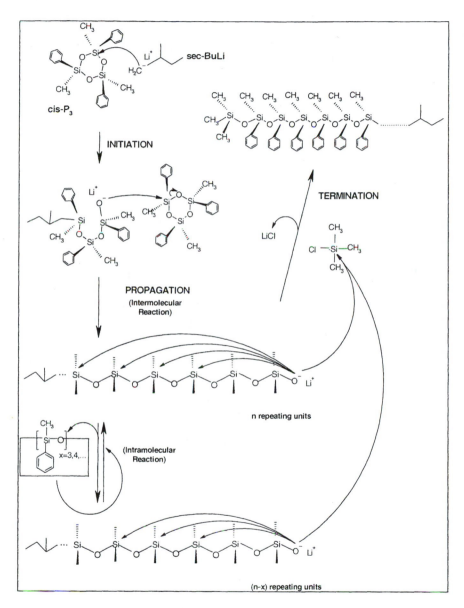

Figure 3. The mechanism of the intramolecular and intermolecular reactions for PMPS in cyclohexane

48

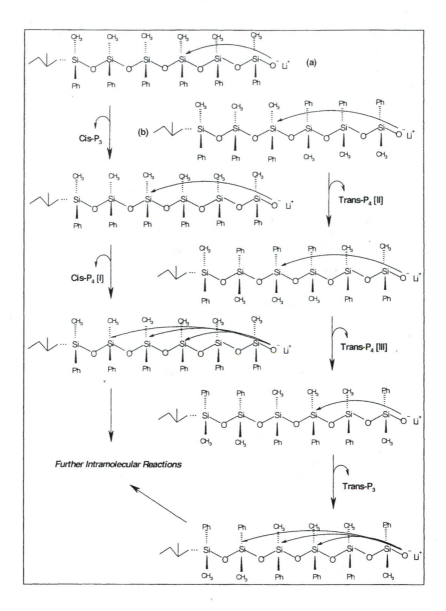

Figure 4. Formation of the cyclic MPS isomers by the intramolecular (back-biting) reaction (Note that we don't see the isomer trans-P$_4$[IV] in this scheme)

11. Boileau, S. In: *Ring Opening Polymerization*, McGrath, J. E. Eds.; American Chemical Society: Washington D. C., 1985, p23.
12. Morton, M.; Bostick, E. E. *J. Polym. Sci.* **1964**, *A2*, 523.
13. Young, C. W.; Servais, P. C.; Currie, C. C.; Hunter, M. J. *J. Amer. Chem. Soc.* **1948**, *70*, 3758.
14. Bresler, L. S.; Mileskevich, V. P.; Yuzhelevskii, Yu.; Timofeeva, N. P. *Zh. Struct. Khim.* **1978**, *19*, 453.
15. Curtis, M. D.; Thanedar, S.; Elshiek, M. *Polym. Prepr. Div. Polym. Chem. Am. Chem. Soc.* **1984**, *25*, 224.
16. Momper, B.; Wagner, Th.; Maschke, U.; Ballauff, M.; Fischer, E. W. *Polym. Comm.* **1990**, *31*, 186.
17. Oishi, M.; Moon, J. Y.; Janvikul, W.; Kawakami, Y. *Polym. Int* . **2001**, *50*, 135.
18. Kuo, C. M.; Saam, J. C. *Polym. Int.* **1994**, *33*, 187.
19. Battjes, K. P.; Kuo, C. M.; Miller, R. L.; Saam, J. C. *Macromolecules* **1995**, *28*, 790.
20. Saam, J. C. *J. Inorg. Organ. Polym.* **1999**, *9(1)*, 3.
21. Llorente, M. A.; de Piérola, I. F.; Saiz, E. *Macromolecules* **1985**, *18*, 2663.
22. Clarson, S. J.; Semlyen, J. A.; Dodgson, K. *Polymer* **1991**, *32*, 2823.
23. Baratova, T. N.; Mileshkevich, V. P.; Guari, V. E. *Pol. Sci. USSR* **1982**, *24*, 27.
24. Baratova, T. N.; Mileshkevich, V. P.; Guari, V. E. *Pol. Sci. USSR* **1983**, *25*, 2889.

Chapter 5

Ru-Catalyzed Hydrosilylation Polymerization

An Overview of RuH$_2$(CO)(PPh$_3$)$_3$-Catalyzed Hydrosilylation Copolymerizations of α,ω-Diketones with α,ω-Dihydrido-oligo-dimethylsiloxanes and Polymerizations of ω-Dimethylsilyloxy Ketones

Joseph M. Mabry, Matthew K. Runyon, Jyri K. Paulasaari, and William P. Weber*

K. B. and D. P. Loker Hydrocarbon Research Institute, Department of Chemistry, University of Southern California, Los Angeles, CA 90089
*Corresponding author: email: wpweber@usc.edu

Activated   dihydridocarbonyltris(triphenylphosphine)ruthen-ium (**Ru**) catalyzes the hydrosilylation copolymerization of both aromatic and aliphatic α,ω-diketones with α,ω-dihy-drido-oligodimethylsiloxanes to yield poly(silyl ether)s. The effect of siloxane chain length on copolymer T$_g$ has been evaluated.   **Ru** also catalyzes the polymerization of both aromatic and aliphatic ω-dimethylsilyloxy ketones to yield poly(silyl ether)s. Chiral centers affect the NMR spectra of these poly(silyl ether)s. The susceptibility of these polymers to hydrolytic degradation, which depends on structure, will be discussed.

Properties of Poly(silyl ether)s

Poly(dimethylsiloxane) (PDMS) is a crystalline polymer, with a melting point (T_m) near -40 °C and a glass transition temperature (T_g) near -125 °C (*1*). Many copolymers, which contain oligodimethylsiloxane (ODMS) units, no longer exhibit detectable T_ms, but still have T_gs close to that of PDMS (*1*). While PDMS is biocompatible (*2*), the strength of the Si-O-Si bond linkage makes it resistant to hydrolysis and thus biodegradation. Poly(silyl ether)s, on the other hand, contain Si-O-C bond linkages, which are susceptible to acid or base catalyzed hydrolysis (*3*). This hydrolytic instability may make them attractive for various applications, such as the controlled release of drugs, or as materials whose degradation will limit their long-term environmental impact (*4*). There is also interest in the possible utility of poly(silyl ether)s as membranes (*5*), sensor materials (*6*), and elastomers (*7*). Poly(silyl ether)s may also be applicable to various space applications because of their stability to high temperatures and ultraviolet radiation (*8*). Some poly(silyl ether)s exhibit excellent flame retardance (*9*).

Previous Synthetic Methods to Prepare Poly(silyl ether)s

A general method to prepare poly(silyl ether)s is the equilibration polymerization of either dialkoxysilanes (*10*), dihalosilanes (*11*), or diamino silanes (*8,12*) with diols. Acid catalyzed ring-opening polymerization of 2-sila-1-oxacyclopentanes has also been reported (*13*). Due to the instability of poly(silyl ether)s to acid or base, methods, which use neutral conditions, are favored. Among these are transition metal catalysis. High molecular weight poly(silyl ether)s have been obtained by rhodium or palladium catalyzed cross-dehydrocoupling of bis(Si-H) compounds with α,ω-diols (*14-17*). Poly(silyl ether)s containing pendant chloromethyl groups have been synthesized by quaternary ammonium chloride catalyzed reaction of bis(oxetane)s or bis(epoxide)s with dichlorosilanes (*18-21*). Palladium catalyzed condensation copolymerization of bis(Si-H) compounds with para-quinones has also been successfully applied (*22*). We have reported the ruthenium catalyzed dehydrogenative silylation condensation copolymerization of ortho-quinones with α,ω-dihydrido-oligodimethylsiloxanes shown in Figure 1 to yield polycyclic aromatic poly(silyl ether)s (*23,24*). We have also reported the ruthenium catalyzed competitive condensation/addition of α-diketones with α,ω-dihydrido-oligodimethylsiloxanes (*25,26*).

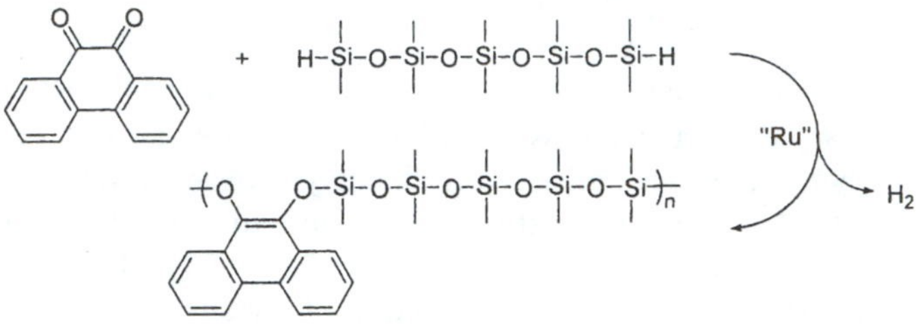

*Figure 1. **Ru** catalyzed condensation copolymerization.*

Ru Catalyzed Hydrosilylation Copolymerization of Aromatic α,ω-Diketones

The catalyst, $RuH_2(CO)(PPh_3)_3$ (**Ru**), prepared from ruthenium trichloride hydrate (*27*), was activated with styrene in toluene at 125 °C for 3 min. The color of the activated catalyst solution is red (*28*). The styrene activates the catalyst by removal of hydrogen and production of ethylbenzene. One triphenylphosphine ligand is also lost. This produces the highly coordinately unsaturated catalyst "$Ru(CO)(PPh_3)_2$" shown in Figure 2 (*29*). Approximately one mole percent of the catalyst was routinely used. Among the advantages of the **Ru** catalyst is that it does not equilibrate the ODMS units. Siloxane equilibration is observed in many cation and anion catalyzed reactions (*30*). This **Ru** catalyzed hydrosilylation copolymerization is unusual in that, while the reacting solution is brown, completion of the reaction is often indicated by the return of the red color associated with the initial activated **Ru** catalyst. In this way, the reaction is comparable to a titration.

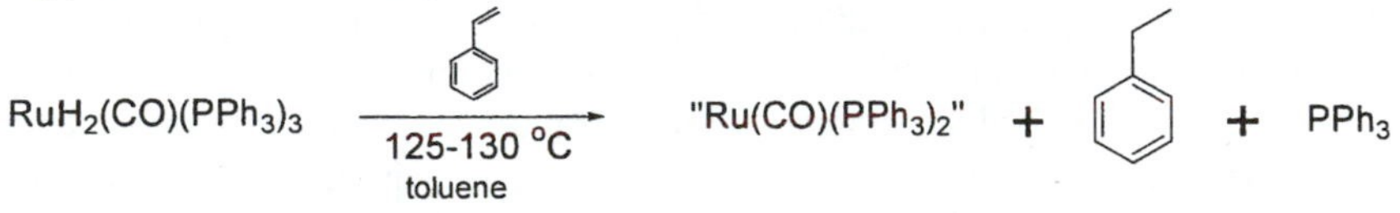

Figure 2. Activation of catalyst.

Aromatic α,ω-diketones, e.g. 1,4-diacetylbenzene and 4,4'-diacetyldi-phenylether, were independently reacted with 1,3-dihydridotetramethyldi-siloxane (**TMDS**), 1,5-dihydridohexamethyltrisiloxane (**HMTS**), 1,7-dihydrido-octamethyltetrasiloxane (**OMTS**), and 1,9-dihydridodecamethylpentasiloxane (**DMPS**). These reactions were carried out in the presence of activated **Ru** to yield poly(silyl ether)s. Copolymerization of 1,4-diacetylbenzene with **DMPS** is shown in Figure 3. **DMPS** was prepared by a triflic acid catalyzed reaction of **TMDS** and hexamethylcyclotrisiloxane (D₃) (*30*).

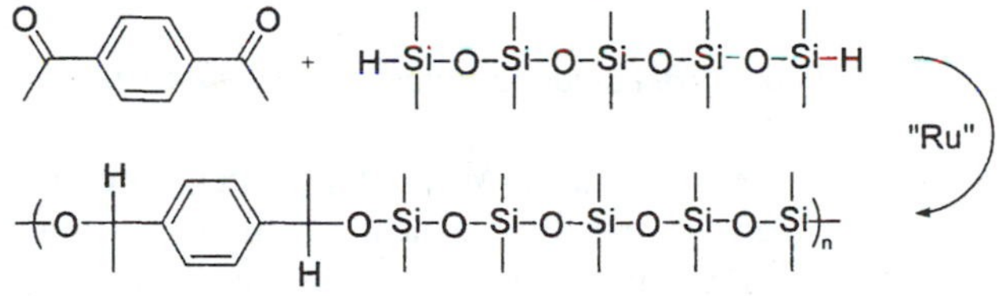

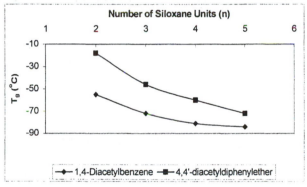

Figure 3. **Ru** *catalyzed reaction of 1,4 diacetylbenzene with* **DMPS**

The T_gs of the copolymers decrease as the length of the ODMS unit between the aromatic units increases. This is expected because of the flexibility of the ODMS segment. The nature of the aromatic unit in the copolymer also affects T_gs. The plot of the T_gs for the copolymers derived from 4,4'-diacetyldiphenylether, while similar to that of those derived from 1,4-diacetylbenzene, has a steeper slope. Both plots are shown in Figure 4.

Figure 4. Plots of T_gs vs. number of siloxane units.

The NMR spectra for these *alt*-copolymers are complicated by the presence of chiral centers (*) as seen in Figure 5. The spectra are similar for both the 1,4-diacetylbenzene and 4,4'-diacetyldiphenylether copolymers. The **Ru** catalyzed addition of the Si-H bond across the C-O double bond of the ketone results in formation of a single chiral center, so each polymer unit contains two chiral centers. This results in two different stereochemical environments for the Si-methyl groups.

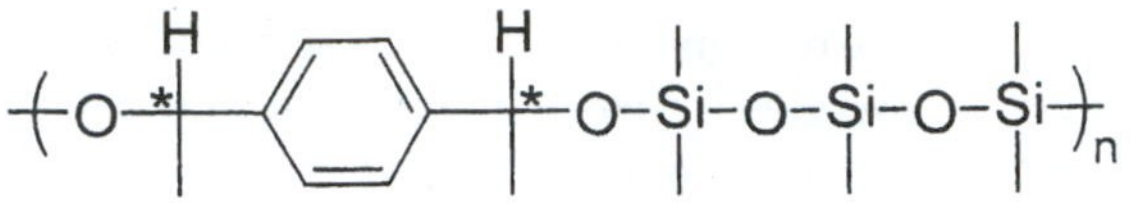

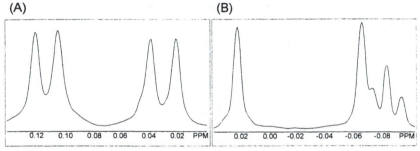

Figure 5. Two chiral centers created in each monomer unit.

The Si-methyl protons in the **TMDS** copolymers are all chemically equivalent. However, each is split by the nearest chiral center and then split again by the more remote chiral center. This produces four diastereotopic environments: RR', SS', RS', and R'S, which are distinct. This results in four singlets of equal intensity as seen in Figure 6A.

In the case of the **HMTS** copolymers, each polymer unit contains three silicon atoms. Two of these are adjacent to a chiral center. Their diastereopic environments are only affected by this adjacent chiral center which can be either R or S. This results in two signals of equal intensity. The central silicon of the trisiloxane copolymer is affected by both chiral centers. This results in four diastereotopic environments: RR, SS, RS and SR. However, RS and SR are meso. This leads to three resonances in a 1:2:1 ratio as seen in Figure 6B.

(A)   (B)

| 0.12   0.10   0.08   0.06   0.04   0.02   PPM | 0.02   0.00   -0.02   -0.04   -0.06   -0.08   PPM |

Figure 6. ^1H NMR spectra of aromatic copolymers.

The **OMTS** copolymers contain four silicons, each of which is affected only by the nearest chiral center, which can be either R or S. The difference between these environments is larger for the Si-methyl groups which are adjacent to chiral center than those which are further removed. This leads to four resonances of equal intensity as seen in Figure 7A.

The **DMPS** copolymers have five Si-methyl groups. Those, which are closest to the chiral center, experience the largest difference in diastereotopic environments. The inner pair of Si-methyl groups experience a smaller difference. Finally, the central Si-methyl groups are not affected by either chiral center because they are too remote. This results in five signals of equal intensity as seen in Figure 7B.

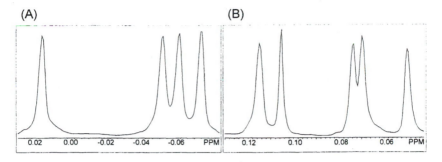

(A)   (B)

| 0.02 0.00 -0.02 -0.04 -0.06 PPM || 0.12 0.10 0.08 0.06 PPM |

Figure 7. ¹H NMR spectra of aromatic copolymers.

Ru Catalyzed Hydrosilylation Copolymerization of Aliphatic Diketones

2,5-Hexanedione and 2,7-octanedione (*31,32*) were independently reacted with the same four α,ω-dihydrido-oligodimethylsiloxanes above (*33*). The reaction is shown in Figure 8. A difference between the two systems is that the color change of the **Ru** catalyst from brown to red at the conclusion of the reaction previously reported with aromatic α,ω-diketones is not observed.

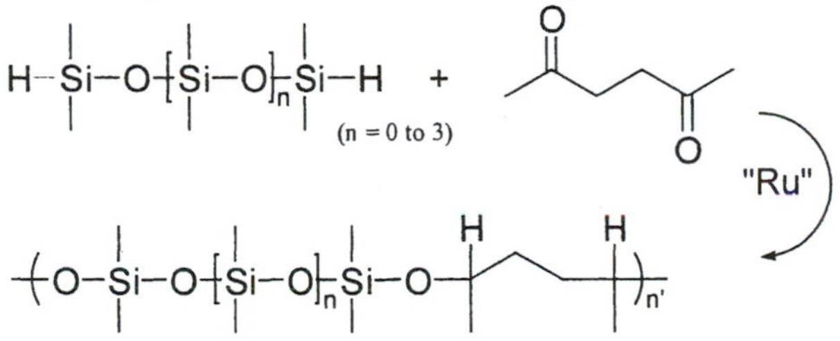

*Figure 8. **Ru** catalyzed reaction of 2,5-hexanedione.*

The NMR spectra of these copolymers are also complicated by the presence of chiral centers. The effect of the chiral centers on Si-methyl groups in ¹H NMR spectra previously discussed is observed to a lesser extent than in aromatic systems. In this system, splitting is clearly visible in the ¹³C NMR spectra. Surprisingly, in the case of the 2,5-hexanedione copolymers, every ¹³C signal from the aliphatic portion of the copolymers is split into two peaks. The terminal Si-methyl signals are also split due to their close proximity to the chiral centers. The internal Si-methyl peaks of copolymers containing more than two silicon atoms are too far removed from the chiral centers for splitting to occur. Therefore, in the **TMDS** copolymer, every peak in the ^{13}C spectrum is split. The copolymer has four nonequivalent carbon atoms, however, there are eight peaks in the ^{13}C spectrum, shown in Figure 9 (*31*). Interestingly, copolymers of 2,7-octanedione only show terminal Si-methyl splitting in ^{13}C NMR spectra (*32*).

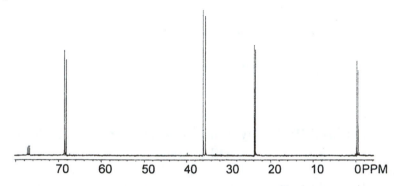

*Figure 9. ^{13}C NMR Spectrum of **TMDS** copolymer.*

The ^{29}Si signal of the **TMDS** copolymer exhibits splitting similar to that of the ^{1}H NMR spectra of the aromatic copolymers above. The signal is split into two peaks by the nearest chiral center. It is then split again by the more remote chiral center, producing four peaks from two chemically equivalent silicon atoms. The ^{29}Si Spectrum of the **TMDS** copolymer is shown in Figure 10. Such splitting is not seen in copolymers with longer siloxane segments. This may be because, as the siloxane chain gets longer, there is less effect from the more remote chiral center.

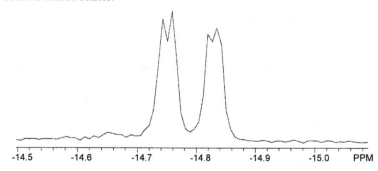

*Figure 10. ^{29}Si NMR Spectrum of **TMDS** copolymer.*

End group analysis by ^{1}H NMR provides a valuable tool to independently determine M_n. End group analysis was carried out before and after the completion of the copolymerization reactions to compare the values of M_n obtained by GPC with those obtained by ^{1}H NMR. For example, the copolymerization of 2,5-hexanedione and **OMTS** was analyzed prior to its completion. The integration of peaks at 2.11 and 2.47 ppm, assigned to the $-CH_2-CO-CH_3$ end group, were compared to the integration of the peak at 3.87 ppm, assigned to the methine protons in the polymer backbone. According to the integration, there should be 6.5 monomer units per oligomer, provided there is a ketone end group at one end of each oligomer. The number of monomer units multiplied by the molecular weight of a single monomer unit gives the M_n of the oligomer. This was calculated to be 2100. Analysis of the integration of the peak at 4.69, from an Si-H end group, and the peak at 3.87 ppm likewise, gives a M_n value of 2100. The M_n determined by GPC (1900) was in good agreement. Comparison of the Si-H to the $CH_2-CO-CH_3$ end groups indicates that proper stoichiometric balance existed at the time of the NMR analysis. Both end group analysis and IR were used to verify that stoichiometric balance was maintained during the copolymerizations.

The T_gs for these copolymers also decrease as the length of the siloxane chain increases, as shown in Figure 11. However, in this system, the approach to the T_g of PDMS seems to be linear, rather than asymptotic. Perhaps, with longer ODMS monomers, the decrease in T_g will be less rapid, as was seen above.

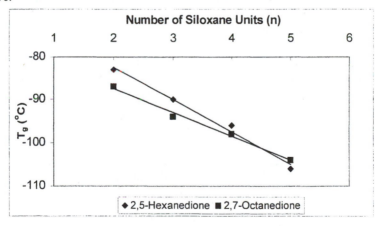

Figure 11. Plots of T_gs vs. number of siloxane units.

Ru Catalyzed Homopolymerization of Aromatic and Aliphatic ω-Dimethylsilyloxy Ketones

4-Hydroxyacetophenone and 4-hydroxybenzophenone, were silylated by reaction with dimethylchlorosilane in THF to yield 4-dimethylsilyloxy-acetophenone and 4-dimethylsilyloxybenzophenone, respectively (*30*). These difunctional monomers were polymerized by the addition of activated **Ru** in toluene. The synthesis and polymerization of 4-dimethylsilyloxyacetophenone is shown in Figure 12. While the length of the siloxane unit is the same in each of the polymers, the size of the aromatic unit differs. As expected, the T_g of poly(4-dimethylsilyloxybenzophenone) (25 °C) is higher than that of poly(4-dimethylsilyloxyacetophenone), which has a T_g of −10 °C. Interestingly, Si-methyl splitting in ^{1}H NMR spectra due to the chiral center is observed only for poly(4-dimethylsilyloxyacetophenone). The ^{1}H NMR of poly(4-dimethylsilyl-oxybenzophenone) shows only a broad peak in the Si-methyl region (*30*).

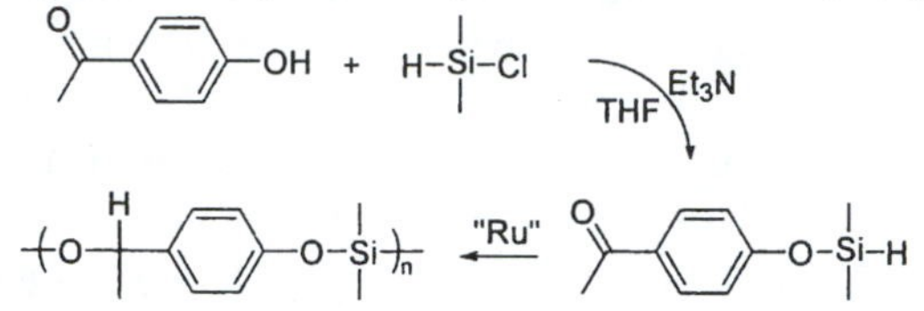

Figure 12. Synthesis and polymerization of 4-dimethylsilyloxyacetophenone.

Aliphatic hydroxy ketones, 5-hydroxy-2-pentanone and 4-hydroxy-2-butan-one, were silylated by reaction with 1,1,3,3-tetramethyldisilazane in THF to yield 5-dimethylsilyloxy-2-pentanone and 4-dimethylsilyloxy-2-butanone, re-spectively. Both monomers undergo **Ru** catalyzed polymerizations to yield poly(5-dimethylsilyloxy-2-pentanone) (T_g = -81 °C) and poly(4-dimethylsilyl-oxy-2-butanone) (T_g = -88 °C), respectively. The synthesis and polymerization of 5-dimethylsilyloxy-2-pentanone is shown in Figure 13. The ^{1}H NMR spectra of both polymers show signal splitting for the methylene group adjacent to the chiral center. Consistent with the aliphatic copolymers above, little Si-methyl splitting is seen. ^{13}C NMR shows only splitting of the Si-methyl peaks.

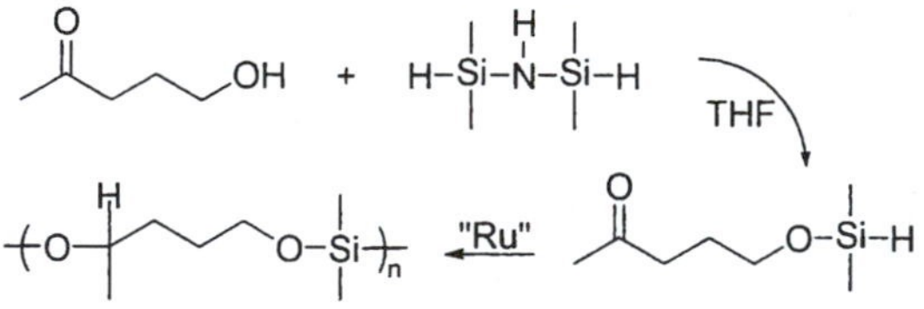

Figure 13. Synthesis and polymerization of 5-dimethylsilyloxy-2-pentanone.

These aliphatic polymers (and monomers) are much less hydrolytically stable than any of the previously mentioned materials. Silyl ethers are well-known to undergo both acid and base catalyzed hydrolysis and methanolysis. In general silyl ethers are more susceptible to acid catalyzed hydrolysis than to basic hydrolysis (*34,35*). The rate of these reactions depends not only on the concentration of acid or base, but also on the nature of the substituents bonded to silicon and on the character of the alkyl or aryl group of the ether. In particular, if the carbon of the Si-O-C bond is secondary, the silyl ether is considerably more resistant to hydrolysis and methanolysis than if it is primary (*36*). These polymers each contain an Si-O-C bond linkage in which the oxygen is bonded to a primary carbon. Even though precipitation is done under neutral conditions, these polymers are quite unstable and can be hydrolyzed at room temperature in a matter of hours. This makes polymer purification by precipitation from methanol impossible due to immediate methanolysis. All other polymers described in this chapter were purified by repeated precipitation from THF and methanol. Instability has been previously observed in polymers produced by the **Ru** catalyzed hydrosilylation of aldehydes, which also produce primary poly(silyl ether) products (*30,37*). A theoretical study of the hydrolysis and condensation of silyl ethers indicates that hydrolysis of the Si-O-C bond linkage, even under neutral conditions, is quite facile (*38*).

References

1. Kennan, J. J. In *Siloxane Polymers;* Clarson, S. J.; Semlyen, J. A., Eds.; Prentice Hall: Englewood Cliffs, NJ, 1993; p. 73.
2. Langer, R. *Science* **1990**, *249*, 1527.
3. Voronkov, M. G.; Mileshkevich, V. P.; Yuzhelevskii, Yu A. *The Siloxane Bond*, Consultants Bureau: New York, 1978; Si-O-Si pp 146-149, Si-O-C pp 323-340.
4. Nagasaki, Y.; Matsukura, F.; Masao, K.; Aoki, H.; Tokuda, T. *Macromolecules* **1996**, *29*, 5859.
5. Stern, S. A.; Shan, V. M.; Hardy, B. J. *J. Polym. Sci., Part B: Polym. Phys.* **1987**, *25*, 1263.
6. Kaganove, S. N.; Grate, J. W. *Polym. Prepr.* **1998**, *39*(1), 556.
7. Dunnavant, W. R.; Markle, R. A.; Sinclair, R. G.; Stickney, P. B.; Curry, J. E.; Byrd, J. D. *Macromolecules* **1968**, *1*, 249.
8. Curry, J. E.; Byrd, J. D. *J. Appl. Polym. Sci.* **1965**, *9*, 295.
9. Webb, J. L.; Nye, S. A.; Grade, M. M. U. S. Patent 5,041,514, 1991.
10. Bailey, D. L.; O'Connor, F. M., DE Patent 1,012,602, 1957.

11. Macfarlane, R.; Yankura, E. S. Contract No. DA-19-020-ORD-5507, Quarterly Report No. &, Naugatuck Division of the U.S. Rubber Co.

12. Nye, A. A.; Swint, S. A. *J. Polym. Sci., Part A: Polym. Chem.* **1994**, *32*, 131.

13. Mironov, V. F.; Kozlkov, V. L.; Fedotov, N. S. *Zh. Obshch. Chim.* **1969**, *39*, 966.

14. Li, Y.; Kawakami, Y. *Macromolecules* **1999**, *32*, 8768.

15. Li, Y.; Kawakami, Y. *Macromolecules* **1999**, *32*, 6871.

16. Li, Y.; Kawakami, Y. *Polym. Prepr.* **2000**, *41*(1), 534.

17. Li, Y.; Seino, M.; Kawakami, Y. *Macromolecules* **2000**, *33*, 5311.

18. Minegishi, S.; Ito, M.; Kameyama, A.; Nishikubo, T. *J. Polym. Sci., Part A: Polym. Chem.* **2000**, *38*, 2254.

19. Itoh, H.; Kameyama, A.; Nishikubo, T. *J. Polym. Sci., Part. A: Polym. Chem.* **1997**, *35*, 3217.

20. Liaw, D. J. *Polymer* **1997**, *38*, 5217.

21. Nishikubo, T.; Kameyama, A.; Kimura, Y.; Nakamura, T. *Macromolecules* **1996**, *29*, 5529.

22. Reddy, P. N.; Chauhan, B. P. S.; Hayashi, T; Tanaka, M. *Chem. Lett.* **2000**, 250.

23. Mabry, J. M.; Teng, C. J.; Weber, W. P. *Polym. Prep.* **2001**, *42*(1), 153.

24. Mabry, J. M.; Runyon, M. K.; Weber, W. P. *Macromolecules* **2001**, *34*, in press.

25. Mabry, J. M.; Weber, W. P. *Polym. Prep.* **2001**, *42*(1), 281.

26. Mabry, J. M.; Teng, C. J.; Weber, W. P. *PMSE* **2001**, *84*, 505.

27. Levison, J. J.; Robinson, S. D. *J. Chem. Soc. A* **1970**, 2947.

28. Guo, H.; Wang, G.; Tapsak, M. A.; Weber, W. P. *Macromolecules* **1995**, *28*, 5686.

29. Noll, W. *The Chemistry and Technology of Silicones,* Academic Press: New York 1968; pp 326.

30. Mabry, J. M.; Paulasaari, J. K.; Weber, W. P. *Polymer* **2000**, *41*, 4423.

31. Brown, H. C.; Geoghegan, P. J.; Kurek, J. T.; Lynch, G. J. *Organometal. Chem. Syn.* **1970**, *1, 7*.

32. Ratcliffe, R.; Rodehorst, R. *J. Org. Chem.* **1970**. *35, 11*.

33. Mabry, J. M.; Weber, W. P. *Polym. Prep.* **2001**, *42*(1), 145.

34. Ackerman, E. *Acta Chem. Scand.* **1957**, *11*, 373.

35. Sommer, L. H. *Stereochemistry, Mechanism and Silicon*; McGraw-Hill: New York, **1965**; pp 132.

36. Burger, C.; Freuzer, F. H. In *Silicon in Polymer Synthesis*; Kricheldorf, H. R., Ed.; Springer: Berlin, Germany, 1996, pp 139.

37. Paulasaari, J. K.; Weber, W. P. *Macromolecules* **1998**, *31*, 7105.

38. Okumoto, S.; Fujita, N.; Yamabe, S. *J. Phys. Chem. A* **1998**, *102*, 3991.

Chapter 6

Synthesis and Characterization of Siloxane-Containing Polymers Obtained by Dehydrocoupling Polymerization

Yusuke Kawakami and Ichiro Imae

Graduate School of Materials Science, Japan Advanced Institute of Science and Technology (JAIST), 1-1 Asahidai, Tatsunokuchi, Ishikawa 923–1292, Japan

Siloxane-containing polymers have been synthesized by catalytic dehydrocoupling polymerization. Their syereochemical structures, diad or triad tacticity, were characterized by NMR spectroscopies. The results showed that the reactivity of bis(silane)s in the dehydrocoupling polymerization is not affected by the structures of the silanes, and that the stereoregularity of the polymer reflected the optical purity of the monomer.

Introduction

Siloxane-containing polymers, such as polysiloxanes and polycarbosiloxanes, have been used as oil, rubber, and insulator due to the unique properties of Si–O bond, *e.g.*, flexibility, high thermal stability, low dielectricity[1]. As a novel synthetic method of siloxane-containing polymers, we reported a catalytic dehydrocoupling polymerization to give disiloxane-phenylene polymer, poly[(oxydimethylsilylene)(1,4-

phenylene)(dimethylsilylene)] from 1,4-bis(dimethylsilyl)benzene with water in the presence of transition metal catalysts under mild conditions[2]. The polymer has high molecular weight and high thermal degradation temperature and high crystallinity. For practical applications of the polymer, it is important to control the thermal properties. In order to modify the morphology of the polymer without losing the high thermal stability, we report here the synthesis of disiloxane-arylene copolymers from bis(silane) derivatives with water in various isomer ratio[3]. The reactivity of Si–H groups of the monomers was investigated by estimating their triad tacticity by NMR. This reaction is applied to the synthesis of stereoregular polysiloxane itself from optically acrtive disiloxane derivatives[4].

Experimental

Analysis

500MHz ^{1}H, 125 MHz ^{13}C, and 99 MHz ^{29}Si NMR spectra were obtained in CDCl$_3$ on Varian 500 MHz Unity *INOVA* spectrometer. Chemical shifts are reported in ppm, relative to CHCl$_3$ (δ 7.26) in ^{1}H NMR, CDCl$_3$ (δ 77.00) in ^{13}C NMR, and tetramethylsilane (δ 0.00) in ^{29}Si NMR. Specific optical rotations were measured with a JASCO DIP-370S digital polarimeter. Size exclusion chromatography (SEC), and HPLC on an optically active stationary phase were performed on a JASCO HPLC on the combination of Shodex KF-803L (exclusion limit: $M_n = 7 \times 10^4$, polystyrene) and KF-804 (exclusion limit: $M_n = 4 \times 10^5$, polystyrene) columns (linear calibration down to $M_n = 100$, polystyrene) using tetrahydrofuran (THF) as an eluent, and on a Daicel CHIRALCEL® OD (cellulose carbamate derivative) and CHIRALCEL® AD (amylose carbamate derivative) with hexane as an eluent, respectively. The differential scanning calorimetry (DSC) and thermogravimetry (TG) were carried out on a Seiko SSC/5200H instrument.

Materials

Tris(dibenzylideneacetone)dipalladium(0)-chloroform adduct (Pd$_2$(dba)$_3$•CHCl$_3$) and chloro(1,5-cyclooctadiene)rhodium(I) dimer ([RhCl(cod)]$_2$) were purchased from Aldrich Chemicals and Kanto Chemicals, respectively. Tetrahydrofuran (THF) and toluene were dried and distilled over calcium hydride and sodium just before use. 1,4-Bis(dimethylsilyl)benzene (*p*-**BSB**), 1,4-bis(dimethylhydroxysilyl)benzene (*p*-**BHB**) were purchased from Shin-Etsu Chemicals and purified by distillation over calcium hydride, and repeated recrystallization from hexane/THF (9/1 v/v) , respectively. 1,3-Bis(dimethylsilyl)benzene (*m*-**BSB**), 4,4'-bis(dimethylsilyl)biphenyl (**BSD**) were synthesized according to the literatures. Synthesis of 9,9'-bis[4-(dimethylsilyl)phenyl]fluorene (**BSF**)[3b], (*S,S*)-1,3-dihydroxy-1,3-dimethyl-1,3-diphenyl-1,3-disiloxane ((*S,S*)-**SiOH**)[5], (*S,S*)-1,3-dihydro-1,3-dimethyl-1,3-diphenyl-1,3-disiloxane (84%o.p., (*S,S*)-**SiH**)[5] is described in our previous reports. Disiloxane-arylene polymers were synthesized by dehydrocoupling polymerization as reported with various molar ratio {[x mmol of *p*-**BSB**]+[(10-x) mmol of *m*-**BSB, BSD,** or **BSF**] with water (0.18g, 10 mmol) or *p*-**BHB** (1.97g, 0.01mol) in THF (4ml) in the presence of Pd$_2$(dba)$_3$•CHCl$_3$(25.9mg, 2.5×10^{-5}mol) at room temperature for 3hr (Scheme 1). Catalyst was removed by passing through a Florisil column with chloroform. The crude polymer was purified by repeated reprecipitation from chloroform into methanol.

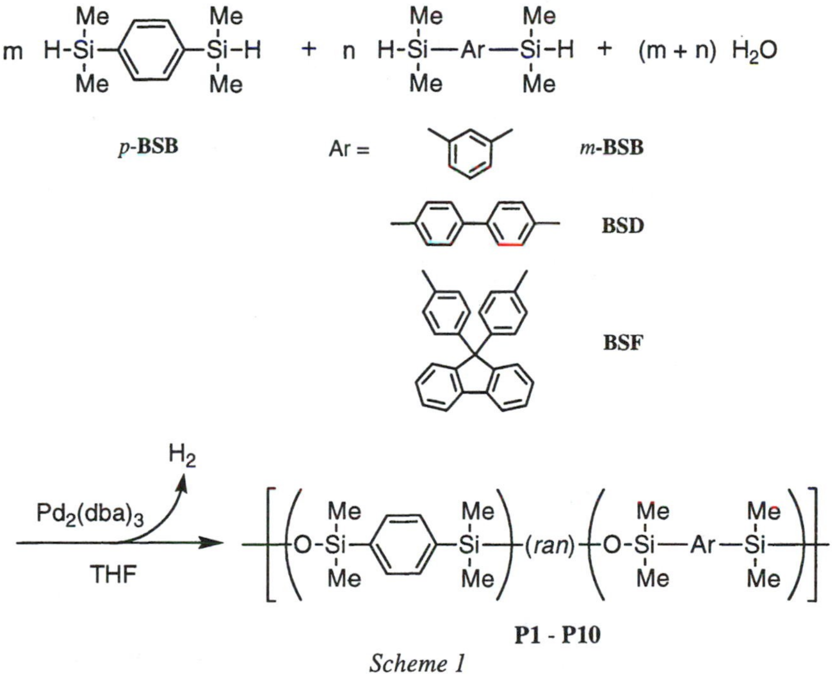

P1 - P10

Scheme 1

Dehydrocoupling polymerization of (*S,S*)-**SiOH** and (*S,S*)-**SiH** was also carried out in toluene at 60°C in the presence of [RhCl(cod)]$_2$ and triethylamine (Scheme 2). Catalyst was removed by passing through a silica-gel column with 5:1 hexane:dichloromethane. The crude polymer was purified by repeated reprecipitation from diethyl ether into methanol.

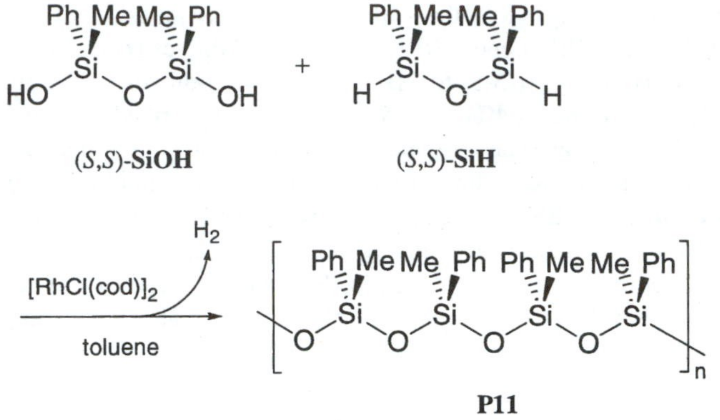
Ph MeMe Ph
HO Si O Si OH
+
Ph MeMe Ph
H Si O Si H
(S,S)-SiOH
(S,S)-SiH
H2
[RhCl(cod)]2
toluene
Ph MeMe Ph Ph MeMe Ph
O Si O Si O Si O Si
n
P11

Scheme 2

Results and Discussion

Synthesis of Disiloxane-Arylene Polymers and Their Sequential Tacticity

Polymers with various compositions of *p*-**BSB** and *m*-**BSB** were synthesized by cross-dehydrocoupling reaction using $Pd_2(dba)_3$•$CHCl_3$ as a catalyst. From SEC chromatograms of the product mixture before purification, it was found that the low molecular weight fraction increased with the increase of *m*-**BSB** in the feed. The oligomer fraction in **P5** was separated by preparative SEC, and proved to contain essentially the cyclic dimer by NMR and MS spectra. Cyclic dimer was almost selectively formed in considerably large quantity in the polymerization of *m*-**BSB** to produce **P5**. **P1** from *p*-**BSB** was produced in higher yield than **P5** from *m*-**BSB**. The kinky structure of *m*-**BSB** apparently made it easy to selectively form cyclic dimer, and resulted in lower polymer yield.

The number average molecular weight of the purified polymer (**P1** – **P5**) tends to decrease with the increase in *m*-**BSB** in the feed. The molecular weight distribution (M_n/M_w) became a little wider in this order.

^{1}H NMR spectra of methyl and aromatic protons (in the region of 7.2-7.8 ppm) of **P1**, **P3**, and **P5** are shown in Figure 1. **P1** from *p*-**BSB** and water showed basically two singlets at 0.323 (H^a) and 7.538 (H^b). **P5** from *m*-**BSB** showed a singlet at 0.301 (H^c) and three peaks at 7.304 (H^e), 7.547 (H^f), and 7.741 (H^d). **P3** showed four signals at 0.304, 0.309, 0.317, 0.322, and two mutiplets at 7.286-7.341, 7.518-7.567, and three singlets at 7.741, 7.754, 7.766.

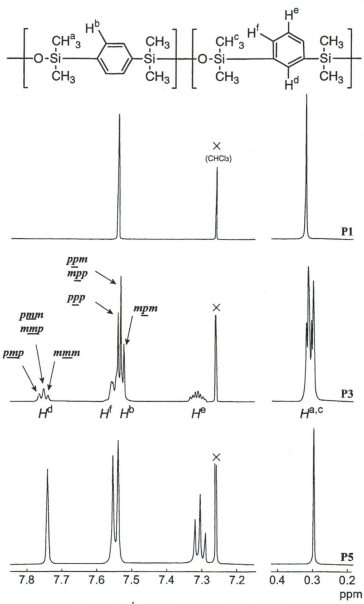

Figure 1. ^1H NMR spectra of P1, P3, P5.

The composition of **P3** was determined by taking the signal around 7.7-7.8 of H^d of *m*-constitutional unit as 1H. Subtraction of 2H (H^f) from the peaks between 7.5-7.6 ppm (which contains $2H^f$ and $4H^b$ protons) will give 4H of *p*-**BSB** unit protons (H^b). Compositions of the other polymers (**P2** and **P4**) were determined similarly and the results of the polymerization are summarized in Table I.

Table I. Cross-dehydrocoupling Polymerization of *p*-BSB and *m*-BSB, BSD, BSF[a]

Polymer	Comonomer	[*p*-BSB / comonomer]		Yield [%][c]	M_n^d
		Feed	Composition[b]		
P1	*m*-**BSB**	100 / 0	100 / 0	87	34800
P2		70 / 30	74 / 26	76	30400
P3		50 / 50	57 / 43	72	26800
P4		30 / 70	38 / 62	62	24200
P5		0 / 100	0 / 100	50	16800
P6	**BSD**	70 / 30	70 / 30	73	13000
P7		50 / 50	48 / 52	85	15700
P8		30 / 70	31 / 69	69	13800
P9		0 / 100	0 / 100	72	24600
P10	**BSF**	0 / 100	0 / 100	73	5000

[a] reaction conditions: 10 mmol of comonomer with above-mentioned feed ratio and 10 mmol of water in 4 mL of THF; 25 μmol of metal; 3 h (48 hr for **P10**) in room temperature. [b] estimated by ^{1}H NMR. [c] after reprecipitations in methanol and pentane. [d] estimated by SEC with polystyrene standard.

It is interesting to note that the apparent *m*-composition is lower than that of feed even at the quantitative conversion of monomers. This could be caused by the difference in the reactivity, or preferential consumption of *m*-monomer in oligomerization.

To estimate the reactivity of each isomer, it is necessary to evaluate the sequence regularity of the polymer. The split methyl proton signals seem to be overlapped reflecting the higher order sequence regularity, and it is difficult to estimate the definite sequence regularity of the polymer from methyl signals. Thus, accurate triad regularity of the sequence was analyzed by the well-separated H^d signals at 7.739, 7.752, and 7.764 of the *m*-centered triad, and H^b signals at 7.522, 7.530, and 7.538 of *p*-centered triad.

The H^d and H^b signals were split in three peaks reflecting the triad sequence regularity **m*mm*, [**m*mp*, **p*mm*], **p*mp*** and **m*pm*, [**p*pm*, **m*pp*], **p*pp***, respectively. Judging from the signal of **P5**, the signal at 7.739 was assigned to **m*mm*** triad. Considering the peak area ratio and feed ratio, the peaks at 7.752 and 7.764 were assigned to [**m*mp*, **p*mm*] and **p*mp*** triads, respectively. Similarly, the peaks at 7.538, 7.530, 7.522 were assigned to **p*pp*, [**p*pm*, **m*pp*], **m*pm*** triads by considering the change in the chemical shifts of the polymers.

Calculated concentration of each triad was obtained by assuming the selection of isomer to form triad is dependent only on the concentration of each isomer. For example, **m*mm*** and **m*pm*** depend on the square of the concentration of *m*-isomer. The triad concentrations of *m*- and *p*-centered signals are almost the same, and simply depend on the feed ratio. This strongly supports the equal reactivity of *m*- and *p*-**BSB** in the cross-dehydrocoupling reaction with water, which, in turn, indicates that the reason of lower content of *m*-unit in the copolymer is the preferential formation of cyclic oligomers from *m*-isomer.

In ^{1}H NMR spectra of the copolymers from **BSD** and *p*-**BSB**, three singlet signals due to the aromatic proton in *p*-**BSB** unit were clearly observed at 7.538, 7.546, and 7.553 ppm reflecting their triad sequential tacticity, although the signals due to the aromatic proton in **BSD** unit were complicated because of their overlapping. From these signals, triad sequence of the copolymers could be estimated similarly to the copolymer of **P2** – **P4**. Furthermore, four singlet signals at 0.323, 0.341, 0.359, and 0.377 ppm due to the methyl proton of *p*-**BSB** and **BSD** were also well-separated. These signals reflected diad sequence, and enabled to estimate diad sequence of the polymers. These results from estimation of diad and triad sequential tacticity support that the reactivity of Si-H on dimethylsilyl group will not be affected by the structure of aromatic groups.

The molecular weight of the polymer from **BSF** with water was low, since the viscosity of the reaction solution became high as the polymerization proceeded.

Thermal properties of the polymers are shown in Table II. T_m was observed at 127 °C for **P1**, not for **P2** – **P10**. The crystallinity of **P1** could be controlled by the incorporation of small amounts of aromatic groups in the polymer structure. The T_{onset} and T_d increased by introduction of rigid **BSD** and **BSF**. T_g decreased from –23 °C to –49 °C with the increase of *m*-**BSB**, but increased to 56 °C with the increase of **BSD**.

P10, which has a bulky, rigid fluorene unit in the polymer, showed the highest T_g, T_d, and T_{onset} among the polymers.

Table II. Thermal Properties of P1 - P5[a]

Polymers	T_g [°C]	T_c^b [°C]	T_m [°C]	T_{onset}^c [°C]	T_d^d [°C]	Char Yield[e] [%]
P1	-23	15, 95	127	463	505	26
P2	-33	---	---	458	505	36
P3	-38	---	---	457	506	33
P4	-44	---	---	460	505	36
P5	-49	---	---	463	505	23
P6	-5	---	---	471	516	46
P7	11	---	---	472	510	57
P8	20	---	---	457	498	56
P9	56	---	---	484	524	62
P10	149	---	---	499	539	58

[a] DSC: determined by second run (5 °C min^{-1}) in air; TG: 5 °C min^{-1} in N$_2$.
[b] crystallization temperature. [c] temperature at 5 % weight loss. [d] temperature at maximum weight loss. [e] char yield at 600 °C.

Synthesis of Stereoregular Poly(methylphenylsiloxane)

It is difficult to obtain highly optically active silanols with a leaving group because of their chemical instability. Stereoregularity of polysiloxanes has been evaluated for polymers obtained by the ring-opening polymerization of cyclic siloxanes[6], such as *cis*- or *trans*-1,3,5-trimethyl-1,3,5-triphenylcyclotrisiloxane (*cis*- or *trans*-D$_3^{Ph}$). However, the stereoregularity was not controlled perfectly by any attempts.

We focused on the polycondensation reaction of optically active disiloxane derivatives.

(1*S*,3*S*)-1-(4-methoxy-1-naphthyl)-1,3-dimethyl-3-(1-naphthyl)-1,3-diphenyl-1,3-disiloxane ((*S*,*S*)-**SiNp**) was synthesized by the reaction of optically pure (*R*)-methyl(1-naphthy)phenylchlorosilane with potassium (*S*)-methyl(4-methoxy-1-naphthyl)phenylsilanoleate. The stereoisomeric ratio of (*S*,*S*)-**SiNp** was found to be (1*S*,3*S*) : (1*S*,3*R*) : (1*R*,3*S*) : (1*R*,3*R*) = 95 : 5 : 0 : 0 by HPLC on optically active stationary phase.

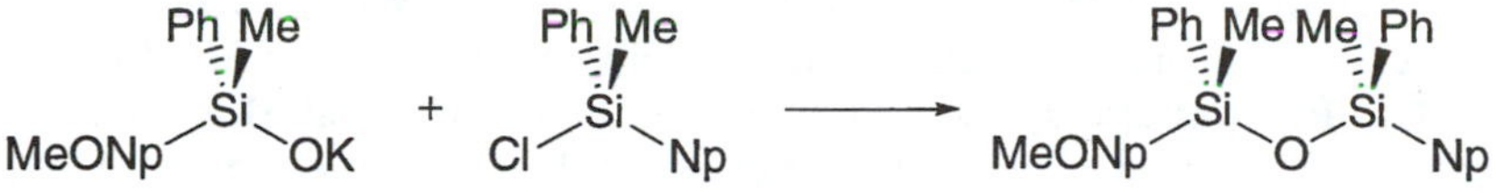

MeONp =
4-methoxy-1-naphthyl

Np =1-naphthyl

(S,S)-SiNp

Scheme 3

The stereochemistry of cleavage reaction of silicon-naphthyl bond in (S,S)-SiNp by bromine was investigated. The stereoisomeric ratio of 1,3-dibromo-1,3-dimethyl-1,3-diphenyl-1,3-disiloxane ((S,S)-SiBr) obtained from (S,S)-SiNp was found to be [(S,S)+(R,R)] : (S,R) = 91 : 9 by ¹H NMR. This result shows that this reaction proceeds with 98% inversion.

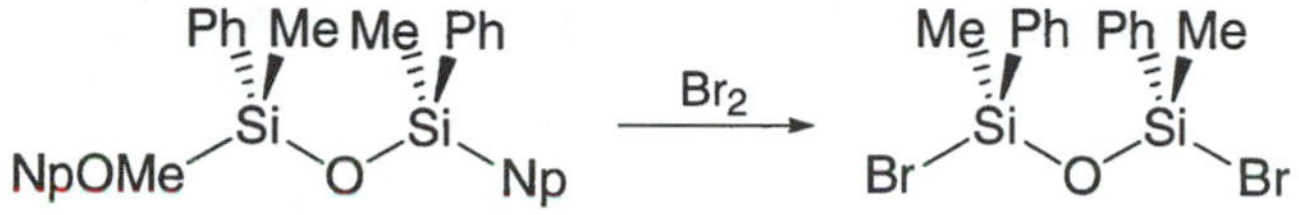

(S,S)-SiNp

(S,S)-SiBr

Scheme 4

(S,S)-SiOH ((S,S) : (S,R) : (R,R) = 86 : 14 : 0) was prepared by hydrolysis of (S,S)-SiBr, and its optical separation by HPLC gave the optically pure (S,S)-SiOH. (S,S)-SiH ((S,S) : (S,R) : (R,R) = 84 : 16 : 0) was also synthesized from (S,S)-SiH by the reduction with LiAlH₄.

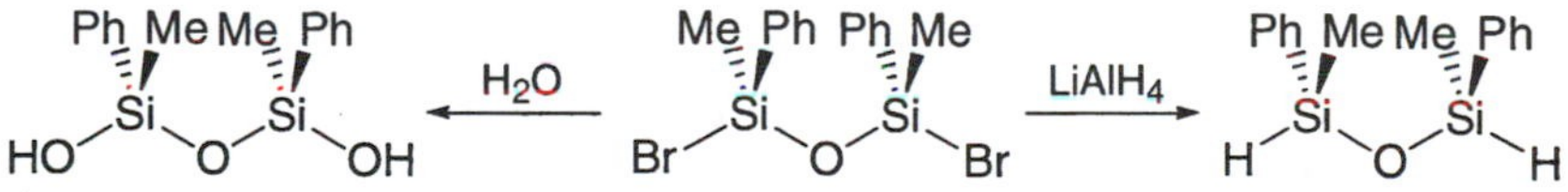

(S,S)-SiOH

(S,S)-SiBr

(S,S)-SiH

Scheme 5

The cross-dehydrocoupling reaction of (S)-methyl(1-naphthyl)phenylsilanol (>99%ee) with (R)-methyl(1-naphthyl)phenylsilane (>99%ee) was carried out in toluene using various rhodium catalysts as a model reaction. Based on this model reaction, dehydrocoupling polymerization of (S,S)-SiOH and (S,S)-SiH was carried out at 60°C in toluene in the presence of [RhCl(cod)]₂ as a datalyst and triethylamine as an additive. The results are shown in Table III.

The triad tacticity of **P11** was assigned by ¹H NMR of the methyl proton (*I* = 0.04, *H* = 0.09, and *S* = 0.14ppm) and ¹³C NMR of the *ipso* carbon of the phenyl group (*S* = 136.7, *H* = 136.9, and *I* = 137.1 ppm). Although the reaction of (*S,S*)-**SiOH** and (*S,S*)-**SiH** gave **P11** with low molecular weight, its triad tacticity was found to be rich in syndiotacticity (*S* : *H* : *I* = 60 : 32 : 8) by ^{13}C NMR. **P11** was also synthesized by the reaction of (*S,S*)-**SiOH** and optically active 1,3-bis(dimethylamino)-1,3-dimethyl-1,3-diphenyl-1,3-disiloxane, but the stereoselectivity of the reaction was not so high (66% retention).

Table III. Rh-catalyzed Cross-dehydrocoupling Polymerization of SiOH with SiH

run	monomers	Yield[a] [%]	$M_n{}^b$	PDI[b]	$S : H : I^c$	$T_g{}^d$ [°C]
1	*meso / dl*	23	7700	1.5	31:42:27 (25:50:25)	−24
2	(*S,S*)[e]	23	2400	1.7	60:32:8 (88:8:4)	−22

[a] isolated yields after reprecipitation into methanol. [b] estimated by SEC with polystyrene standard. [c] estimated by ^{13}C NMR (in parenthesis: calculated triad tacticity from the stereoisomer concentration). [d] determined by DSC with heating rate at 5 °C min^{-1} on the second scan. [e] (*S,S*) : (*S,R*) : (*R,R*) = 25 : 50 : 25.
[f] (*S,S*)-**SiOH**: (*S,S*) : (*S,R*) : (*R,R*) = 100 : 0 : 0; (*S,S*)-**SiH**: (*S,S*) : (*S,R*) : (*R,R*) = 84 : 16 : 0

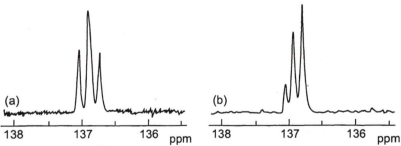

Figure 2. ^{13}C NMR spectra of ipso-carbon of phenyl group of poly(methylphenylsiloxane): (a) atactic; (b) rich in syndiotacticity

References

1 *Silicon-Containing Polymers;* Jones, R. G.; Ando, W.; Chojnowski, J., Eds.; Kluwer Academic Publishers: Dordrecht, Netherlands, 2000.

2 (a) Li, Y.; Kawakami, Y. *Macromolecules* **1999**, *32*, 3540, (b) Li, Y.; Kawakami, Y. *Macromolecules* **1999**, *32*, 6871, (c) Li, Y.; Kawakami, Y. *Macromolecules* **1999**, *32*, 8768, (d) Li, Y.; Seino, M.; Kawakami, Y. *Macromolecules* **2000**, *33*, 5311.

3 (a) Kawakita, T.; Oh, H.-S.; Moon, J.-Y.; Liu, Y.; Imae, I.; Kawakami, Y., *Polym. Int.* **2001**, in press, (b) Moon, J.-Y.; Miura, T.; Imae, I.; Park, D.-W.; Kawakami, Y., *Silicon Chem.* **2001**, in press.

4 (a) Oishi, M.; Moon, J.-Y.; Shirakawa, E.; Kawakami, Y., *Polym. J.* **2000**, *32*, 980, (b) Oishi, M.; Moon, J.-Y.; Janvikul, W.; Kawakami, Y., *Polym. Int.* **2001**, *50*, 135.

5 Oishi, M.; Kawakami, Y., *Org. Lett.* **1999**, *1*, 549.

6 Tsvetkov, V. N.; Andrianov, K. A.; Vinogradov, E. L.; Yakushkina, S. E.; Vardosanidze, T. N. *Vysokomol. Soedin.* **1967**, *9B*, 983. (b) Bostick, E. E. *Polym. Prepr.* **1969**, *10*, 877.

Chapter 7

Synthesis of Linear Copolysiloxanes with Regular Microstructures and Synthesis of Soluble Polysiloxanes with Trifunctional $RSiO_{3/2}$ (T) or Tetrafunctional $SiO_{4/2}$ (Q) Units

William P. Weber[1*], Jyri Kalevi Paulasaari[1], and Guoping Cai[2]

[1]K. B. and D. P. Loker Hydrocarbon Research Institute, Department
of Chemistry, University of Southern California,
Los Angles, CA 90089–1661
[2]Department of Polymer Science and Engineering, Zhejiang
University, Hangzhou, China
*Corresponding author: email: wpweber@usc.edu

Anionic ring-opening polymerization (AROP) of substituted cyclotrisiloxanes permits the preparation of copolysiloxanes with regular microstructures as a result of chemo- and regio-selective polymer propagation. Soluble linear polysiloxanes containing trifunctional $RSiO_{3/2}$ (T) or tetrafunctional $SiO_{4/2}$ (Q) units have been prepared by AROP of substituted cyclotrisiloxanes.

Background

Linear polysiloxanes made up of difunctional $RR'SiO_{2/2}$ (D) units, in which R is methyl, phenyl, and R' is methyl, phenyl, 3-cyanopropyl, 3,3,3-trifluoropropyl, hydrogen or vinyl are commercially the most important type of inorganic/organic polymeric materials. Copolysiloxane properties depend on both the molar ratio and sequence of the components. Random, regular, and block copolymers prepared from the same monomers can have distinctly different properties. Two approaches to the synthesis of copolysiloxanes, which have regular microstructures, have been reported. These are condensation polymerization of difunctional monomers such as vinylmethyldichlorosilane with tetramethyldisiloxane-1,3-diol (1) and anionic ring-opening polymerization (AROP) of substituted cyclotrisiloxane (2-4). Trifunctional $RSiO_{3/2}$ (T) units are found at branch points while tetrafunctional $SiO_{4/2}$ (Q) units may be located at the center of star burst polymers or at the junction point of two linear polysiloxane chains. Siloxane polymers that contain large numbers of T or Q units are typically not soluble. Only a few Q resins are commercially available (5).

Chemo- and Regioselectivity

We have found that anionic ring-opening polymerization (AROP) of mono-substituted and 1,1-disubstituted cyclotrisiloxanes permits, in some cases, the preparation of copolysiloxanes that have regular microstructures. In these polymers, the microstructure of the chain is controlled by the propagation step. Regioselectivity is observed when nucleophilic attack by the propagating silanolate anion occurs selectively on the substituted silyl center of the cyclotrisiloxane monomer. Ring-opening of the hypervalent silanolate thus formed leads to an ordered structure of the siloxane units in the polymer chain.

For example, AROP of 1-hydrido-pentamethylcyclotrisiloxane is initiated by dilithio diphenylsilanediolate (6). This reaction yields poly(1-hydrido-pentamethyltrisiloxane) with a highly regular microstructure (2). Similar AROP of both 1-vinyl-pentamethylcyclotrisiloxane and 1,1-divinyl-tetramethylcyclo-trisiloxane proceeds in a highly chemo- and regioselective manner to yield poly-(1-vinylpentamethyltrisiloxane) (3,4) and poly(1,1-divinyltetramethyltri-siloxane), (7) respectively. Each of these polymers have highly regular micro-structures. Mechanism of AROP and ^{29}Si NMR specta of poly(1,1-divinyltetra-methylcyclotrisiloxane) are shown in Figures 1 and 2. Chemical modification of such polysiloxanes that have either Si-H or Si-vinyl functional groups regularly arranged along the polymer backbone has been accomplished by Pt or Ru catalyzed hydrosilylation reactions.

Incorporation of T Units

Herein, we report several examples of the synthesis of soluble linear polysiloxanes, which regularly incorporate a number of T units. For instance,

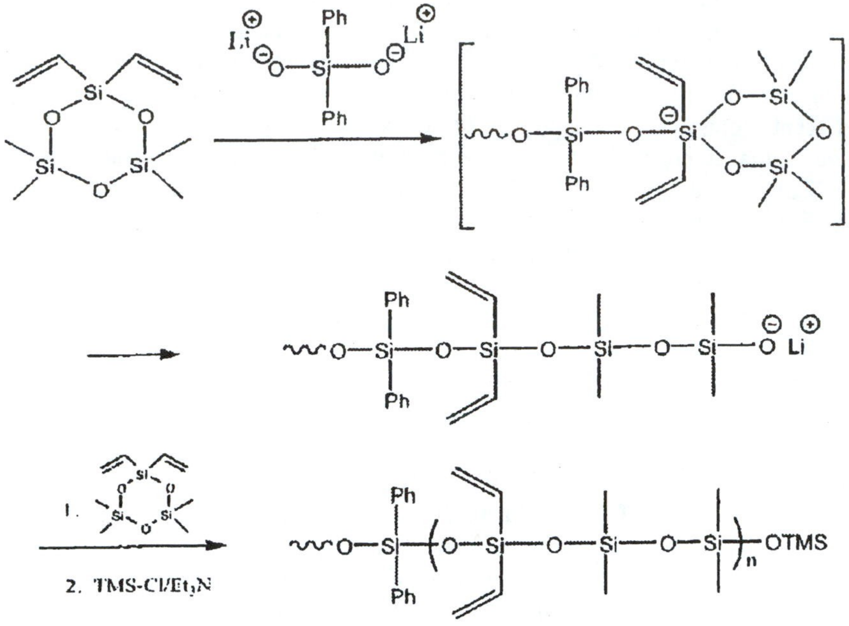

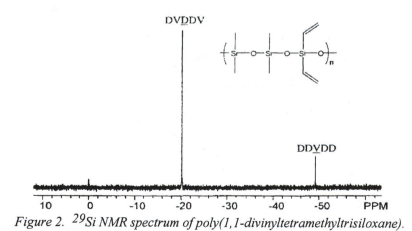

Figure 1. Synthesis of poly(1,1-divinyltetramethyltrisiloxane).

DVDDV

DDVDD

10   0   -10   -20   -30   -40   -50   PPM

Figure 2. ^{29}Si NMR spectrum of poly(1,1-divinyltetramethyltrisiloxane).

AROP of a mixture of cis and trans 1,3,5-trimethyl-1,3,5-tris(trimethylsiloxy)-cyclotrisiloxane leads to high molecular weight poly[methyl(trimethylsiloxy)-siloxane] [M_w/M_n = 25,000/16,000; T_g = -95 °C]. In this polymer, each silyl center of the polymer backbone is a T unit (8). The required cyclotrisiloxane has been prepared by reaction of 1,1-dichlorotetramethyldisiloxane with DMSO (9-11). 1,1-Dichlorotetramethyldisiloxane was, in turn, prepared by a superacid (FeCl$_3$/HCl) catalyzed equilibration reaction of methyltrichlorosilane with hexamethyldisiloxane (12). This reaction is shown in Figure 3. In this way, methyltrichlorosilane, an often-unwanted by-product of the direct synthesis reaction of methyl chloride with silicon metal, can be converted into a linear polysiloxane that has significant thermal stability and a low T_g.

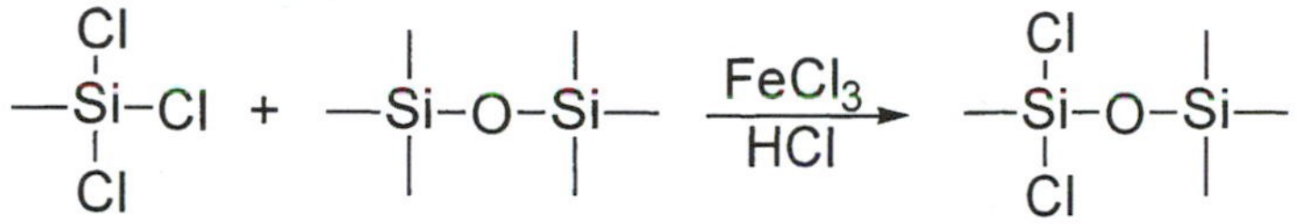

Figure 3. Synthesis of 1,1-dichlorotetramethyldisiloxane.

Isomeric polysiloxanes with regular microstructures that contain both Si-H and T units have been prepared. Thus, AROP of 1-dimethylsiloxypentamethyl-cyclotrisiloxane gives poly(1-dimethylsiloxypentamethyltrisiloxane) [M_w/M_n = 6060/4120; T_g -131 °C] whereas AROP of 1-hydrido-1-trimethylsiloxytetra-methylcyclotrisiloxane gave poly(1-hydrido-1-trimethylsiloxytetramethyltri-siloxane) [M_w/M_n = 17,900/11,800; T_g = -120 °C]. Both of these have highly regular microstructures as determined by the ^{29}Si NMR spectra, Figures 4 and 5. Competition between propagation and chain transfer apparently limits the molecular weight of poly(1-dimethylsiloxypentamethyltrisiloxane) (13,14).

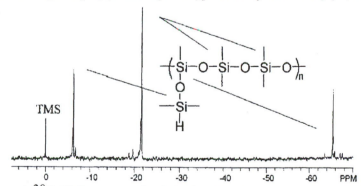

Figure 4. ^{29}Si NMR spectra of poly(1-dimethylsiloxypentamethyltrisiloxane).

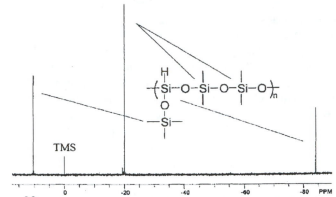

Figure 5. ^{29}Si NMR spectra of poly(1-hydrido-1-trimethylsiloxytetramethyl-trisiloxane).

The T groups of poly(1-hydrido-1-vinyldimethylsiloxytetramethyltri-siloxane) are substituted with both vinyldimethylsiloxy and Si-H functional groups. This polysiloxane has been prepared by AROP of 1-hydrido-1-vinyldi-methylsiloxy-tetramethylcyclotrisiloxane. While AROP of this monomer is chemoselective, it is not regioselective. Poly(1-hydrido-1-vinyldimethyl-siloxytetramethyltrisiloxane) undergoes platinum catalyzed hydrosilylation crosslinking (*15*).

Incorporation of Q Units

Finally, we would like to report the synthesis of a polysiloxane in which every third siloxane unit is a Q unit. AROP of 1,1-bis(trimethylsiloxy)tetra-methylcyclotrisiloxane undergoes chemoselective, but not regioselective, ring-opening. This produces three triads, Figure 6. The monomeric cyclotrisiloxane was prepared by reaction of DMSO/triethylamine with 3,3-dichlorohexa-methyltrisiloxane. Poly[1,1-bis(trimethylsiloxy)tetramethyltrisiloxane] [M_w/M_n = 25,000/16,000; T_g = -121 °C] undergoes catastrophic decomposition in nitrogen beginning at ~ 400 °C. In air, on the other hand, decomposition commences at ~ 250 °C. A char yield of almost 40 % is stable between 500 and at least 800 °C (*16,17*).

While AROP has proved successful in preparing polysiloxanes which contain both T and Q units, attempts to prepare a polysiloxanes in which each silyl center of the backbone is a Q unit by AROP of hexakis(trimethylsiloxy)-cyclotrisiloxane failed (*18*).

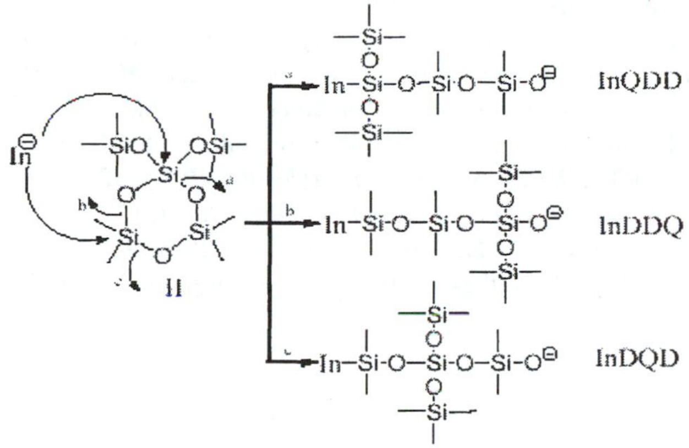

Figure 6. Triads from AROP of 1,1-bis(trimethylsiloxy)tetramethyltrisiloxane.

Conclusion

AROP of trimethylsilyloxy substituted cyclotrisiloxanes allows the preparation of unusual polysiloxanes. These polymers may have highly regular microstructures as a result of the chemo- and regioselectivity of polymer propagation. Soluble linear polysiloxanes containing trifunctional $RSiO_{3/2}$ (T) or tetrafunctional $SiO_{4/2}$ (Q) units have been prepared.

References

1. Harris, G. I. *J. Chem. Soc.* **1963**, 5978.
2. Paulasaari, J. K.; Weber, W. P. *Macromolecules*, **1999**, *32*, 6574.
3. Hempenius, M. A.; Lammertink, R. G. H.; Vancso, G. J. *Macromolecules* **1997**, *30*, 266
4. Rozga-Wijas, K.; Chojnowski, J.; Zundel, T.; Boileau, S. *Macromolecules* **1996**, *29*, 2711.
5. Gelest Inc. Tullytown, PA.
6. Battjes, K.; Kuo, C.-M.; Miller, R. L.; Saam, J. C. *Macromolecules* **1995**, *28*, 790.
7. Weber, W. P.; Cai, G. *Macromolecules* **2001**, *34*, 4355.
8. Cai, G. P.; Weber, W. P. *Macromolecules* **2000**, *33*, 6310.
9. Lu, P.; Paulasaari, J. K.; Weber, W. P. *Organometallics* **1996**, *15*, 4649

10. Goossens, J. C. General Electric Co., French Patent, 1,456,981, Oct. 1, 1964. *Chem. Abstr.* **1967**, *67*, 54259.
11. Voronkov, M. G.; Basenko, S. V. *J. Organomet. Chem.* **1995**, *500*, 325.
12. Paulasaari, J. K.; Weber, W. P. *J. Organomet. Chem.* **1999**, *584*, 376.
13. Cai, G.; Weber, W. P. *Polymer Prep.* **2001**, *42(1)* 198.
14. Cai, G. P.; Weber, W. P. *Macromolecules* **2000**, *33*, 8976.
15. Paulasari, J. K.; Weber, W. P. *Macromolecules,* **1999**, *32*, 5217.
16. Cai, G.; Weber, W. P. *Polym. Preprints* **2001**, *42(1)*, 171.
17. Cai, G.; Weber, W. P. *Macromol. Chem. Phys.* **2000**, *201*, 2234.
18. Paulasaari, J. K. Ph.D. thesis, University of Southern California, 1999.

Chapter 8

Polysiloxanes in Compressed Carbon Dioxide

Sarah L. Folk[1] and Joseph M. DeSimone[1,2,*]

[1]Department of Chemistry, University of North Carolina at Chapel
Hill, Chapel Hill, NC 27599
[2]Department of Chemical Engineering, North Carolina State
University, Raleigh, NC 27695

Compressed carbon dioxide (CO_2) is an environmentally
friendly solvent alternative with potential for extensive use in
industrial processes. Polysiloxanes are one of only a few
classes of polymers that are soluble in this medium. This
chapter reviews the properties and synthesis of polysiloxanes
in liquid and supercritical CO_2, as well as their areas of
utilization.

As the negative impact of extensive water and traditional organic solvent
use in industrial processes becomes more apparent, researchers look for
environmentally friendly solvent alternatives. Carbon dioxide (CO_2) is an
excellent choice because it is environmentally benign, nontoxic, nonflammable,
and relatively inexpensive. Additionally, CO_2 is readily available from natural
reservoirs and as a byproduct of current industrial processes as well as being
recyclable in many applications. As a replacement solvent, CO_2 is used in its
compressed liquid or supercritical fluid (SCF) phases. SCFs have gas-like
viscosities and liquid-like densities; the tuning of these properties, and hence the
solvency of the medium, can be accomplished with minimal adjustments to
temperature and pressure. A convenient advantage of the utilization of $scCO_2$ as
compared to other SCFs, is its readily accessible critical temperature and
pressure ($T_{critical} = 31$ °C; $P_{critical} = 73.8$ bar) (1).

Solubility in CO_2

Because liquid and supercritical CO_2 (liq, scCO_2) have low dielectric constants, low polarizability per volume, and minimal Van der Waals interactions, many highly polar species (i.e. water, salts) and most polymeric species are insoluble in CO_2. Only two main classes of polymers were shown to be appreciably soluble in CO_2: amorphous fluoropolymers and polysiloxanes (2). Poly(ether-carbonates) (3) and oligomeric poly(propylene oxide) (PPO) (4) were also found to be soluble in CO_2 but to a lesser extent. Although polysiloxanes are less soluble than most fluoropolymers, they are easier to characterize and significantly less expensive to use, especially on a large scale. Their relatively low cost combined with characteristics such as low glass transition temperature, low surface tension, good thermal and oxidative stability, and optical transparency have resulted in widespread attention in the field.

In order for polysiloxanes to be successfully employed in compressed CO_2 processes, basic research on the solution properties of the most elementary system, poly(dimethylsiloxane)/CO_2, is necessary.

Macroscopic Solubility

The most commonly used method to determine the macroscopic solubility of poly(dimethylsiloxane) (PDMS) in liq and scCO_2 is with cloud point measurements. PDMS forms a clear, colorless solution in CO_2. The transition from a one-phase to a two-phase solution occurs by varying the temperature or pressure. The point of precipitation is determined visually, or through more advanced techniques such as turbidimetry or light scattering. The characteristic upper and lower critical solution behavior of PDMS homopolymer in CO_2, including the upper and lower critical solution temperatures, was demonstrated (5).

The addition of a polar group to PDMS homopolymers usually results in a distinct decrease in CO_2 solubility (Figure 1). However, an increased solubility of PDMS with an addition of propyl acetate side chains was shown to be due to interactions between the side chain carbonyl and CO_2 (6). The solubilities of PDMS analogs of conventional hydrocarbon surfactants were also determined in CO_2 (7).

Determination of solubility properties of PDMS in liq and scCO_2 is important for processes such as fractionation of high polydispersity samples which depend upon differences in solubility (8). The rate of pressure quench on the pressure-induced phase separation (PIPS) of PDMS in CO_2 (9) was investigated along with the mechanistic change in phase separation from nucleation and growth to spinodal decomposition as a function of the depth of pressure quench (*10*). The ability to determine the binodal and spinodal envelopes, and hence the change in growth mechanism, is beneficial for determining the final particle morphology. PIPS is an important step in many SCF-based processes such as spray coating, textile dyeing, and the rapid expansion of supercritical solutions (RESS). A recent paper exploiting the RESS process explored spraying PDMS from a solution to coat a chemical sensor (*11*). The chemical sensor coated with uniform microspheres of PDMS showed an increased sensitivity and a rapid and reversible response.

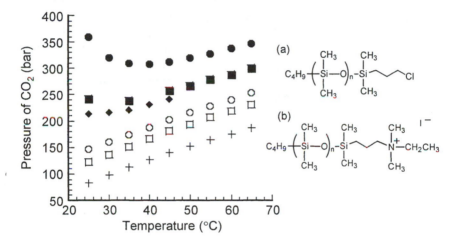

Figure 1. Solubility curves before (a: +, □, ○) and after (b: ♦, ■, ●) polar functionalization at 5 w/w% and 0.5 w/w%, respectively, of 2 kg/mol (+, ♦), 3 kg/mol (□, ■), and 5 kg/mol(○,●) PDMS.

Microscopic Solubility

Although the solubilities of homopolymers in CO_2 were found to correlate strongly with the cohesive energy density and surface tension of the polymers in the bulk phase (*4*), the extent of solubility of individual polymer chains in any solvent is dependent upon the relative strengths of the polymer-polymer, solvent-solvent, and polymer-solvent interactions. The net interaction between two polymer chains serves to classify the solvent as a "good," "theta (θ)," or "poor" solvent for a particular polymer at specified conditions. For systems where the net interaction is repulsive, the chains are swollen (larger than the ideal Flory size) and the solvent is deemed a "good solvent" for the polymer. Conversely, if the net interaction is attractive, then the polymer is in a "poor solvent" environment and either insoluble, or soluble and smaller than ideal size. When the net interaction is zero (or very nearly zero), the polymer is in "θ solvent" conditions and assumes an ideal conformation.

Small angle neutron scattering (SANS) studies probed these interactions for solutions of PDMS in CO_2. Although PDMS was reasonably soluble in CO_2, CO_2 was designated as a poor solvent for PDMS, according to the above solvent definitions, based on experimental results wherein the individual chains experienced aggregation except in very dilute solutions (12). Additional work with this system at a constant CO_2 density of 0.95 g/cm^3 showed the existence of a theoretically predicted theta temperature (T_θ) of 65 °C at which the chains exhibited ideal behavior (13, 14). Perhaps more interesting was that the compressible nature of CO_2 permitted the researchers to determine a theta pressure (P_θ) of 52MPa (Figure 2) along with the more traditional T_θ (13). This is believed to be the first P_θ on record.

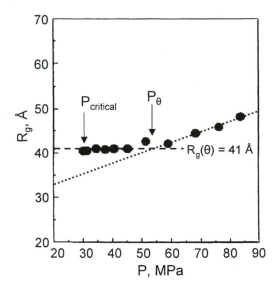

Figure 2. Variation of the radius of gyration (Rg) for PDMS with M_W = 22,500 g/mol in a solution of (h + d) PDMS in $scCO_2$ vs pressure at T = 70 °C.

SANS in combination with dynamic light scattering more recently was used to measure the dynamic and static correlation lengths of PDMS chains in poor solvent conditions (T < T_θ) (15). Additionally, fluorescence techniques were used to determine the mean-free distance between termini of pyrene end-labeled PDMS as a function of density (16).

Swelling with CO_2

The swelling of both cross-linked and uncross-linked PDMS are areas of current research. The swollen cross-linked PDMS system is beneficial to modeling stationary phase behavior of SCF chromatography (17, 18). Uncross-linked PDMS swollen with CO_2 at ambient temperatures, on the other hand, is an analog to thermoplastic melts at their processing temperatures. The sorption of gases into polymer melts is important to polymer processing with SCFs, especially processes such as extrusion and injection molding in the manufacture of foams and composites. Dissolved CO_2 modifies the rheological properties of the melt by lowering the viscosity. This decrease in viscosity is due primarily to the increase in free volume upon swelling although also due to the decreased concentration of polymer chains (19). The use of $scCO_2$ as an additive enables well-defined tuning of the processing systems as well as ease of additive removal. This is in contrast to traditional additives which present challenges in achieving desired properties of the system both during and after processing.

The effects of pressure, temperature, and molecular weight on the extent and kinetics of swelling of the uncross-linked PDMS system were determined (20-22). Swelling, monitored in situ as a function of time, showed an initial region of dramatic increase in volume followed by a relatively small increase to reach an equilibrium value as seen in Figure 3 (21). The overall change in volume increased with an increase in pressure or molecular weight, but exhibited pressure-dependent temperature effects. In situ ATR (Attentuated Total Reflectance)-IR spectroscopy was used to simultaneously measure CO_2 sorption and polymer swelling as well as to provide insight into the molecular interactions (22). The effect of CO_2 swelling on the dynamics of a non-polar solute in PDMS was also determined using fluorescence measurements (23).

Additionally, the first report of the swelling of thin films (<200-nm) was conducted on uncross-linked PDMS (24). The swelling and sorption values were measured by in situ ellipsometry based on the change in refractive index and thickness of the approximately 100-nm exposed films. The uncross-linked thin films were reported to swell more than bulk films due to excess CO_2 at the polymer interfaces and a change in chain conformation due to the influence of the solid substrate and CO_2. The actual comparison, however, was with a partially cross-linked bulk sample. Examination of the swelling of a purely uncross-linked bulk sample with the thin film reveals nearly identical swelling behavior (*21, 22, 24*).

On the opposite end of the spectrum, researchers measured the increase in viscosity of CO_2 upon addition of small percentages of PDMS (*25*). These results can be of practical value, as thickeners are formulated for control of CO_2 in oil recovery.

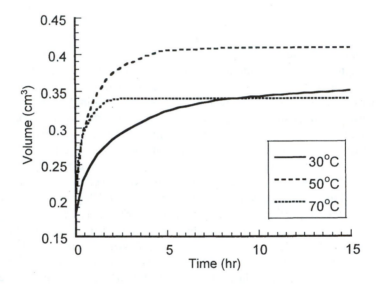

Figure 3. Swelling behavior as a function of time for a 95 kg/mol PDMS sample exposed to CO_2 at 20.7 MPa and specified temperatures.

Adapted from reference 13. Copyright 1999 American Chemical Society.

The Sanchez-Lacombe and Panayiotou-Vera equations of state were used extensively to model changes in viscosity and results of swelling measurements of PDMS in CO_2 with (*18*) and without (*17, 20, 21, 25, 26*) added cosolvents. In most cases, these models are in reasonable agreement with experimental data, although the development of improved thermodynamic models that are able to more accurately describe systems over extended ranges of temperature and pressure is important.

Synthesis of Polysiloxanes in CO_2

Researchers have recognized the advantages of using CO_2 as a polymerization medium. A recent review discussed these benefits (27) and they therefore will not be addressed here. The synthesis of CO_2-philic siloxane-based homopolymers and copolymers in liq and $scCO_2$ has received attention. Two monomers used for the synthesis of PDMS are hexamethylcyclotrisiloxane (D_3) and octamethylcyclotetrasiloxane (D_4). Anionic polymerization of these monomers in $scCO_2$ was shown to be unsuccessful due to the prominent side reaction of the anionic species with CO_2 (28). Cationic polymerizations of D_4 with triflic acid and methyl triflate initiators, however, formed PDMS in yields as high as 80% depending upon the CO_2 pressure (29). Homopolymers of PDMS were also synthesized using methacryloxypropyl terminated poly(dimethylsiloxane) (M-PDMS) (30). In this case, the vinyl termini underwent free radical polymerization and achieved a degree of polymerization of 5 to 6. Higher molecular weight polymer formation was limited due to the bulkiness of M-PDMS. Copolymerizations of 1,1-dihydroperfluorooctyl methacrylate (FOMA) with M-PDMS in $scCO_2$ were investigated (Scheme 1) and the reactivity ratio r_1 of FOMA determined (30).

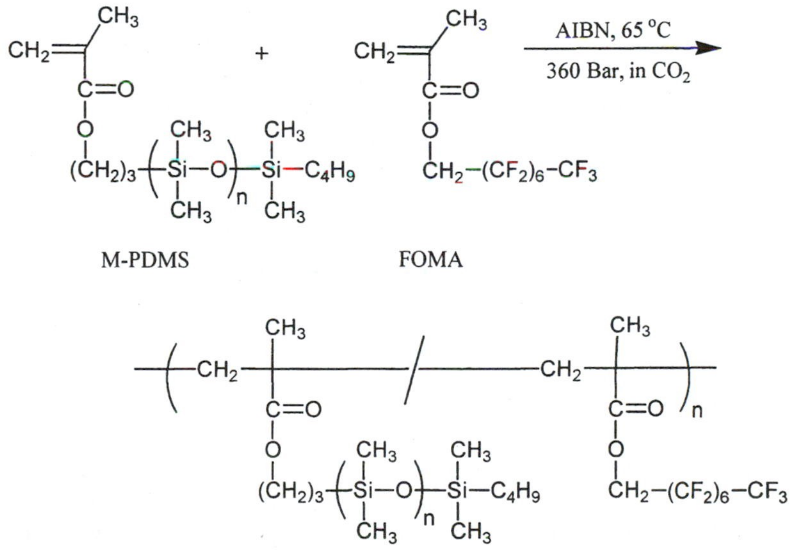

Scheme 1. Copolymerization of M-PDMS and FOMA in scCO2. (Reproduced with permission from reference 30. Copyright 2000 John Wiley & Sons.)

These fluorinated organosiloxanes have potential for use as high temperature lubricants, elastomers, or low surface energy and fouling release coatings. An alternative route to fluorinated organosiloxanes is through hydrosilation of poly(methylhydrosiloxane) by a fluorinated olefin in scCO$_2$ with Karstedt's catalyst as shown in Scheme 2 (*31*). The resulting products of these reactions possessed varying solubilities in CO$_2$ based on molecular weight and composition.

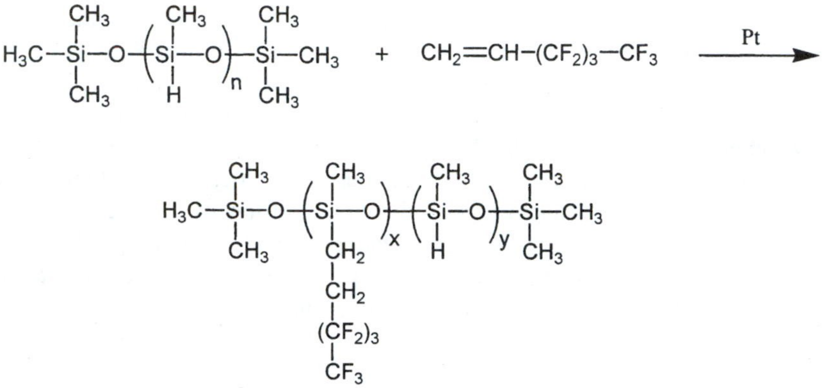

Scheme 2. A hydrosilation reaction in scCO$_2$ between poly(methylhydrosiloxane) and nonafluorohexene to form a fluorinated polysiloxane. (Reproduced with permission from reference 31. Copyright 2001 Wiley-VCH.)

Polysiloxanes as Stabilizers

Dispersion Polymerizations

Unlike PDMS, most polymers, even at low molecular weights, are insoluble in CO$_2$. In contrast to the resultant polymer, monomers and initiators for dispersion polymerizations are CO$_2$-soluble. To prevent the growing polymer particles from precipitating out of solution at a critical molecular weight, stabilizers are used which either physically adsorb to the particle surface or are chemically grafted into the growing particle. Effective stabilizers must provide sufficient anchorage to the growing particle as well as sufficient extension and solubility of the CO_2-soluble moiety in CO_2 to prevent coagulation. These dispersion polymerizations can result in uniform particles ca. 0.1 to 10 μm (*32*) in size which can be collected as dry, free-flowing white powders upon venting of CO_2 after polymerization. Researchers have used PDMS as the CO_2-soluble portion of the stabilizers.

PDMS homopolymer

The first account in the literature of a siloxane-based stabilizer for dispersion polymerizations in CO_2 was for the polymerization of methyl methacrylate (MMA) using poly(dimethylsiloxane)-monomethacrylate (PDMS-mMA) as the stabilizer (*33*). This macromonomer was chemically grafted to the particle during polymerization to afford stabilization and the formation of uniform particles. Rates of reaction, kinetic rate constants, loci of polymerization, and the mechanism of particle formation were reported for this system (*34, 35*). Because CO_2 is in the poor solvent regime for PDMS during the polymerizations, the presence of monomer in the continuous phase is required to prevent flocculation of the dispersions. To ensure that the polymerization was a dispersion as compared to a precipitation polymerization, a minimum pressure (~207 bar) and stabilizer concentration (~2 wt% stabilizer/monomer) were also necessary to achieve sufficient solvent quality and surface coverage, respectively. Gamma radiation was uniquely employed to polymerize MMA with PDMS-mMA as a stabilizer in addition to commercially available biacryloxy, biepoxy, and biacetoxy terminated PDMS (*36*). PDMS-mMA was used successfully along with other vinyl-terminated PDMS-based macromonomers for the polymerization of styrene (*33*), vinyl acetate (*37*), and random copolymers of MMA and ethyl methacrylate (*38*).

PDMS copolymer

Chemical incorporation of the stabilizer into the polymerized particles may not always be desirable and the amount of stabilizer able to be incorporated may be insufficient for stabilization. An alternative is to use diblock copolymer stabilizers such as polystyrene-*b*-PDMS (PS-*b*-PDMS) where the CO_2-phobic block adheres to the polymer particle by means of physical adsorption. Dispersion polymerizations of styrene were conducted with this diblock stabilizer. The ratio of CO_2-phobic to CO_2-philic blocks (the anchor-soluble balance) dramatically effected the progress of the reactions and the resulting particle morphologies (*39*). Particle stabilization as related to interfacial activity was also looked at for this system (*32*). In addition to the polymerization of styrene, PS-*b*-PDMS was also used as a successful stabilizer for the polymerization of the water soluble monomer, vinylpyrrolidone, when stabilizer concentrations were less then 1.9 wt% (*40*). Higher concentrations resulted in a plasticized outer shell of PDMS which allowed the poly(vinylpyrrolidone) particles to fuse.

"Ambidextrous" and unsaturated stabilizers

"Ambidextrous" block copolymer stabilizers were investigated to enable colloidal PMMA to be synthesized in CO_2 and then redispersed in water without further modification (*41, 42*). This was first accomplished by use of the stabilizer PDMS-*b*-poly(methacrylic acid) (PDMS-*b*-PMA) where the PDMS blocks afforded stabilization during polymerization in CO_2 and collapsed onto the surface of the particles when placed in water. In contrast, the PMA block sufficiently adsorbed to the particle while of neutral charge in CO_2, and provided electrostatic stabilization of the colloids when transferred into water (Figure 4). This system supported a stable aqueous latex on the order of one hour with subsequent agglomeration due to insufficient stabilization of the large colloids.

To solve this problem, new stabilizers were used that had higher molecular weight PDMS tails for increased steric stabilization, and a CO_2-phobic portion

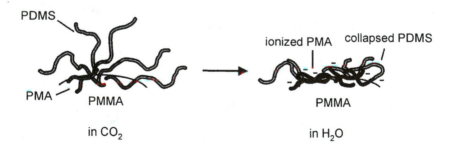

Figure 4. PDMS-b-PMA effectively stabilizes PMMA particles through steric stabilization in CO₂ and electrostatic stabilization in H₂O. (Reproduced with permission from reference 43. Copyright 2001 Elsevier Science Ltd.)

composed of either methacrylic or acrylic acid in combination with MMA or *t*-butyl acrylate for improved surface coverage by increased adsorption to the particles. The resultant colloids could be redispersed to form up to 40 wt% aqueous latexes with no evidence of agglomeration (*42*). It was also shown that colloids synthesized with PDMS-mMA could be redispersed in aqueous solution upon addition of sodium dodecyl sulfate (*42*).

The only example thus far of an unsaturated stabilizer for dispersion polymerizations in scCO$_2$ is for the polymerization of MMA with the polymer stabilizer synthesized from (bicyclo[2.2.1]hept-5-en-2-yl)triethoxysilane by ring opening metathesis polymerization using Grubbs catalyst. The unsaturated backbone of the stabilizer functions as the CO$_2$-phobic surface active moiety and the silicon functionalities as the CO$_2$-philic portion (*44*). A disadvantage of unsaturated stabilizers is the possibility for chain transfer to the stabilizer and cross-linking of the developing polymer particle. The solubility of the PMMA resulting from the polymerization of MMA with the stated stabilizer indicates that no extensive cross-linking occurred. However, it is proposed that only 0.4 wt% stabilizer actually participated in the polymerization, and therefore the existence of minimal cross-linking is possible.

Cy = cyclohexyl

Water/CO$_2$ systems

CO$_2$-in-water (C/W) and water-in-CO$_2$ (W/C) emulsions are useful for many functions such as interfacial reactions, dry cleaning applications, and synthesis. PDMS-based ionic and nonionic copolymer surfactants were used to form both C/W and W/C emulsions. The first study compared a series of PDMS-*g*-poly(ethylene oxide) (PDMS-*g*-PEO), PDMS-*g*-(PEO-*b*-PPO), and PDMS-*g*-PPO with a fluorinated anionic surfactant and a series of pluronics and showed that the decrease in interfacial tension between CO$_2$ and water caused by the surfactant correlated with the amount of solvated water in CO$_2$ (*45*). The most effective surfactant, PDMS$_{24}$-*g*-PEO$_{22}$, lowered the interfacial tension from ca. 20 to 0.2 mN/m and was able to uptake approximately 10 w/w% water upon W/C emulsion formation.

More recent work for 50/50 W/C emulsions focused on PDMS-*b*-PMA and PDMS-*b*-poly(acrylic acid) surfactants which have varying hydrophilicities dependent upon the pH of the buffered water phase (*46*). Flocculation and coalescence of the emulsions could be tailored as a function of pH, CO_2 pressure, temperature, and the hydrophilic-CO_2-philic balance (HCB). Emulsions were stable for more than seven days in some cases, and gelation of flocculated droplets stabilized the emulsions whereas addition of a cosolvent solvated the PDMS chains and reduced flocculation and stability.

C/W emulsions with nonionic polymeric surfactants at 50/50 w/w% CO_2 and water were also targeted (*47*). Although in some cases W/C emulsions were expected based on Bancroft's rule, C/W emulsions were formed instead. This unanticipated result was attributed to limited solvation of the CO_2-philic tails, rapid diffusion of droplets and surfactant through CO_2, and rapid drainage of CO_2 films between droplets - all of which disfavor stabilization of W/C emulsions (*47*).

Particle Stabilization

Building upon the successful stabilization of colloidal polymer particles in CO_2, the steric stabilization of hydrophobic and hydrophilic silica gel particles in CO_2 was also investigated with both physically absorbed (*48*) and surface-grafted (*49*) PDMS chains. In addition, unfunctionalized PDMS homopolymer was used for steric stabilization of the sol-emulsion-gel polymerization of phenolic/furfural gel microspheres in scCO$_2$ (*50*).

Summary

The potential for compressed CO_2 to be utilized in industrial processes is enhanced by the application of polysiloxanes in this medium. The research advances of polysiloxanes in compressed CO_2 demonstrate the utility of this binary system alone, as a model system for thermoplastic melts at their processing temperatures, and as a tool to provide suitable environments for other chemistries. Determining a feasible method to reduce the quantity of water and organic solvents in industrial wastestreams is a challenge, and the replacement of traditional solvents with liq or scCO$_2$ is a viable solution.

Acknowledgments

We thank the National Science Foundation Science and Technology Center for Environmentally Responsible Solvents and Processes for support under Agreement No. **CHE-9876674** and the Alfred P. Sloan Foundation for a research fellowship to J.M.D.

References

1. Quinn, E. L.; Jones, C. L. *Carbon Dioxide;* Reinhold: New York, 1936.
2. DeSimone, J. M.; Maury, E. E.; Menceloglu, Y. Z.; McClain, J. B.; Romack, T. J.; Combes, J. R. *Science* **1994,** *265,* 356-359.
3. Sarbu, T.; Styranec, T.; Beckman, E. J. *Nature (London)* **2000,** *405,* 165-168.
4. O'Neill, M. L.; Cao, Q.; Fang, M.; Johnston, K. P.; Wilkinson, S. P.; Smith, C. D.; Kerschner, J. L.; Jureller, S. H. *Ind. Eng. Chem. Res.* **1998,** *37,* 3067-3079.
5. Bayraktar, Z.; Kiran, E. *J. Appl. Polym. Sci.* **2000,** *75,* 1397-1403.
6. Fink, R.; Hancu, D.; Valentine, R.; Beckman, E. J. *J. Phys. Chem B* **1999,** *103,* 6441-6444.
7. Fink, R.; Beckman, E. J. *J. Supercrit. Fluids* **2000,** *18,* 101-110.
8. Zhao, X.; Watkins, R.; Barton, S. W.; *J. Appl. Polym. Sci.* **1995,** *55,* 773-778.
9. Li, J.; Zhang, M.; Kiran, E. *Ind. Eng. Chem. Res.* **1999,** *38,* 4486-4490.
10. Liu, K.; Kiran, E. *J. Supercrit. Fluids* **1999,** *16,* 59-79.
11. Tepper, G.; Levit, N. *Ind. Eng. Chem. Res.* **2000,** *39,* 4445-4449.
12. Chillura-Martino, D.; Triolo, R.; McClain, J. B.; Combes, J. R.; Betts, D. E.; Canelas, D. A.; DeSimone, J. M.; Samulski, E. T.; Cochran, H. D.; Londono, J. D.; Wignall, G. D. *J. Mol. Struct.* **1996,** *383,* 3-10.
13. Melnichenko, Y. B.; Kiran, E.; Wignall, G. D.; Heath, K. D.; Salaniwal, S.; Cochran, H. D.; Stamm, M. *Macromolecules* **1999,** *32,* 5344-5347.
14. Melnichenko, Y. B.; Kiran, E.; Heath, K.; Salaniwal, S.; Cochran, H. D.; Stamm, M.; Van Hook, W. A.; Wignall, G. D. *ACS Symp. Ser.* **2000,** *739,* 317-327.
15. Melnichenko, Y. B.; Brown, W.; Rangelov, S.; Wignall, G. D.; Stamm, M. *Phys. Lett. A* **2000,** *268,* 186-194.
16. Kane, M. A.; Pandey, S.; Baker, G. A.; Perez, S. A.; Bukowski, E. J.; Hoth, D. C.; Bright, F. V. *Macromolecules* **2001,** *34,* 6831-6838.

17. Xu, S.; Wells, P. S.; Tao, Y.; Yun, K. S.; Parcher, J. F. *ACS Symp. Ser.* **2000,** *748,* 96-118.
18. West, B. L.; Bush, D.; Brantley, N. H.; Vincent, M. F.; Kazarian, S. G.; Eckert, C. A. *Ind. Eng. Chem. Res.* **1998,** *37,* 3305-3311.
19. Gerhardt, L. J.; Manke, C. W.; Gulari, E. *J. Polym. Sci., Polym. Phys. Ed.* **1997,** *35,* 523-534.
20. Garg, A.; Gulari, E.; Manke, C. W. *Macromolecules* **1994,** *27,* 5643-5653.
21. Royer, J. R.; DeSimone, J. M.; Khan, S. A. *Macromolecules* **1999,** *32,* 8965-8973.
22. Flichy, N. M. B.; Kazarian, S. G.; Lawrence, C. J.; Briscoe, B. J. *J. Phys. Chem. B* **2002,** *106,* 754-759.
23. Niemeyer, E. D.; Bright, F. V. *Macromolecules* **1998,** *31,* 77-85.
24. Sirard, S. M.; Green, P. F.; Johnston, K. P. *J. Phys. Chem. B* **2001,** *105,* 766-772.
25. Xiong, Y.; Kiran, E. *Polymer* **1995,** *36,* 4817-4826.
26. Gerhardt, L. J.; Garg, A.; Manke, C. W.; Gulari, E. *J. Polym. Sci., Polym. Phys. Ed.* **1998,** *36,* 1911-1918.
27. Kendall, J. L.; Canelas, D. A.; Young, J. L.; DeSimone, J. M. *Chem. Rev.* **1999,** *99,* 543-563.
28. Mingotaud, A.-F.; Cansell, F.; Gilbert, N.; Soum, A. *Polym. J. (Tokyo)* **1999,** *31,* 406-410.
29. Mingotaud, A.-F.; Dargelas, F.; Cansell, F. *Macromol. Symp.* **2000,** *153,* 77-86.
30. Shiho, H.; DeSimone, J. M. *J. Polym. Sci., Polym. Chem. Ed.* **2000,** *38,* 1139-1145.
31. Mera, A. E.; Morris, R. E. *Macromol. Rapid Commun.* **2001,** *22,* 513-518.
32. Harrison, K. L.; da Rocha, S. R. P.; Yates, M. Z.; Johnston, K. P.; Canelas, D.; DeSimone, J. M. *Langmuir* **1998,** *14,* 6855-6863.
33. Shaffer, K. A.; Jones, T. A.; Canelas, D. A.; DeSimone, J. M.; Wilkinson, S. P. *Macromolecules* **1996,** *29,* 2704-2706.
34. O'Neill, M. L.; Yates, M. Z.; Johnston, K. P.; Smith, C. D.; Wilkinson, S. P. *Macromolecules* **1998,** *31,* 2848-2856.
35. O'Neill, M. L.; Yates, M. Z.; Johnston, K. P.; Smith, C. D.; Wilkinson, S. P. *Macromolecules* **1998,** *31,* 2838-2847.
36. Filardo, G.; Caputo, G.; Galia, A.; Calderaro, E.; Spadaro, G. *Macromolecules* **2000,** *33,* 278-283.
37. Canelas, D. A.; Betts, D. E.; DeSimone, J. M.; Yates, M. Z.; Johnston, K. P. *Macromolecules* **1998,** *31,* 6794-6805.
38. Giles, M. R.; Hay, J. N.; Howdle, S. M. *Macromol. Rapid Commun.* **2000,** *21,* 1019-1023.

39. Canelas, D. A.; DeSimone, J. M. *Macromolecules* **1997,** *30,* 5673-5682.

40. Berger, T.; McGhee, B.; Scherf, U.; Steffen, W. *Macromolecules* **2000,** *33,* 3505-3507.

41. Yates, M. Z.; Li, G.; Shim, J. J.; Maniar, S.; Johnston, K. P.; Lim, K. T.; Webber, S. *Macromolecules* **1999,** *32,* 1018-1026.

42. Li, G.; Yates, M. Z.; Johnston, K. P.; Lim, K. T.; Webber, S. E. *Macromolecules* **2000,** *33,* 1606-1612.

43. DeSimone, J. M.; Keiper, J. S. *Curr. Opin. Solid State Mater. Sci.* **2001,** *5,* 333-341.

44. Giles, M. R.; Howdle, S. M. *Eur. Polym. J.* **2001,** *37,* 1347-1351.

45. da Rocha, S. R. P.; Harrison, K. L.; Johnston, K. P. *Langmuir* **1999,** *15,* 419-428.

46. Psathas, P. A.; da Rocha, S. R. P.; Lee, C. T., Jr.; Johnston, K. P.; Lim, K. T.; Webber, S. *Ind. Eng. Chem. Res.* **2000,** *39,* 2655-2664.

47. da Rocha, S. R. P.; Psathas, P. A.; Klein, E.; Johnston, K. P. *J. Colloid Interface Sci.* **2001,** *239,* 241-253.

48. Calvo, L.; Holmes, J. D.; Yates, M. Z.; Johnston, K. P. *J. Supercrit. Fluids* **2000,** *16,* 247-260.

49. Yates, M. Z.; Shah, P. S.; Johnston, K. P.; Lim, K. T.; Webber, S. *J. Colloid Interface Sci.* **2000,** *227,* 176-184.

50. Lee, K.-N.; Lee, H.-J.; Kim, J.-H. *J. Supercrit. Fluids* **2000,** *17,* 73-80.

Characterization

Chapter 9

Neutron Scattering Studies of Cyclic and Linear Poly(dimethylsiloxanes)

A. C. Dagger[1,5], V. Arrighi[2], S. Gagliardi[2], M. J. Shenton[3], S. J. Clarson[4], and J. A. Semlyen[1,6]

[1]Department of Chemistry, University of York, Heslington, York YO10 5DD, United Kingdom
[2]Chemistry Department, Heriot-Watt University, Riccarton, Edinburgh EH14 4AS, United Kingdom
[3]Polymer Research Centre, School of Physics and Chemistry, University of Surrey, Guildford, Surrey GU2 7XH, United Kingdom
[4]Polymer Research Center, College of Engineering, University of Cincinnati, Cincinnati, OH 45221–0012
[5]Current address: Smith and Nephew Group Research Center, York Science Park, Heslington, York YO10 5DF, United Kingdom
[6]Tony Semlyen left us in this world shortly before the oral presentation of this work that was given in San Diego by Tony Dagger.
We shall all miss his wonderful company.

Small angle neutron scattering (SANS) has been used to investigate the conformations of linear and cyclic poly(dimethylsiloxanes) (PDMS) in chemically identical undiluted blends over a range of molar mass and composition. Symmetrical (molar masses of the two components closely matched) and asymmetrical (molar masses of the two components significantly different) isotopic blends of linear hydrogenated PDMS (H-PDMS) and linear deuterated PDMS (D-PDMS) were shown to have radii of gyration, which agreed with theoretical predictions for random-coil linear polymers obeying Gaussian statistics. Blends of cyclic H-PDMS and cyclic D-PDMS showed that the PDMS cyclics adopt more expanded average conformations in such melts.

Introduction

Small angle neutron scattering (SANS) has been used to investigate the conformations of linear and cyclic poly(dimethylsiloxanes) (PDMS) in chemically identical undiluted blends. Here we present the results from a series of SANS experiments studying linear and cyclic PDMS over a range of molar mass and composition. Cyclic polymers offer the ideal conditions for studying the statistics of linear polymers. Ring polymers consist of identical monomers, free from the defects introduced in linear chains by terminal ends. These features make cyclic polymers the subject of interesting, yet relatively scarce investigations from both the theoretical and experimental point of view. Symmetrical (where the molar masses of the two components were closely matched) and asymmetrical (where the molar masses of the two components were significantly different) isotopic blends of linear hydrogenated PDMS (H-PDMS) and linear deuterated PDMS (D-PDMS) were shown to have radii of gyration, which agreed with theoretical predictions for random-coil linear polymers obeying Gaussian statistics. Blends of cyclic H-PDMS and cyclic D-PDMS have also been studied for the first time. The results obtained showed that the PDMS cyclics adopt more expanded average conformations in such melts. This is in line with some conjectures and computer simulations published in the literature. The findings presented in this paper are compared with our previous studies of cyclic and linear PDMS in dilute solution and the interaction parameters involved are derived from our new experimental data for the undiluted systems. Despite the range of experimental data that are available describing the behaviour of cyclic polymers in solution, the static and dynamic properties of ring polymers in the melt remain largely unexplored. As stated by Müller, Wittmer and Cates [1] :

"No experimental study of the radius of gyration of rings in the melt appears to have been made. A good understanding of the static properties of the system is of course an indispensable starting point for a reasonable description of the dynamics, so this is unfortunate".

Experimental

Materials

The preparation and characterisation of sharp fractions of hydrogenated cyclic and linear poly(dimethylsiloxanes) (PDMS) have been described in previous papers [2-4] . Many sharp fractions have been obtained, each on a scale of several grams, using preparative gel permeation chromatography (GPC)[5] . The fractions consist of ring molecules $((CH_3)_2SiO)_x$, with number-average numbers of skeletal bonds, n_n, up to ~700. A number of studies of cyclic PDMS have been made, comparing their properties with those of the corresponding linear polymers [6-8] . For example, GPC retention volumes [5,9] , dilute solution viscosities [10] , neutron scattering behaviour and radii of gyration in dilute solution [11] have all been investigated experimentally.

Both per-deuterated cyclic and linear PDMS have been prepared in one reaction from bis(trideuteriomethyl)diphenylsilane $((CD_3)_2SiPh_2)$ by treating with a catalytic amount of trifluoromethanesulphonic (triflic) acid (CF_3SO_3H) at high dilution in dichloromethane in the presence of water. Random copolymers of hydrogenated and deuterated PDMS have also been prepared in a similar fashion [12] .

The polydisperse per-deuterated linear and cyclic PDMS recovered from the catalytic reaction was fractionated into sharp fractions using our preparative GPC instrument [5] . The molar masses of the cyclic fractions show a typical spread for a ring/chain equilibration reaction. The dispersities (M_w/M_n) of the samples were also typically low, all being less than 1.2.

The purity of the fractions were verified using infrared absorption, mass spectrometry and ^1H, ^{13}C and ^{29}Si nuclear magnetic resonance spectroscopies[2-4].

Neutron Scattering

Blends of the polymers have been studied using the LOQ small angle neutron scattering diffractometer at the ISIS Spallation Neutron Source, Rutherford Appleton Laboratory (RAL), Didcot, UK [13] , and the D22 SANS instrument at the Institut Laue-Langevin, Grenoble, France [14] . The blends were investigated at different compositions (typically 5%, 25% and 50% H-PDMS by volume). Only information obtained over the scattering vector range $0.008 < q < 0.25$ Å^{-1} is examined here. The instruments were used under the normal operating conditions at all times. The samples were examined at 298K with the use of a temperature controlled cell holder. For the purpose of subtracting the background scattering, scattered intensities were recorded for pure H-PDMS and D-PDMS samples for each blend. All of the scattered intensities recorded from the blends were radially averaged about the incident beam direction and were corrected for transmission, thickness and normalised by standard procedures. Subtraction of the background was performed using the appropriately volume fraction-weighted sum of the scattered intensities of the pure H-PDMS and D-PDMS samples, shown to be accurate through the use of our randomly labeled copolymers [12].

Results and Discussion

Scattering profiles such as those shown in Figure 1 were obtained for a range of blends of PDMS and fitted using he random-phase approximation (RPA) equation, as introduced by de Gennes [15] :

$$d\Sigma(q)/d\Omega = k_n[(z_A v_A \Phi_A g_{DA}(q))^{-1} + (z_B v_B \Phi_B g_{DB}(q))^{-1} - 2\chi/v_0]^{-1}$$

-where $k_n = N_A(b_A/v_A - b_B/v_B)^2$, z_i is the degree of polymerisation, Φ_i is the volume fraction for a chain of type "i", v_i is the monomer molar volume, bi is the coherent scattering length of a monomer unit of type "i", $v_0 = (v_A v_B)^{1/2}$, related to the thermodynamical interaction between chains of different natures. The RPA describes the total scattered intensity $\delta\Sigma(q) / \delta\Omega$ in terms of the scattering vector, $q = 4\pi/\lambda \sin(\theta / 2)$, where λ is the neutron wavelength and θ is the scattering angle, the radius of gyration of the polymer coil, R_g, and the site-site interaction parameter, χ. For Gaussian random-coil polymers, the single chain structure factor, $g_{Di}(q)$, is given by the Debye function [16]:

$$g_{Di}(q) = 2(\exp(-x) - 1 + x)/x^2$$

-where $x = q^2 (R_{gi\ (chain)})^2$. The structure factor of a ring was first calculated by Cassassa [17] and has been proven to be consistent by several other independent studies [18-20] . It is expressed as the well-known function:

$$g_{Di}(q) = \left(\frac{2}{\sqrt{u}}\right)\exp-\frac{u}{4}\int_{0}^{\frac{\sqrt{u}}{2}} \exp(x^2)dx = \left(\frac{2}{\sqrt{u}}\right)D\left(\frac{\sqrt{u}}{2}\right)$$

-where $u = (q^2 2(R_{gi\ (ring)})^2)$ and $D(x)$ denotes the Dawson integral. All of the profiles were fitted via a least squares operation using only R_g and χ as fitting variables. Only details of the R_g values are presented here.

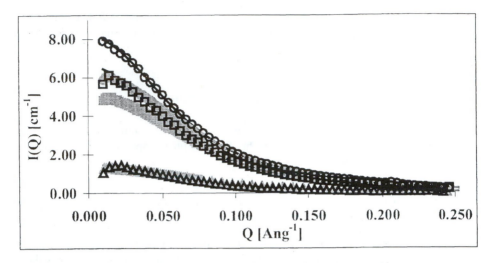

Figure 1. SANS data of a H-linear/D-linear PDMS blend (black points) and a H-cyclic/D-cyclic PDMS blend (grey points) after subtraction of the incoherent background. Both blends are symmetrical of molar mass $M_w \sim 10000$. Data has been recorded at three different volume fractions of H-PDMS (approximately 0.05 (triangles), 0.25 (squares) and 0.50 (circles)). Lines indicate fits using the Random Phase Approximation (see text).

Figure 2 shows the results of the SANS experiments as a plot of Log R_g versus log M_w for both the linear/linear blends and the cyclic/cyclic blends over a range of molar mass. It is apparent, as has been shown for many other properties of cyclic polymers when compared with their linear counterparts, that there is a difference in behaviour. In both cases linear dependencies are observed, meaning the results can be described by a power law ($R_g \sim M_w^{\nu}$). The radii of gyration of the linear blends were found to scale with molar mass with a critical exponent (ν) of 0.53 ± 0.02, while the cyclic blends were found to scale with a critical exponent of 0.39 ± 0.06.

Conclusions

The mean square radius of gyration for linear chains is twice the corresponding quantity for rings under theta solvent conditions [6,8]. In good solvents, several models are available for the description of chain swelling of rings and linear polymers. They do not predict similar tendencies. Perturbation calculations, the

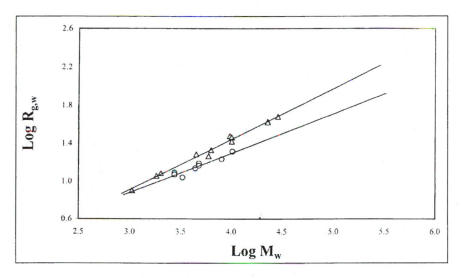

Figure 2. Log weight average radii of gyration versus log M_w for linear (Δ) and cyclic (o) H/D blends investigated in this work.

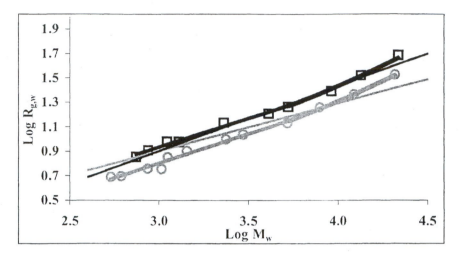

Figure 3. Log weight average radii of gyration versus log M_w for linear PDMS in dilute benzene (open black squares and thick black line), cyclic PDMS in dilute benzene (open grey circles and thick grey line), blends of H/D linear PDMS (thin black line) and blends of H/D cyclic PDMS (thin grey line).

Flory equation of swelling and field theoretical calculations predict that the ratio $\beta = R_{gl}^{\,2}/R_{gc}^{\,2}$ decreases under good solvent conditions and becomes lower than 2, the value obtained under theta conditions in the absence of excluded volume interactions. Bloomfield-Zimm, Yu-Fujita and Bensafi-Benmouna models predict higher values [21]. Exisiting data obtained from scattering experiments and computer calculations are not sufficient to draw a definite conclusion about the real tendencies although they favour a reduction of β in good solvents.

The linear blends we have examined display radii of gyration which scale with molar mass according to Gaussian statistics ($v = 0.5$) and are similar to those observed by neutron scattering at molar masses below the onset of excluded volume effects in dilute solution (see Figure 3)[11]. The cyclic PDMS polymers appear to be more expanded in a cyclic host melt than under dilute solution conditions at low molar masses becoming similar at the higher molar masses investigated. The ratio β is seen to increase with increasing molar mass in the melt. The critical exponent v has been examined extensively by computer simulations looking at the variation of the radius of gyration versus molar mass in the bulk state [21]. Exponents between 0.4 and 0.45 have been reported indicating deviations from Gaussian statistics. This is the first time the critical exponent has been measured for a blend of purely cyclic polymers in the bulk state and the exponent measured is in excellent agreement with the values obtained by computer simulations.

Acknowledgement

The authors would like to acknowledge grants awarded by the Engineering and Physical Research Council to A.C.D. (GR/M00596) and S.G. (GR/M00725), which enabled this research to be carried out. We also thank Dow Corning, Barry, S. Wales for beneficial discussions during the synthesis of the deuterated PDMS. We would also like to acknowledge the instrument scientists, Drs. Steve King and Richard Heenan, responsible for the LOQ instrument at the ISIS facility, Didcot, Oxon, UK, and Dr. Isabelle Grillo, responsible for the D22 instrument at the ILL, Grenoble, France, for their assistance with the neutron scattering measurements and data handling.

References

1. Müller, M.; Wittmer, J.P.; Cates, M.E. *Phys. Rev. E* **1996**, *53*, 5063.
2. Dagger, A.C.; Semlyen, J.A. *Polymer* **1998**, *39*, 2621.
3. Dagger, A.C.; Semlyen, J.A. *Polym. Prepr. (Am. Chem. Soc., Div. Polym. Chem.)* **1998**, *39*, 579.
4. Dagger, A.C.; Semlyen, J.A. *Polym. Commun.* **1999**, *40*, 3243.
5. Dodgson, K.; Sympson, D.; Semlyen, J.A. *Polymer* **1978**, *19*, 1285.
6. Semlyen, J.A., (Ed.), *"Cyclic Polymers"*, Elsevier Applied Science Publishers Ltd., London, **1986**.
7. Clarson, S.J. and Semlyen, J.A., (Eds.), *"Siloxane Polymers"*, PTR Prentice Hall, New Jersey, **1993**.
8. Semlyen, J.A., (Ed.), *"Cyclic Polymers (second edition)"*, Kluwer Academic Publishers, Dordrecht, **2000**. .
9. Semlyen, J.A. and Wright, P.V., *"Gel Permeation Chromatography of Cyclic Siloxanes"*, Chapter 23 of Epton, R., (Ed.), *"Chromatography of Synthetic and Biological Polymers. Column Packings, GPC, GF and Gradient Elution."* Volume 1, Ellis Horwood Ltd., Chichester, **1978**.
10. Dodgson, K. and Semlyen, J.A., *Polymer*, **1977**, *18*, 1265-1268.
11. Higgins, J.S., Dodgson, K. and Semlyen, J.A., *Polymer*, **1979**, 20, 553-588.
12. Dagger, A.C.; Elkin, A.; Semlyen, J.A. *to be published.*
13. King, S.M.; Heenan, R.K. *"The LOQ Handbook"*, RAL Technical Report, June **1996**.
14. *"D22 Manual"*; Scientific Secretariat, Institut Laue Langevin, 6 rue Jules Horowitz, BP 156, 38042 Grenoble Cedex 9, France.
15. De Gennes, P.-G. *"Scaling Concepts in Polymer Physics"*; Cornell University Press: New York, **1979**.

16. Debye, P. Tech. Rep. 637 to Rubber Reserve Co., **1945**; reprinted in: Debye, P. *The Collected Papers of Peter J. W. Debye*; Interscience: New York, **1954**.
17. Cassasa, E. *J. Poly. Sci. A* **1965**, 3, 605.
18. Burchard, W.; Schmidt, M. *Polymer* **1980**, 21, 745.
19. Edwards, C.J.C.; Richards, R.W.; Stepto, R.F.T.; Dodgson,K.; Higgins, J.S.; Semlyen, J.A. *Polymer* **1984**, 25, 365.
20. Kosmas, M; Benoît, H.; Hadziioannou, G. *Colloid Polym. Sci.* **1994**, 272, 1466.
21. See Benmouna, M. and Maschke, U., *"Theoretical Aspects of Cyclic Polymers: Effects of Excluded Volume Interactions"*, Chapter 16 of Semlyen, J.A., (Ed.), *"Cyclic Polymers (second edition)"*, Kluwer Academic Publishers, Dordrecht, **2000** for a discussion of these models and predictions.

Chapter 10

Photoexcitation and Photoemission Spectra of Phenyl-Substituted Cyclosiloxanes

Michael W. Backer[1] and Udo C. Pernisz[2]

[1]Dow Corning Limited, New Ventures R & D, Chemical Synthesis Research Group, Cardiff Road, Barry, South Glamorgan CF63 2YL, United Kingdom
[2]Dow Corning Corporation, New Ventures R & D, Physics Research Group, 2200 West Salzburg Road, Midland, MI 48686

The unique interaction between conjugated π-electron systems and the Si atom in phenyl substituted silanes and siloxanes gives rise to chromophore behavior of these compounds. Variously substituted cyclotrisiloxanes and a cyclotetrasiloxane have been synthesized, and complete excitation-emission maps were obtained in the ultraviolet to visible spectral range. The position of the emission maxima was analyzed from intensity contour plots, and two different types of behavior were identified: Well defined absolute emission maxima and local maxima shifting with increasing excitation wavelengths. A model for this phenomenon and evidence for the effects of methyl versus phenyl substituents are discussed.

Silanes and siloxanes are generally well characterized due to their importance as precursors in industrial products. Despite the broad knowledge about their physical and material properties (*1*), little is known about the photoluminescence behaviour of these classes of compounds. In recent years, first results of investigations on phenyl substituted disilanes (*2,3*) and polysilanes (*4*) as well as on variously substituted silsesquioxanes (*5*) have been published. Besides that, first observations were made of the dependence of photoluminescence features on substitution pattern in silanes (*6*) and on ring size in cyclosiloxanes (*7*). The use of specific chromophores, e.g. stilbene–analogous silacyclobutenes, resulted in the successful preparation of highly luminescent silanes and siloxanes (*8-10*). In the following, the positions of emission maxima in photoluminescence spectra of cyclotrisiloxanes measured in the solid state at room temperature are discussed with respect to the number of phenyl substituents.

Compounds

Hexamethylcyclotrisiloxane **1**, the permethylated representative of the investigated series of cyclic siloxanes, is a commercially available compound and can be purified by sublimation. Mixed methyl and phenyl substituted cyclotrisiloxanes **2-5** (with **2**: R^1=Me, R^2=R^3=Ph; **3**: R^1=Ph, R^2=R^3=Me; **4**: R^1=R^2=Ph, R^3=Me; **5**: R^1=R^2=Ph, R^3=H) have been generated via condensation reaction of accordingly substituted disiloxanediols with dichlorosilanes in diethylether in the presence of trialkylamines as described in Figure 1. The raw materials can be recrystallized from cold alkanes (*10, 11*). The perphenylated cyclotrisiloxane **6** and cyclotetrasiloxane **7** are generated by condensation reaction of diphenylsilanediol in acidic or caustic alcohol, respectively. Recrystallization can be carried out from warm butanone/chloroform (**6**) or hot glacial acetic acid (**7**) (*12*).

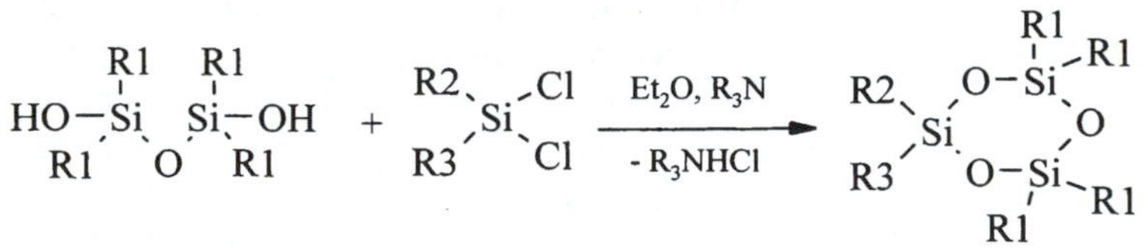

Figure 1. Preparation of Cyclosiloxanes 2- 5

Instrumentation

The photoluminescence spectra were obtained at room temperature with a Spex Fluorolog 2 spectrophotometer made by Jobin/Yvon (see Figure 2). The single grating excitation and emission monochromators are equipped with 1200 lines/nm gratings blazed at 330 nm and 500 nm, respectively. The focal length of the system is $f = 0.22$ m which results in a dispersion of about 3 nm/mm.

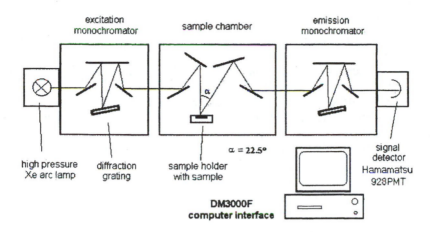

Figure 2. Schematic instrumentation (Jobin/Yvon SPEX Fluorolog 2)

The light source was a 150 W high pressure Xe arc lamp. Measurements were carried out using a circular aperture that resulted in a spectral bandwith of about 0.8 nm. The powdered samples were measured in a flat sample holder (with recessed area of ca. 1.0 cm x 0.5 cm) covered with a quartz glass. The sample holder was placed normal to the excitation beam, and the fluorescence light was collected at an angle of 22.5°. The spectra were recorded with a SPEX DM3000 and evaluated with the software program GRAMS32 (*13*). Intensity contour plots have been obtained by measuring the emission intensity in excitation wavelength's blocks of 10 nm with an increment of 2 nm for which the range of the emission wavelength was kept constant. In order to prevent excitation light from reaching the emission detector, each subsequent excitation block was run with the emission wavelength range also incremented by 10 nm.

Results and Discussion

The initial photoluminescence measurements of the cyclotrisiloxane series $[Ph_{2n}Me_{2(3-n)}SiO_3]$ **1-3** and **6** with n = 0,1,2,3, recorded at an excitation wavelength of λ_{xc} = 320 nm (*9*), appeared to suggest a simple correlation by which the increasing number of phenyl substituents causes an increase in luminescence emission intensity. However, a comparison between the emission intensities of hexaphenylcyclotrisiloxane **6** and octaphenylcyclotetrasiloxane **7** raised doubts about this concept as the hexaphenyl substituted ring exhibited, again irradiated at the wavelength λ_{xc} = 320 nm, an emission intensity about 100

108

times higher than the octaphenyl substituted ring. When similar measurements were carried out at different excitation wavelengths, completely new features were discovered, e.g. large shifts of the emission maxima position within the spectra of cyclotetrasiloxane 7 from the far UV into the blue visible region, and varying intensity ratios within the cyclotrisiloxane series as a function of the excitation wavelength. These observations prompted the decision to generate complete excitation–emission maps of the phenyl substituted cyclosiloxanes in order to obtain a basic understanding of the relationship between the intensity maxima positions and the substitution pattern of the ring systems.

Hexamethylcyclotrisiloxane [D₃] 1

The photoemission spectrum of hexamethylcyclotrisiloxane **1** recorded at the excitation wavelength $\lambda_{xc} = 320$ nm is shown in Figure 3. Small spikes at the emission wavelengths $\lambda_{ms} = 350$ and 525 nm have been identified as artefacts caused by the equipment settings. There occurs only a small increase in photoluminescence emission below 360 nm which may be assigned to the Si-Me group; the further increase in intensity below 340 nm is dominated by stray light from the excitation monochromator. Since no particular emission features could be observed in this spectrum, the creation of a contour plot was omitted.

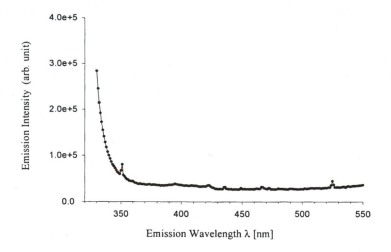

*Figure 3. Photoemission spectrum of hexamethylcyclotrisiloxane [D₃] 1;
spectrum measured in solid state at room temperature with spectral bandwidth
of 1.0 nm; excitation wavelength $\lambda_{xc} = 320$ nm.*

2,2-Diphenyl-4,4,6,6-tetramethylcyclotrisiloxane [$D_2D^{Ph_2}$] 2

In the spectrum of 2,2-diphenyl-4,4,6,6-tetramethylcyclotrisiloxane **2** presented as intensity contour plot in Figure 4, luminescence is observed to occur in the high energy region at λ_{ms} = 293 nm when compound **2** is irradiated with light of λ_{xc} = 276 nm. This transition can be assigned to the dimethylsiloxy groups of the molecule. The insertion of a diphenylsiloxy group into the ring system leads to two major transitions at the excitation wavelength λ_{xc} = 290 nm with emission at λ_{ms} = 356 and 368 nm (ΔE = 0.80 and 0.91 eV), respectively. In the lower energy regions, the spectrum shows a shift of the emission maxima towards longer wavelengths indicative of a vibrational structure. This feature with $\Delta \nu$ = 1300 – 1350 cm^{-1}, observed in the wavelength range of λ_{xc} = 290 – 320 nm cannot be assigned, however, to one of the major vibrational stretching modes such as Si-C$_{aryl}$ with $\Delta \nu$ = ~1430 cm^{-1}, or Si-O-Si at $\Delta \nu$ = 1020 – 1080 cm^{-1}(*1*).

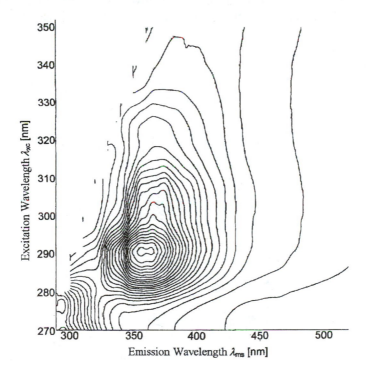

Figure 4. Photoexcitation-emission map of diphenyltetramethylcyclotrisiloxane [$D_2D^{Ph_2}$] 2; presented as intensity contour plot; spectrum measured in solid state at room temperature with spectral bandwidth 1 nm.

110

2,2-Dimethyl-4,4,6,6-tetraphenylcyclotrisiloxane [DD$^{Ph2}_2$] 3

In the spectrum of 2,2-dimethyl-4,4,6,6-tetraphenylcyclotrisiloxane **3**, presented in Figure 5, luminescence emission maxima are observed at wavelength λ_{ms} = 385 nm with excitation wavelengths of λ_{xc} = 292 and 302 nm (ΔE = 0.89 and 1.03 eV). At these excitation wavelengths, small shoulders on the emission maxima are visible at a wavelength of λ_{ms} = 368 nm with slightly reduced transition energies of ΔE = 0.73 and 0.88 eV, respectively. Surprisingly, the absolute emission maximum of compound **3** is detected in the blue region at λ_{ms} = 459 nm when the sample is irradiated with light of λ_{xs} = 388 nm (ΔE = 0.50 eV). This enormous bathochromic shift of the luminescence into the visible range gives rise to a special conjugation of the molecular orbitals of the phenyl groups within this cyclotrisiloxane. Further investigations of the molecular structure of **3** (crystal structure: monoclinic, space group P2$_1$/n; in comparison to **2**: ortho-rhombic, space group Pnma, with orthogonally positioned phenyl rings (*10, 14*)) will provide the basis for quantum mechanical calculations of this specific phenomenon.

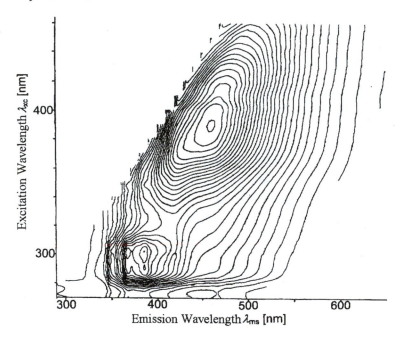

Figure 5. Photoexcitation-emission map of dimethyltetraphenylcyclotrisiloxane [DD$^{Ph_2}_2$] 3; presented as intensity contour plot; spectrum measured in solid state at room temperature with spectral bandwidth 1 nm.

Methylpentaphenylcyclotrisiloxane [D^{Ph}D^{Ph_2}_2] 4

The insertion of a fifth phenyl substituent in methylpentaphenylcyclotrisil-oxane **4** leads to nearly the same positions of emission maxima as observed in compound **3**. The spectrum, shown in Figure 6, exhibits also maxima at wavelength λ_{ms} = 385 nm and shoulders at λ_{ms} = 368 nm with excitation wavelengths of λ_{xc} = 292 and 302 nm. However, a third maximum is observed in the high energy range of the spectrum at λ_{ms} = 385 nm when the sample is irradiated with light of λ_{xc} = 288 nm (ΔE = 1.09 eV). The most pronounced difference between the spectra of compounds **3** and **4** is the disappearence of the maximum in the blue region (Figure 5) and the appearance of a new lower intensity maximum, also at λ_{ms} = 385 nm (E = 3.22 eV) when the compound is irradiated with light of wavelength λ_{xc} = 339 nm (ΔE = 0.45 eV). Obviously the suggested conjugation in compound **3** is effectively disturbed by the additional aryl substituent.

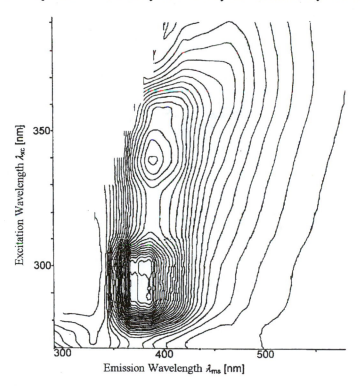

Figure 6. Photoexcitation-emission map of methylpentaphenylcyclotrisiloxane [D^{Ph}D^{Ph_2}_2] 4; presented as intensity contour plot; spectrum measured in solid state at room temperature with spectral bandwidth 1 nm.

Pentaphenylcyclotrisiloxane [$D^{Ph,H}D^{Ph}{}_2$] 5

The replacement of the methyl substituent in molecule **4** by a hydrogen atom to form pentaphenylcyclotrisiloxane **5** results in a slight bathochromatic shift ($\Delta\lambda$ = 6 nm) of the emission maxima and the shoulders to λ_{ms} = 394 and 374 nm, respectively, whereby the shoulders here form well defined local maxima as presented in Figure 7. More interesting, however, is the large shift of the corresponding excitation to longer wavelengths at λ_{xc} = 324 and 339 nm. To these belong the transition energy pairs ΔE = 0.34, 0.51 eV, and 0.51, 0.68 eV which are both shifted by δE = 0.17 eV. The highest oscillator strength is observed for the transition at λ_{xc} = 339 (with λ_{ms} = 394 nm). Such a bathochromatic effect for the excitation wavelength (cf. [$D^{Ph}D^{Ph}{}_2$] **4**) in conjunction with a hyperchromic effect for the emission intensity has been described for a series of triphenylsilanes Ph$_3$SiR (**6**) where for R = H the photoluminescence efficiency was also observed to be largest.

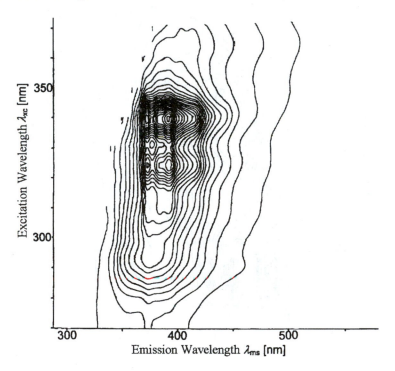

Figure 7. Photoexcitation-emission map of pentaphenylcyclotrisiloxane [$D^{Ph,H}D^{Ph}{}_2$] 5; presented as intensity contour plot; spectrum measured in solid state at room temperature with spectral bandwidth 1 nm.

Hexaphenylcyclotrisiloxane 6

The contour map of hexaphenylcyclotrisiloxane **6**, shown in Figure 8, exhibits two main intensity peaks at the emission wavelength of λ_{ms} = 356 nm with excitation at λ_{xc} = 292 and 327 nm (ΔE = 0.77 and 0.31 eV), respectively. This is an exceptional example for the behavior that the emission peak does not shift its position as the excitation wavelength is changed; thus it correlates with a single electronic transition. It is, however, quite surprising that in the spectra of the triclinic and orthorhombic modifications of this perphenylated cyclotrisiloxane (*15*) no difference in the position of the emission maxima could be observed. Besides, it also cannot be explained why linear tetraphenyldisiloxanediol exhibits emission maxima at exactly the same excitation and emission wavelengths (*6*).

Octaphenylcyclotetrasiloxane 7

In contrast to the spectrum of compound **6**, the spectrum of octaphenylcyclotetrasiloxane **7** (*16,17*), see Figure 9, shows well-defined maxima only at excitation wavelengths shorter than 300 nm, at λ_{xc} = 286 and 292 nm, with emission at λ_{ms} = 322 and 356 nm (ΔE = 0.49 and 0.77 eV), respectively. In the remaining parts of the spectrum, the second, minor emission maximum, or rather shoulder, is bathochromatically shifted with increasing excitation wavelengths. For instance, a local emission maximum is observed at λ_{ms} = 368 nm with an excitation at λ_{xc} = 302 nm (ΔE = 0.74 eV). This behavior, which is different from that of the stationary maxima in cyclotrisiloxane **6**, could possibly be explained with the similarity between cyclotetrasiloxanes or larger cyclic systems and linear molecules (*1*), although this hypothesis has to be proven by luminescence investigations on polydiphenylsiloxanes.

Summary and Conclusions

The investigation of the photoluminescence behavior of a series of cyclotrisiloxanes and cyclotetrasiloxanes variously substituted with methyl and phenyl has demonstrated that each single compound exhibits specific emission and excitation features with respect to the positions or relative intensities of the maxima, depending on its substitution pattern. Figure 10 shows the collected data set of the intensity maxima of compounds **2** to **7** in the near-ultraviolet region in an emission-excitation plot. The perphenylated cyclotrisiloxane provides a basis of two peaks (see dashed line) from which the individual substitutents appear to shift to longer excitation and emission   wavelengths

114

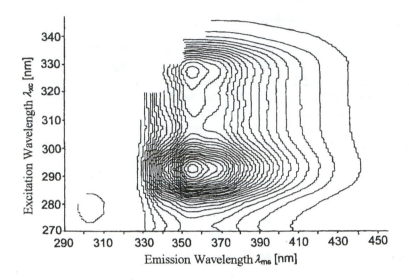

Figure 8. Photoexcitation-emission map of hexaphenylcyclotrisiloxane [$D^{Ph_2}_3$] 6; presented as intensity contour plot; spectrum measured in solid state at room temperature with spectral bandwidth 1 nm.
(Reproduced from reference 6. Copyright 2001 American Chemical Society)

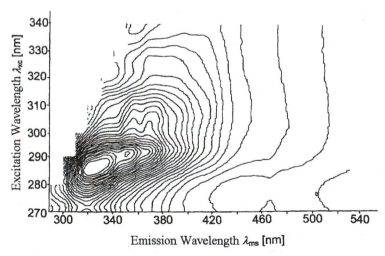

Figure 9. Photoexcitation–emission map of octaphenylcyclotetrasiloxane [$D^{Ph_2}_4$] 7; presented as intensity contour plot; spectrum measured in solid state at room temperature with spectral bandwidth 1 nm.

(arrows), D^{Me2} producing a small, D^{Ph} a large, and $D^{Ph,H}$ an intermediate bathochromatic excitation shift in the 385 nm emission region with a split of the maxima for the H-containing compound. One conclusion from the observations is that higher symmetry of the diphenylsiloxy and dimethylsiloxy groups results in the simpler photoluminescence structure.

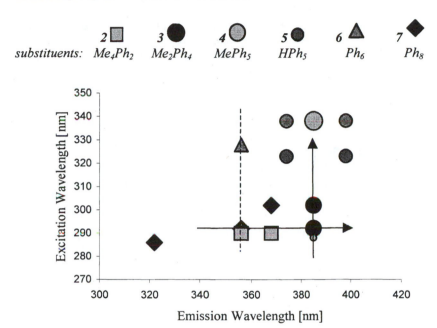

Figure 10. Overview plot of the positions of the photoluminescence intensity maxima in an emission–excitation wavelength plane for compounds 2 to 7. The substitutent groups are identified in the legend above.

Acknowledgements

The authors thank Dr. J. Weis of the Wacker-Chemie GmbH, Munich, for the donation of pure hexamethylcyclotrisiloxane **1**. They also acknowledge the support of Prof. Dr. N. Auner, Johann-Wolfgang Goethe Universität, Frankfurt, during M.W.B.'s doctoral thesis work and thank him for many stimulating discussions. Helpful comments on the data analysis by G. Zank and the computational support by T. Lauer are gratefully acknowledged. Some photo-

116

luminescence spectra were taken by A. Hart (all Dow Corning Corporation) with great skill and expert handling of the equipment.

References

1. *The Analytical Chemistry of Silicones*; Smith, A. L., Ed.; Chemical Analysis; John Wiley & Sons, Inc.: New York, 1991; Vol. 112.
2. Shizuka, H.; Obuchi, H.; Ishikawa, M.; Kumada, M. *J. Chem. Soc., Chem. Comm.* **1981**, 405.
3. Shizuka, H.; Obuchi, H.; Ishikawa, M.; Kumada, M. *J. Chem. Soc., Faraday Trans. 1* **1984**, *80*, 383.
4. Michl, J.; West, R. In Silicon-Containing Polymers: The Science and Technology of their Synthesis and Applications; Jones, R. G.; Ando, W.; Chojnowski, Eds.; Kluwer Academic Publisher: Dordrecht, The Netherlands, 2000; 449-529.
5. Ossadnik, C.; Veprek S.; Marsmann, H. C.; Rikowski E. *Monatsh. Chem.* **1999**, *130 (1)*, 55.
6. Pernisz, U.; Auner, N.; Backer, M. *Polym. Prepr.* **2001**, *42(*1), 122.
7. Pernisz, U.; Auner, N.; Backer, M. In *Silicones and Silicone-Modified Materials*; Clarson, S. J.; Fitzgerald, J. J.; Owen, M. J.; Smith S. D., Eds.; ACS Symposium Series 729, American Chemical Society: Washington, DC, 2000; 115-127.
8. Pernisz, U.; Auner, N.; Backer, M. *Polym. Prepr.* **1998**, 39(1), 450.
9. Pernisz, U.; Auner, N. In *Organosilicon Chemistry – From Molecules to Materials*; Auner, *N.*; Weis, J., Eds.; VCH-Wiley: Weinheim, Germany / New York, 2000; Vol. 4, 505-520.
10. Backer, M. *Silacyclobutene - Synthese, Struktur, Reaktivität und Materialien*; Mensch und Buch Verlag: Berlin, Germany, 1999; PhD thesis, Humboldt-Universität zu Berlin, Berlin, Germany, 1999.
11. Sporck, C. R. U.S. Patent 3340287, 1967.
12. Burkhard, C. A. *J. Am. Chem. Soc.* **1945**, *67*, 2173.
13. GRAMS32 Spectral Notebase v.4.02, Lev. II, Galactic Industries Corp..
14. Auner, N.; Backer, M.; Herrschaft, B.; Holl, S.; in preparation.
15. Ovchinnikov, Yu. E.; Struchkov, Yu. T.; Buzin, M. I.; Papkov, V. S. *Vysokomol. Soedin Ser. A* **1997**, 39 (3), 430; *Polymer Science (engl.)* **1997**, *39 (3)*, 273.
16. Ovchinnikov, Yu. E.; Shklover, V. E.; Struchkov, Yu. T.; Dement'ev, V. V.; Frunze, T. M. *Metalloorg. Khim.* **1988**, *1*, 1117; *Organomet. Chem. USSR (engl.)* **1988**, *1*, 613.
17. Hossain, M. A.; Hursthouse, M. B.; Malik, K. M. A. *Acta Cryst.* **1979**, *B 35*, 522.

Chapter 11

Measuring the Size and Shape of Silicones: The Utilization of Chemometrics and Spectroscopy

Martyn J. Shenton[1,3], Henryk Herman[1,*], and Anthony C. Dagger[2]

[1]Polymer Research Centre, School of Physics and Chemistry, University
of Surrey, Guildford, Surrey GU2 7XH, United Kingdom
[2]Smith and Nephew GRC, York Science Park, Heslington, York YO10 5DF,
United Kingdom
[3]Current address: AG Fluoropolymers, P.O. Box 4, Thornton, Lancashire
FY5 4QD, United Kingdom

When describing a polymer, two fundamental parameters often
used are its molar mass and topology, *i.e.* what is its size and
shape?   Traditional routes for measuring these properties,
although generally accurate, are often time consuming and
unsuitable for real-time on-line processing.   However it is
possible to use molecular spectroscopy and chemometrics to
probe molar mass and molecular topology information of some
polymers. In the case of linear polydimethylsiloxanes (PDMS)
the number average molar mass ($<M_n>$) may be measured by
Raman spectroscopy using this approach in a few seconds
once chemometric models have been constructed. In addition,
distinguishing between PDMSs of the same $<M_n>$ but different
topology, *i.e.*, linear versus cyclic, is also possible.   In this
chapter an account of the preparation of approximately
monodisperse linear and cyclic PDMS, their spectroscopic
analysis and how the chemometric models to predict their size
and shape are developed is given. Finally, some possibilities
of this approach of polymer analysis are discussed.

Introduction

When describing a polymer, two fundamental parameters that are often quoted include its size (molar mass) and shape. Traditional routes for measuring these properties, (e.g., viscometry (1), gel permeation chromatography (GPC) (2) and trapping (3) etc.) although generally accurate, are often time consuming and unsuitable for real-time on-line processing. In addition, environmentally unfriendly solvents at elevated temperatures may be required to dissolve the polymer. Hence our desire to develop a rapid, remote, precise and accurate methodology to address these questions.

We outline a methodology that uses Raman spectroscopy and chemometrics to yield molar mass and molecular shape information (4). Spectroscopy is a widely used, rapid and non-destructive analytical tool that can probe structural information about materials; when coupled to the powerful data-handling characteristics of chemometrics (5) (multivariate data analysis), correlations between spectral features and property data may be modeled. Previous examples where spectroscopy coupled to chemometrics has been used includes gasoline blending (6,7) real-time quantitative monitoring of filler amount in extruded polymer (8) and condition monitoring of insulating materials (9,10).

In this paper, we demonstrate that spectroscopy coupled to chemometrics can give molar mass and topology information for polydimethylsiloxanes (PDMS) within a few seconds, once a model has been constructed. This makes on-line monitoring of molar mass characteristics of siloxanes, and potentially other polymers possible.

Experimental

Preparing a series of pure, well characterized and approximately monodisperse polymeric materials can be a difficult task. However, the ability to use preparative GPC (11) to fractionate a polydisperse polymer can make this task at least manageable; this was the approach used in these studies with silicones.

Preparation of Linear PDMS Fractions

A polydisperse PDMS with trimethyl end groups supplied by Dow Corning Ltd. (DC 200 series) was fractionated by preparative GPC. After purification, this resulted in a series of fifteen approximately monodisperse PDMS fractions with molar masses between 1000 and 31000; see Table 1. These clear liquid samples were purified and stored in glass vials.

Table 1. Molar Mass Data from GPC Analysis for the Fractionated Linear PDMS

Linear Fraction	$<M_n>$ /g mol^{-1}	$<M_w>/<M_n>$
L F01	1070	1.01
L F02	1430	1.02
L F03	2285	1.04
L F04	3370	1.08
L F05	5135	1.05
L F06	7090	1.10
L F07	9430	1.03
L F08	10525	1.04
L F09	10935	1.07
L F10	12400	1.05
L F11	14005	1.08
L F12	16146	1.06
L F13	22029	1.05
L F14	25396	1.07
L F15	30510	1.03

Adapted with permission from reference 4. Copyright 2000 John Wiley

Preparation of Cyclic PDMS Fractions

As with the linear PDMS, a polydisperse cyclic PDMS melt was fractionated by preparative GPC. However, as high $<M_n>$ polydisperse cyclic PDMS is not commercially available, a ring-chain equilibration reaction was performed (12). The formation of large cyclic polymers by the ring-chain equilibration reaction relies on the random cleavage and reformation of backbone bonds. When this is performed in dilute solution, a distribution of polymers may be formed. One starting material for preparing cyclic PDMS is octamethylcyclotetrasiloxane (D_4). In the presence of a strong acid (e.g. trifluoromethanesulfonic acid), hydrolysis of the Si-O backbone occurs to form silanol and silyl ester end groups, see step (i) in Figure 1. The silyl ester is readily hydrolysed to form another silanol-terminated chain, step (ii). The random recombination of silanol terminated chains results in an increase in the average molar mass of the silicones in solution, step (iii). If silanol end groups on the same molecule condensate, which is relatively likely in dilute solution, then a cyclic molecule is formed. These reactions result in a mixture of linear and cyclic PDMS that may be separated by solution fractionation. Full details of the preparation are available elsewhere (13,14). After fractionation and purification, a series of fifteen approximately monodisperse cyclic PDMS fractions with molar masses between 1100 and 29000 were available; see Table 2. These clear liquid samples were purified and stored in glass vials. Small cyclics may be obtained by vacuum distillation of the ring fraction prior to fractionation.

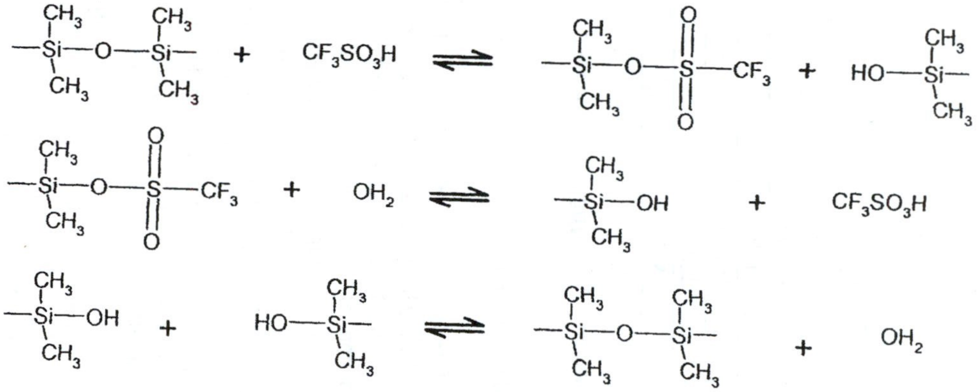

(i)
(ii)
(iii)

Figure 1. *Mechanism of polymerization of silicones during the Chojnowski and Wilczek ring-chain equilibration reaction (12)^*

FT-Raman Spectroscopy

The PDMS samples were analyzed in the liquid phase using a Perkin-Elmer System 2000 FT-Raman instrument fitted with a Nd:YAG laser (λ = 1064 nm). Spectra were recorded from 200 to 3500 cm^{-1} at a resolution of 4 cm^{-1} with 16 scans co-added. Analysis time was typically 30 seconds per sample. Glass is effectively invisible to Raman spectroscopy; hence spectra were recorded through the glass vials negating further sample preparation. The range normalized Raman spectra of the linear and cyclic PDMS fractions listed in Tables 1 and 2 are illustrated in Figures 2 and 3 respectively. Note that the spectra of each fraction are not distinguished; by eye they are all similar. The main spectroscopic features were assigned according to Table 3 (*15*).

There are two notable features in the spectra in Figures 2 and 3. Firstly, and most importantly is that all of the main spectral features have similar positions and shapes. Hence by eye, it is impossible to allocate a spectrum to a molar mass, let alone a topology. However, this is where one of the strengths of chemometrics is realized, as the mathematics is able to model small but significance variances between the data sets. The second feature of the spectra is the appearance of a baseline variation at lower wavenumbers. This is due to sample (and impurity) fluorescence. Taking the first differential can negate this or, as in this study, use of a spectral region where the fluorescence background is negligible, *i.e.* above 2500 cm^{-1}.

Results

Chemometric Analysis

When using chemometrics, a well-characterized calibration sample set is required. In this case, a series of linear and cyclic PDMS fractions with a $<M_n>$ in the range of 1000 to 31000 were accessible. In all cases, the polydispersities were less than 1.25. These molar masses and whether the PDMS was linear or cyclic were used as the property data. The chemometrics can extract "hidden" information about the spectra and seek correlations with the property data; this process is known as principal components regression (PCR) and results in the formation of a model. The chemometric analysis was performed using commercially available software called Unscrambler v7.5 from Camo Ltd. (*16*).

Table 2. Molar Mass Data from GPC Analysis for the Fractionated Cyclic PDMS

Cyclic Fraction	$<M_n>$ /g mol^{-1}	$<M_w>/<M_n>$
C F01	1119	1.23
C F02	1526	1.20
C F03	1800	1.14
C F04	2290	1.15
C F05	2890	1.18
C F06	3732	1.12
C F07	4832	1.19
C F08	6223	1.12
C F09	7911	1.15
C F10	10331	1.15
C F11	14007	1.11
C F12	17169	1.19
C F13	21143	1.13
C F14	24619	1.13
C F15	29036	1.15

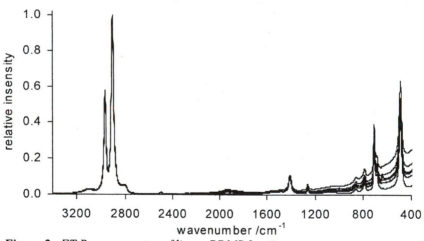

Figure 2. FT Raman spectra of linear PDMS fractions

Adapted with permission from reference 4. Copyright 2000 John Wiley

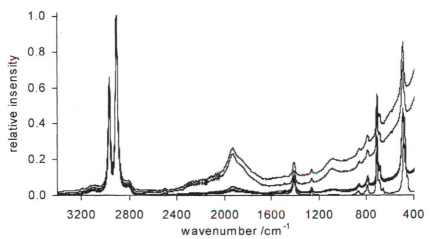

Figure 3. *FT Raman spectra of cyclic PDMS fractions*

Table 3. Peak Assignments for the FT-Raman Spectra of Linear and Cyclic PDMS (4). (Asym. - asymmetric; sym. - symmetric; def. - deformation; str. - stretch and r. -rock)

Peak /cm^{-1}	Assignment	Peak /cm^{-1}	Assignment
2966	CH$_3$ asym. str.	863	[Si(CH$_3$)$_2$] r.
2906	CH$_3$ sym. str.	845, 792	Asym. [Si(CH$_3$)$_2$] r.
2501	CH$_2$ def. overtone	757	Sym. [Si(CH$_3$)$_3$] r.
1412	CH$_3$ asym. def.	710	Si-C sym. str.
1264	CH$_3$ sym. def.	646	Asym [Si(CH$_3$)$_3$] r.
1200-1000	Si-O-Si asym. str.	491	Si-O-Si sym. str.

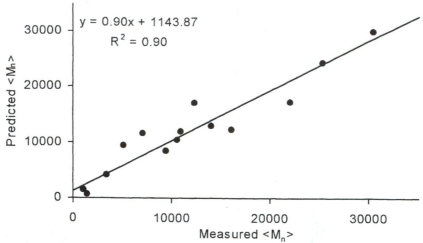

Figure 4. *Predicted vs. measured plot for linear PDMS (adapted from*
Adapted with permission from reference 4. Copyright 2000 John Wiley

The C-H region (2750 - 3250 cm^{-1}) of the spectra of the linear PDMS was used in the PCR analysis. This resulted in a model that had 4 significant principal components (PCs). The PCs describe the variance across the data set; a more detailed description of chemometrics and the role of PCs see the text by Esbensen *et al.* (5). Figure 4 contains the output from the chemometric modeling in the form of a predicted versus measure plot for $<M_n>$. In a perfect model, the slope and the regression coefficient would both be unity. However, if both of these values are about or in excess of 0.9, then a good model is generally formed, as is the case here.

The model must be validated with independent samples; in this case, further DC200 siloxanes were used. It was found that for siloxanes with a $<M_n> > 5k$, then the model was good. However, for the smaller siloxanes, masses from this method tended to be over-estimated. We believe this is due to end-group effects that can be accounted for in further iterations of the modeling process.

By generating a model containing all of the data for the linear and cyclic PDMS fractions, a cluster type analysis may be performed as illustrated in Figure 5.

From this figure, it is clear that the linear and cyclic fractions form different clusters as the weighting of principal components differ. On the left-hand side of the figure, the linear siloxanes are clustered in a triangle; the cyclic species are clustered in an arc on the right-hand side. This means that topologically different, yet chemically and even mass identical PDMS fractions may be distinguished by spectroscopy when processed by this model.

More recently, preliminary investigations on commercial silicone elastomers have been performed. Correlations between the spectra of a silicone containing medical device and the number of sterilization cycles it has been subjected to, have been found, thereby making a rapid and non-destructive method of condition assessment for this device available. Furthermore, in

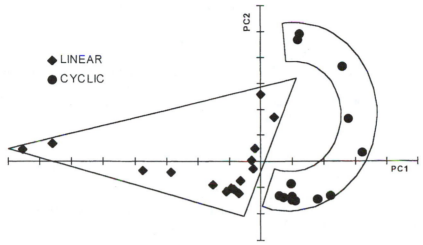

Figure 5. *PC plot showing linear and cyclic grouping*

another study, correlations between mechanical properties of silicones and their spectra have also been found. These results complement our earlier findings on cellulose containing insulating materials where correlations between spectral features and age (*17*) and mechanical strength (*9,10*) were modeled.

The powerful combination of spectroscopy and chemometrics has been used for many years in the oil and gas (*6-8*) industry and for pharmaceutical (*18*) and medical analysis (*19,20*). and with improvements in computer processing speeds and the availability of commercial chemometric software, the authors hope and expect many other applications to develop.

Conclusions

Molar mass and topology of PDMS can be determined using spectroscopy when coupled to the powerful multivariate data processing algorithms that encompass chemometrics. The use of chemometrics has enabled these correlations between spectral features and molar mass to be obtained from the CH region of the Raman spectrum, even though by eye, no significant variation between the spectra of siloxanes of differing molar masses can be seen.

These preliminary investigations suggest that extracting 'hidden' information from easy to measure data, in this case Raman spectroscopy is possible when robust chemometric models are generated. The authors are currently trying to gain a greater understanding of the scientific meanings of these correlations before applying this approach to other polymeric systems.

126

References

1. Dodgson, K.; Sympson, D.; Semlyen, J.A., *Polymer* **1977**, *18*, 1265-1268.
2. Moore, J.C., *J. Polym. Sci., Part A* **1964**, *2*, 835-843.
3. Garrido, L.; Mark, J.E.; Clarson, S.J.; Semlyen, J.A., *Polym. Commun.* **1985**, *26*, 53-55.
4. Shenton, M.J.; Herman, H.; Stevens, G.C., *Polym. Int.* **2000**, *49*, 1007-1013.
5. Esbensen, K.; Schönkopf, S.; Midtgaard, T., "Multivariate Analysis - In Practice." Camo ASA, Oslo, Norway, 1994.
6. Lambert, D.; Descales, B.; Llinas, J.R.; Espinosa, A.; Osta, S.; Sanchez, M.; Martens, A. *Analusis Magazine* **1995**, *23*, 4, M9-M13.
7. Espinosa, A.; Sanchez, M.; Osta, S.; Boniface, C.; Gil, J.; Martens, A.; Descales, B.; Lambert, D. *Oil and Gas Journal* **1994**, *October*, 49-56.
8. Fischer, D.; Bayer, T.; Eichhorn, K.-J.; Otto, M. *Fresenius J. Anal. Chem.* **1997**, *359*, 74-77.
9. Shenton, M.J.; Herman, H.; Stevens, G.C.; Heywood, R.J., *Proc. 8th Int. Conf. Dielect. Mat. Meas. Appl.* **2000**, *IEE Conf. Publication 473*, 346-351.
10. Shenton, M.J.; Herman, H.; Stevens, G.C.; Heywood, R.J. In *Ageing Studies and Lifetime Extension of Materials;* Mallinson, L.G. Ed.; Kluwer Academic, New York, 2000; 359-365.
11. Dodgson, K.; Sympson, D.; Semlyen, J.A., *Polymer* **1978**, *19*, 1285-1289.
12. Chojnowski, J.; Wilczek, L., *Makromol. Chem.* **1979**, *180*, 117-130.
13. Shenton, M.J. *D.Phil. Thesis*, University of York, UK, 1995.
14. Dagger, A.C. *D.Phil. Thesis*, University of York, UK, 1999.
15. Lipp, E.D.; Smith, A.L. In *The Analytical Chemistry of Silicones*; Smith, A.L. Ed.; John Wiley & Sons Ltd., New York, 1991; Ch. 11.
16. http://www.camo.no
17. Ali, M.; Emsley, A.M.; Herman, H.; Heywood, R.J., *Polymer* **2001**, *42*, 2893-2900.
18. Lee, K R; Zuber, G; Katrincic, L., *Drug Development and Industrial Pharmacy* **2000**, *26*, 2, 135-147.
19. Abuzaruraloul, R.; Gjellan, K.; Sjolund, M.; Lofqvist, M.; Graffner, C.; *Drug Development And Industrial Pharmacy* **1997**, *23*, 8, 749-760.
20. Gambart, D.; Cardenas, S.; Gallego, M.; Valcarcel, M.; *Analytica Chimica Acta* **1998**, *366*, 1-3, 93-102.

Chapter 12

Thermally Stable Siloxane Polymers for Gas Chromatography

G. M. Day, A. I. Hibberd, J. Habsuda, and G. J. Sharp

SGE International Pty. Ltd., 7 Argent Place, Ringwood, Victoria 3134, Australia

Several novel polysiloxane copolymers containing various aryl substituents in the backbone and side chains were prepared and evaluated as possible stationary phases for GC capillary columns. All copolymers prepared were examined using TGA and were found to be thermally stable above 400°C. The new copolymers were evaluated as GC stationary phases and while all were perfectly suitable as such, only two copolymers would be suitable for commercial use due to their superior robustness.

High temperature gas chromatography (GC) is a widely used technique for analyzing mixtures of volatile organic compounds (*1,2*). The separation system involves coating a thin film of polymeric material (stationary phase) onto a fused silica capillary column wall, which is then continuously swept by a stream of carrier gas (mobile phase) containing the mixture of organics to be separated. Since the invention of the open tubular column for gas chromatography several decades ago, much work has been performed on developing stationary phases to improve the application and performance of capillary GC (*1,2*). The most successful and commonly used stationary phases to date have all been based on polysiloxanes, with poly(dimethylsiloxane) (PDMS) being the most frequently phase used for general analysis (*1,2*).

Stationary phase materials must meet certain criteria to be considered satisfactory for use, and each is outlined below.

- *Thermal stability:* The phase must be able to withstand high temperatures (200-400°C) for long periods of time without thermal decomposition, the main cause of column bleed, as well as being able to withstand many heating and cooling cycles.
- *Physical stability:* The phase must be able to uniformly coat the inner wall of the capillary, and be robust enough to withstand column manipulation.
- *Immobilization:* Although not an absolute necessity, it is desirable that the stationary phase be designed such that a cross-linking or immobilization process can be undertaken, to impart greater thermal stability and render them non-extractable by solvents.
- *Partitioning capability:* The phase material must allow vaporized molecules to be able to move freely between the mobile (gas) phase and stationary phases at a range of temperatures (0-300°C).
- *Chemical inertness:* The phase must be chemically inert to the analytes, the solvent and any impurities in the system.
- *Selectivity:* For good retention there must be a strong interaction between the analytes and the stationary phase. Possible modes of interaction are dispersive interactions (Van Der Waals forces), dipole-dipole interactions, dipole-induced interactions, acid-base interactions, and molecular shape interactions (in chiral or liquid crystal phases).
- *Reproducibility:* Synthesis of the stationary phase must be reproducible so that the same separations are possible from different batches of material.

Polysiloxanes are ideally suited as GC stationary phases owing to their thermal and physical stability, chemical inertness, excellent film forming properties and their ability to facilitate separation of a variety of organic substances. It has been well documented that polysiloxanes incorporating aryl substituents have higher resistance to thermal and oxidative degradation when compared to the non-aryl analogues (2,3). Numerous high temperature polysiloxanes with aryl substituents either as side chains or incorporated into the polymer backbone have previously been successfully prepared (3). The considerable stiffening effect observed upon incorporation of 1,4-bis(dimethylhydroxy-silyl)benzene (silphenylene) into polysiloxane backbones has led to this material being used to enhance the thermal stability of a variety of siloxanes for a number of applications (3-5). However, the wide-spread use of arylene siloxanes as GC capillary column stationary phases is a relatively recent occurrence (1,2). The excellent thermal stability (low bleed) and good partitioning properties of existing silphenylene based capillary columns stationary phases are good examples (1,6). With the ever increasing sensitivity of GC instrumentation, there is a constant need to improve the performance of high temperature stationary phases in order to minimize capillary column bleed levels. New silphenylene polysiloxane analogues are excellent candidates for further investigation in this area.

In this study, new polysiloxane materials incorporating aryl substituents in the polymer backbone have been prepared (Figure 1). These random copolymers have been characterized and the inherent thermal stability has been studied by thermogravimetric analysis (TGA). Their suitability as stationary phase materials will be discussed.

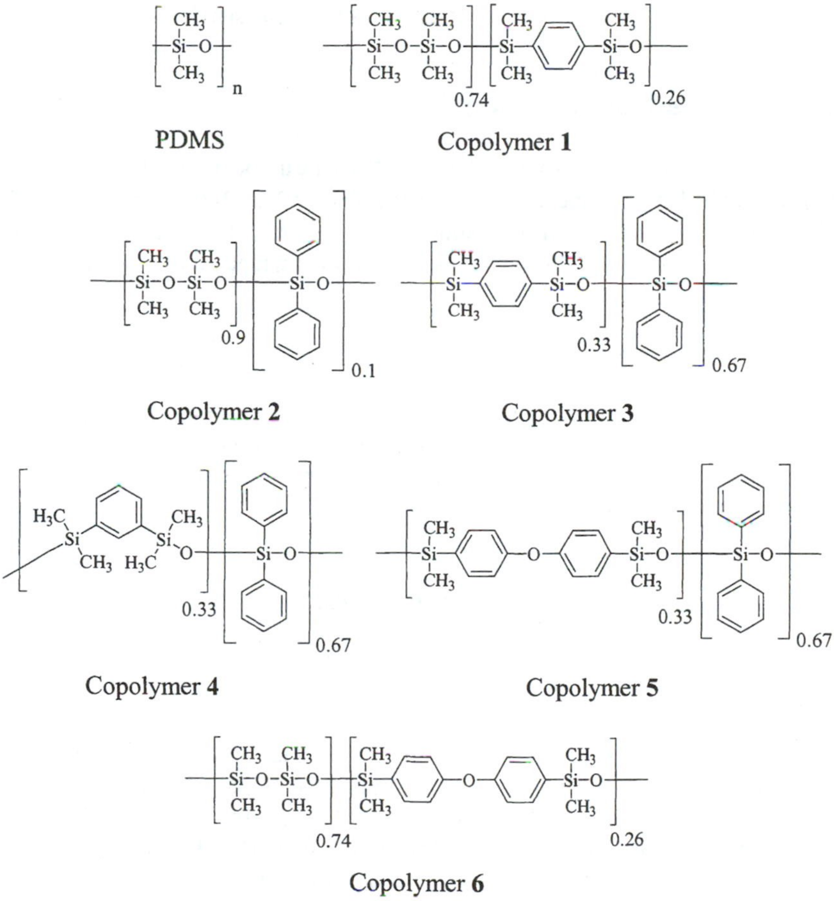

Figure 1 Structures of PDMS and the copolymers studied in this work.

Experimental

Copolymer synthesis

Copolymers **1**, **2**, and **6** (*7,8*) were prepared as previously described. Copolymers **3-5** were prepared according to the procedure outlined by Zhang et al. (*5*). All materials were obtained as opaque, viscous gums. The copolymers were dried at *ca.* 25°C under vacuum for more than a week before being studied. Monomer incorporation ratios were confirmed using NMR and FTIR spectroscopy. Copolymer molecular weights were determined via gel permeation chromatography (GPC) relative to polystyrene standards.

Thermogravimetric analysis (TGA)

TGA was carried out using a Setaram TGA instrument. Each copolymer sample weighed *ca.* 30 mg and was heated from 50 to 1000°C at a rate of 10°C/minute while under a nitrogen atmosphere. The samples were tared at zero time and the weight loss was recorded as a function of temperature measured.

Capillary column preparation

Fused silica tubing was prepared in-house. The methods for preparing capillary columns using polysiloxane stationary phases have been well documented (*9*). Column specifications and run conditions for individual chromatograms are listed in the Figures.

Results and Discussion

Characterization

Characterization of the copolymers by FTIR and ^{1}H NMR spectroscopy confirmed the repeat unit structures and incorporation ratios. Molecular weights determined by GPC were as follows: copolymer **1**, $M_w = 160,000$; copolymer **2**, $M_w = 126,000$; copolymer **3**, $M_w = 250,000$; copolymer **4**, $M_w = 180,000$; copolymer **5**, $M_w = 200,000$; copolymer **6**, $M_w = 145,000$.

Thermal stability

   All copolymers were examined using TGA under nitrogen (Figures 2-4). TGA studies of the inherent thermal stability of the polymeric materials in air were not undertaken since nearly all chromatographic analyses are performed under an inert atmosphere. Comparison of the thermal properties of copolymers 1-3 (Figure 2) clearly shows that that the polymers incorporating arylene units in the polymer backbone (copolymer **1** and **3**) are more thermally stable than the polymer with aryl substituents solely as pendant groups (copolymer **2**). Weight loss for copolymer **2** began at around 480°C, with 98% decomposition having occurred well before the temperature reached 700°C. By contrast, copolymers **1** and **3**, both having backbone silphenylene substituents, exhibited weight loss starting at approximately 540°C. The small weight loss exhibited at around 100°C for copolymer **3** was attributed to the presence of moisture and/or solvent still present in the sample. At 650°C, copolymer **1** had lost 68% of its weight, while copolymer **3** has lost only 46%, with no further decomposition evident at 1000°C. All copolymers compare favorably with poly(dialkylsiloxane) based materials currently used as stationary phase materials, which typically decompose at temperatures less than 400°C.

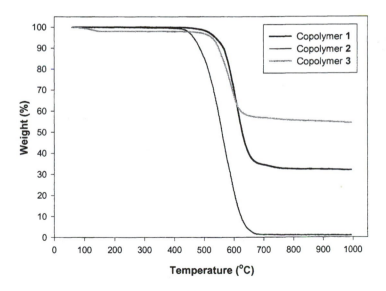

Figure 2   TGA of copolymers **1-3**.

Two analogues of copolymers **3** were prepared, with variations in the type of backbone aryl substituent used; *meta*-substituted silphenylene analogue (copolymer **4**) and a phenyl ether silphenylene analogue (copolymer **5**). Their inherent thermal stability was examined and compared to that of copolymer **3** (Figure 3). It is evident that the silphenylene based copolymer **3** is more thermally stable than copolymers **4** and **5**, with the onset of degradation at a much higher temperature (540°C). At 1000°C, copolymer **3** had exhibited a 46% weight loss while copolymers **4** and **5** exhibited weight losses of 73% and 69% respectively. While it was not surprising that a polymer having a *meta*-substituted arylene analogue in the backbone (copolymer **4**) would be less thermally stable that a polymer containing a *para*-substituted silphenylene analogue (copolymer **3**), previous work had shown that phenyl ether based siloxanes were more thermally stable than the corresponding silphenylene analogues (*3*). This was shown not to be the case for these particular siloxane analogues and indicates that it may not be true for all siloxane types.

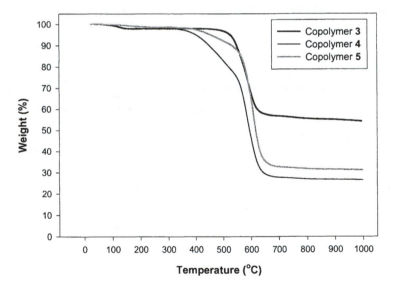

Figure 3   TGA of copolymers **3- 5**.

Figure 4 compares the inherent thermal degradation of copolymer **1** (having a silphenylene moiety in the backbone) and copolymer **6** (having a phenyl ether based moiety in the backbone). It is clear that the silphenylene based copolymer **1** has better thermal stability than the phenyl ether based copolymer **6**. The minor weight loss exhibited by copolymer **6** before true degradation onset could indicate the presence of volatile low molecular weight materials remaining in the polymer matrix, even after purification and prolonged drying.

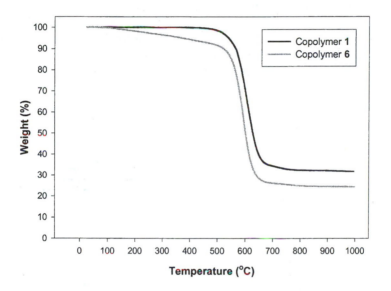

Figure 4   TGA of copolymers **1** and **6**.

Evaluation of polymers as GC stationary phases

The copolymers **1-6** were coated onto the internal surface of fused silica capillary columns, and after undergoing an immobilization process at high temperature, the bleed levels were recorded for each column under the same conditions.   As expected, all columns coated with copolymers containing aryl substituents had significantly lower bleed levels at higher temperatures (> 300°C) under an inert atmosphere when compared to columns coated with PDMS. Capillary columns coated with copolymers **1** and **3**, both containing the silphenylene moiety in the polymer backbone, exhibited lower bleed levels than columns coated with copolymer **2** containing pendant aryl groups only. This is consistent with the TGA results shown earlier.

The combination of high temperature and an oxygenated atmosphere will severely damage the stationary phase of most GC capillary columns. Figure 5 shows a typical example of GC partitioning and bleed testing of a capillary column coated with copolymer **1**. From the bleed levels shown, it is evident that incorporation of an aryl substituent in the backbone of the polysiloxane stationary phase remarkably aids thermal stability after contact with air at high temperatures. In addition, the copolymer **1** stationary phase also retains its inertness to the test mix constituents after contact with air at high temperature.

An example chromatogram for copolymer **3** is shown in Figure 6. Excellent partitioning performance was achieved with this material as stationary phase using the polynuclear aromatic hydrocarbon test mix.

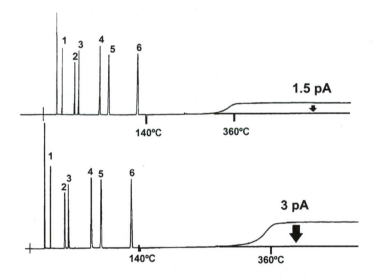

Figure 5 GC chromatograms for a capillary column coated with copolymer **1** (12 m, 0.22 mm ID, 0.25 μm film thickness), shown with bleed levels. **Top:** Before the use of air as carrier gas. **Bottom:** After 80 hours with air as carrier gas at 200°C. **Peaks:** (1) Decane, (2) 4-Chlorophenol, (3) Decylamine, (4) Undecanol, (5) Biphenyl, (6) Pentadecane.

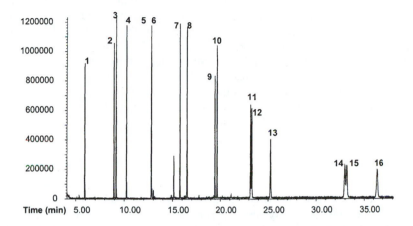

Figure 6 Separation of polynuclear aromatic hydrocarbons using a fused silica capillary column coated with the copolymer **3** (30 m, 0.25 mm ID, 0.25 µm film thickness). **Peaks:** (1) Naphthalene, (2) Acenaphthylene, (3) Acenaphthene, (4) Fluorene, (5) Phenathrene, (6) Anthracene, (7) Fluoranthene, (8) Pyrene, (9) Benzo(a)anthracene, (10) Chrysene, (11) Benzo(b)fluoranthene, (12) Benzo(k)fluoranthene, (13) Benzo(a)pyrene, (14) Indeno(1,2,3-c,d)pyrene, (15) Dibenzo(a,h)anthracene, (16) Benzo(g,h,i)perylene.

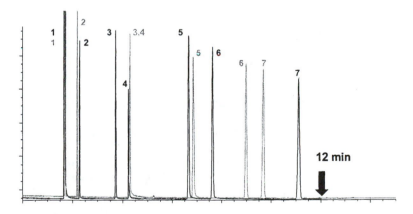

Figure 7 Partitioning capability of fused silica capillary columns coated with copolymer **1** and copolymer **6** (30 m, 0.25 mm ID, 0.25 µm film). **Conditions:** isothermal, 140°C, 15 min. **Peaks:** (1) Solvent, (2) Decane, (3) 4-Chlorophenol, (4) Decylamine, (5) Undecanol, (6) Biphenyl, (7) Pentadecane.

A small change to the chemical architecture of a stationary phase polymer can have a pronounced effect on the chromatography of coated capillary columns. This is illustrated in Figure 7 for GC columns coated with copolymers **1** and **6**. While the phase materials vary only in the type of aryl substituent present, the chromatography is markedly different due to the dissimilar polarities of the two copolymers. In addition, bleed testing of the columns prepared from the two copolymers showed that the column prepared using copolymer **1** as stationary phase exhibited a much lower bleed level at high temperature than the column prepared using copolymer **6** as stationary phase. This bleed comparison result is consistent with the TGA results for these two copolymers (Figure 4).

Conclusion

All copolymers prepared in this work are suitable for use as high temperature GC stationary phases since all materials are thermally stable and chemically inert under standard GC conditions. In all cases the monomer ratios could be modified to suit particular GC applications. However, TGA and bleed testing results indicate that only two of the copolymers tested (**1** and **3**) could be commercially useful since they are the most robust materials. This work has also shown that for GC applications, it is desirable to include arylene moieties into the backbone of potential stationary phase materials for maximum thermal stability. This is highlighted by the strong resistance to degradation of the silphenylene based copolymer **1** stationary phase after exposure to air at elevated temperature.

References

1. Jennings, W.; Mittlefehldt, E.; Stremple, P. *Analytical Gas Chromatography* (2nd Edition); Academic Press, NY, 1997.
2. Blomberg, L. *LCGC Europe*, **2001**, *14 (2)*, 106.
3. Dvornic, P. R.; Lenz, R. *High Temperature Siloxane Elastomers*; Hulthig and Wepf, Basel, 1990.
4. Zhang, R.; Pinhas, A. R.; Mark, J. E. *Macromolecules* **1997**, *30*, 2513.
5. Zhang, R.; Pinhas, A. R.; Mark, J. E. *Polym. Prep.* **1998**, *39(1)*, 575.
6. Aichholz, R., Lorbeer, E. *J. High Resol. Chromatogr.* **1998**, *21*, 363.
7. Zhu, H. D.; Kantor, S. W.; MacKnight, W. J. *Macromolecules*, **1998**, *31*, 850.
8. Merker, R. L.; Scott, M. J. *J. Polym. Sci.: Part A* **1964**, *2*, 15.
9. Grob, K. *Making and Manipulating Capillary Columns for Gas Chromatography*; Huethig, Heidelberg, 1986.

Elastomers and Reinforcement

Chapter 13

The Effect of Thermal Aging on the Non-Network Species in Room Temperature Vulcanized Polysiloxane Rubbers

Mogon Patel and Anthony R. Skinner

Atomic Weapons Establishment, Aldermaston, Reading RG7 4PR, United Kingdom

RTV siloxane rubbers have been thermally aged in inert gas atmospheres and in the presence of moisture at temperatures up to 190°C. Toluene extractable matter measurements showed an increase in the levels of non-network fragments with age for those samples aged in sealed containers. Samples aged open to air did not show this behaviour. The chemical composition and distribution in dimethylsiloxane species in the non-network fragments was analysed using Gas chromatography-mass spectrometry (GC-MS). This showed that cyclic polydimethylsiloxane species within the range D_4 to D_{18} dominate the extractable matter. Thermal ageing causes a significant build up of octamethyltetrasiloxane (D_4) within the polymer matrix. This has been attributed to thermally activated degradation processes involving depolymerisation reactions leading to monomer formation from random points along the chain and from sites containing reactive functional species.

Introduction

Silicone rubbers have in general the following advantageous characteristics: a wide service temperature range (-100°C to 250°C); excellent non-stick, non-adhesive properties; low toxicity; possible optical transparency; low chemical reactivity; and excellent resistance to attack of oxygen, ozone & sunlight. One of the major drawbacks with siloxanes that have been reported in the literature is 'hydrolysis' on thermal ageing in closed systems in the presence of trace amounts of water and the formation of polymeric silanols. A siloxane hydrolysis mechanism, which contributes to the build-up of low molecular weight cyclic fragments in the rubber network, has been proposed (1). This can lead to loss of crosslink density if the cyclic fragments contain a long segment of the polymer backbone. Furthermore, fillers such as iron oxide, being a Lewis acid, have been reported to catalyse the hydrolysis reaction and accelerate cleavage of the siloxane backbone (2). An increase in low molecular weight cyclic siloxane species on ageing silicones at elevated temperatures has also been observed (3). In this paper, an assessment of the non-network fragments associated with virgin and thermally aged RTV polysiloxane rubbers is reported and potential degradation mechanisms discussed.

Experimental

Materials

The RTV5370 polysiloxane rubber pads were prepared by adding 1.5g of XY-70 catalyst (Tin Octoate) to 30g of RTV5370 gum and the mixture vigorously stirred for approx. 15 seconds. 20g of this mixture were added to a 15cm diameter (2 mm deep) mould. The filled mould was transferred to a Moors Press and the polymer was allowed to cure under ambient conditions for 20 minutes. The mould was then dismantled, the cured rubber removed and postcured in an oven at 70°C for 16 hours.

Accelerated Ageing Trials

The rubber samples were thermally aged in an inert gas (argon) atmosphere in sealed containers at temperatures up to 80°C. The containers were periodically removed for sample analysis over a six-month period. Samples were also over-tested at temperatures up to 190°C in both sealed and open containers sometimes in the presence of moisture. In order to carry out a preliminary empirical comparison of data obtained at different temperatures and thereby investigate trends in behaviour, arbitrary simulated ages were calculated based on the assumption that the rate constant of the underlying degradation mechanism doubled per 10°C rise in temperature. These simulated ages have been used to assess the data on a common time axis.

Extractable matter measurements

Samples of rubber of known weight (approximately 5g) were extracted by immersing in analytical grade toluene at 70°C for 96 hours. The specimens were then dried to constant weight and the amount of non-network polymer calculated from the weight loss.

Gas chromatography-mass spectrometry

Measurements on the toluene extracted rubber solutions were made using a HP 5890 series 2 Gas Chromatograph (GC) with a 5971 mass selective detector (MSD).

Results and Discussion

Extractable matter

A plot of the extractable matter (% by wt) against simulated age is shown in Figure 1. An initial plateau is indicated after which a general increase in extractable matter (non-network species) is evident. The sample aged at 190°C for 48 hours in sealed containers shows extractable matter levels of up to 5.5 % ± 0.5. In comparison, the sample heated at 190°C for 48 hours open to the air shows extractable matter levels of 2.7 % ± 0.5. This is slightly lower than those observed for the virgin sample (3%) suggesting a possible loss of low molecular weight fragments from volatilisation effects when the material is thermally aged in open systems. In a sealed system these fragments are presumably retained.

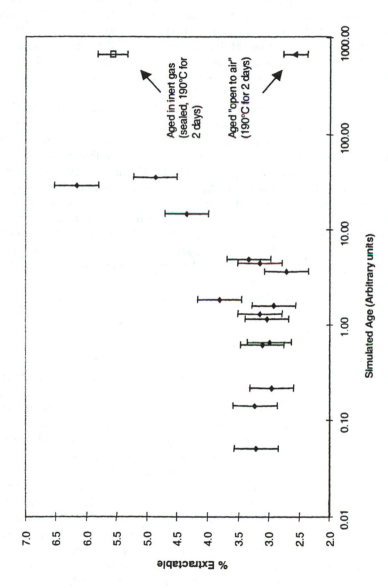

Figure 1: Toluene extractable matter measurements

Chemical Analysis

Figure 2 shows the total ion chromatogram (TIC) of the toluene extractable fraction derived from virgin RTV5370. The area of each peak is indicative of the relative concentration of the corresponding components. A blend of four different plasticisers including, dibutyl phthalate (retention time 39.9min) and diethyl phthalate (retention time 31.5min) were identified by the characteristic 149 m/z (phthalate cation) peak. Evidence for the presence of 2-ethylhexanoic acid (from curing reactions) was found from mass spectra peaks corresponding to McLafferty fragmentation ions (m/z of 73 & 88).

Cyclic polydimethyl siloxanes were found to dominate the extractable matter, representing almost 50% by weight of the non-network material as shown in Figure 3. Further detailed examination of the mass spectra reveals that the distribution of the cyclic siloxane species was D_4 to D_{18} with a molecular mass range of 296 to 1300 g/mol (see Figure 4). The D_9, D_{10} and D_{11} cyclics dominate the distribution with almost 40% of the relative concentration. The D_4 cyclic (octamethylcyclotetrasiloxane) eluted at a retention time of 13.4 min and was identified by the characteristic $[D_4-30]^{2+}$ doubly charged ion with an m/z of 133 and a $[D_4-CH_3]$ ion with an m/z of 281. Tetradecamethylcycloheptasiloxane (D_7) eluted at a retention time of 28.1min and showed fragmentation ions at m/z of 415 (corresponding to $[D_7-103]^+$), and 503 (corresponding to $[D_7-CH_3]^+$). There was no evidence of hexamethylcyclotrisiloxane (D_3) possibly because it is eluting from the GC at a similar retention time as the toluene solvent.

Effect of Thermal Ageing

Figure 5 shows that open to air ageing of the rubber causes significant changes to the distribution of the cyclics with loss of the lower molecular weight species (presumably from volatilisation) and an increase in the relative intensities of the remaining high molecular weight fragments. In comparison, the samples heated in sealed containers (closed system aged) showed little change (see Figure 6) presumably because in a sealed system, the low molar mass fragments are retained. In both open to air and in sealed ageing conditions, a significant build up of octamethyltetrasiloxane (D_4) within the polymer matrix was found. An increase in the relative D_4 intensity of almost 25% is indicated on ageing the polymer at 180°C for 48 hours in an inert atmosphere. In comparison, an approx. relative increase of 50% in the D_4 cyclic is indicated on ageing the polymer open to air at 180°C for 48 hours. The fact that ageing in sealed systems causes an increase in the toluene extractable fraction (see Figure 1) suggests that the D_4 cyclics are being produced from degradation processes associated with the polymer network rather than breakdown of the higher

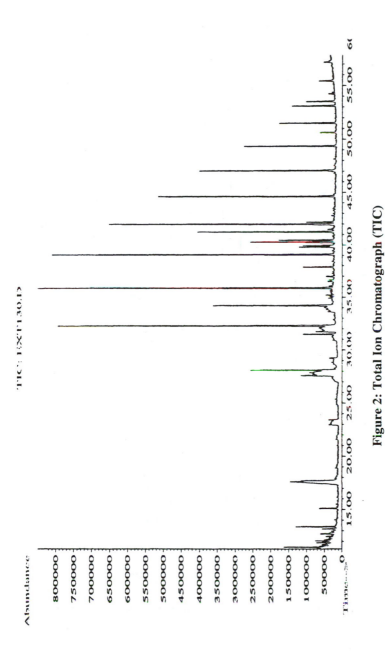

Figure 2: Total Ion Chromatograph (TIC)

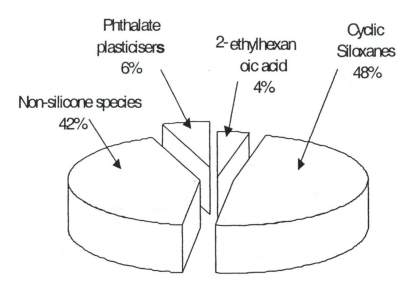

Figure 3: Relative intensity of constituents in the toluene extractable matter

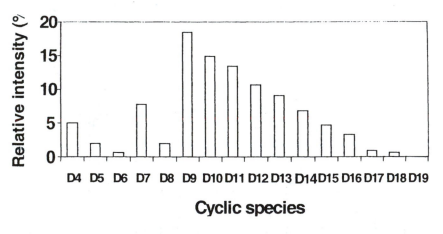

Figure 4: Distributión of cyclic siloxane species in virgin rubber

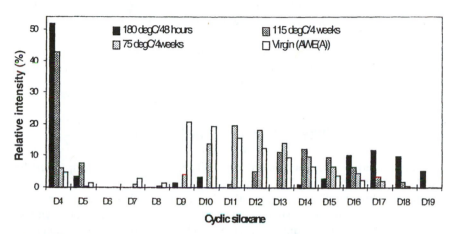

Figure 5: Effect of open to air ageing on the distribution of cyclic siloxanes

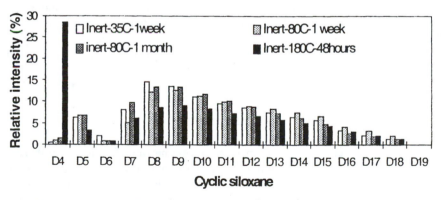

Figure 6: Effect of ageing in sealed systems on the cyclic siloxane distribution.

molecular weight cyclic fragments within the polymer. The elimination of this cyclic from the polymer network rather than other siloxane species may be explained by the fact that RTV5370 polysiloxane gums tend to be manufactured from ring opening polymerisation of low molecular weight cyclics such as the D_4 cyclic monomer (4). Figure 7 shows the ring opening polymerisation reaction of octamethylcyclotetrasiloxane leading to linear polysiloxane diols used in silicone resins. Analysis of the RTV5370 silicone gum, using GC-MS reveals a distribution in cyclic species with the lower molecular weight (D_4 and D_5) species in higher concentration than the other cyclics, see Figure 8. This distribution appears to be significantly different to that shown by the virgin rubber (see Figure 4) and may be explained by the loss of cyclic species during the postcure process for the rubber sample.

The degradation mechanism

The underlying degradation process likely to result in a build up of cyclic siloxane fragments within the polymer matrix is thermally activated depolymerisation reactions along the siloxane chain (5). These reactions may be initiated at active sites (e.g. silanols, tin-siloxane sites, etc.) or at any random points along the polymer chain (see Figure 9 and Figure 10). Active sites may be envisaged as weak links, which are present either residual from the cure process or created from acid accelerated hydrolysis effects. At these sites, thermally activated depolymerisation reactions are more likely than those along the polymer backbone. Recent studies (6) have suggested that these reactions are not initiated by direct siloxane bond scission and that the mechanisms are complex, involving cyclic complexes that lead to a weakened Si-O bond and chain scission.

Conclusions

RTV5370 rubbers show an increase in non-network fragments on thermal aging in sealed inert gas conditions. Those samples heated open to air show extractable matter levels slightly lower than those observed for the virgin sample due to volatilisation of low molecular weight fragments. In each case, thermal ageing causes significant increases in octamethylcyclotetrasiloxane (D4) and is indicative of depolymerisation processes within the polymer network.

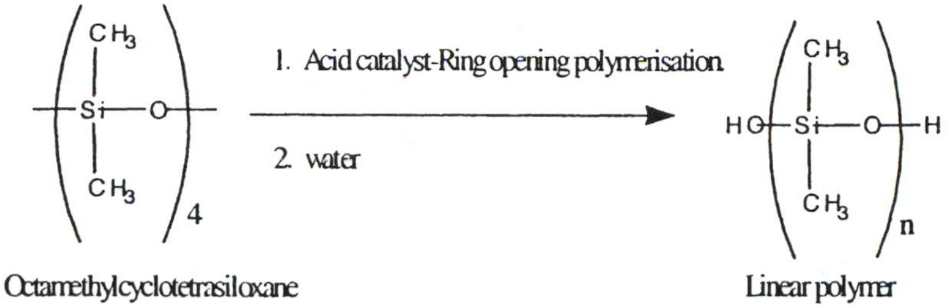

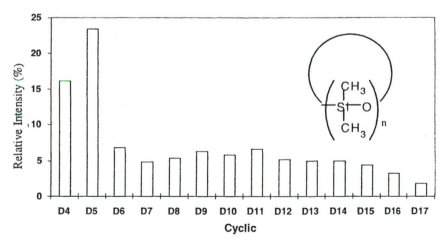

Octamethylcyclotetrasiloxane   Linear polymer

Figure 7: Silicones from ring-opening polymerisation.

Figure 8: Distribution of cyclic siloxane species in the siloxane gum

148

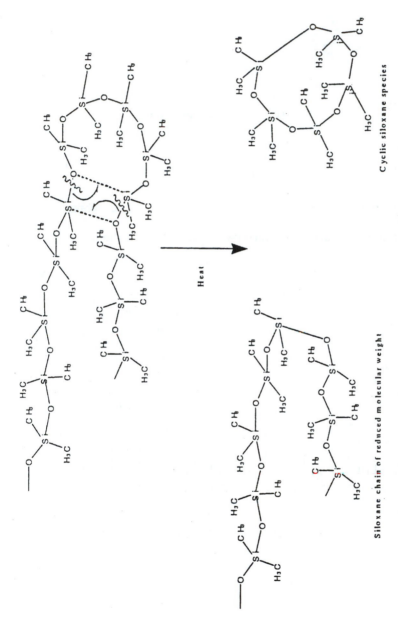

Siloxane chain of reduced molecular weight

Cyclic siloxane species

Figure 9: Typical depolymerisation mechanism initiated at random points along the chain

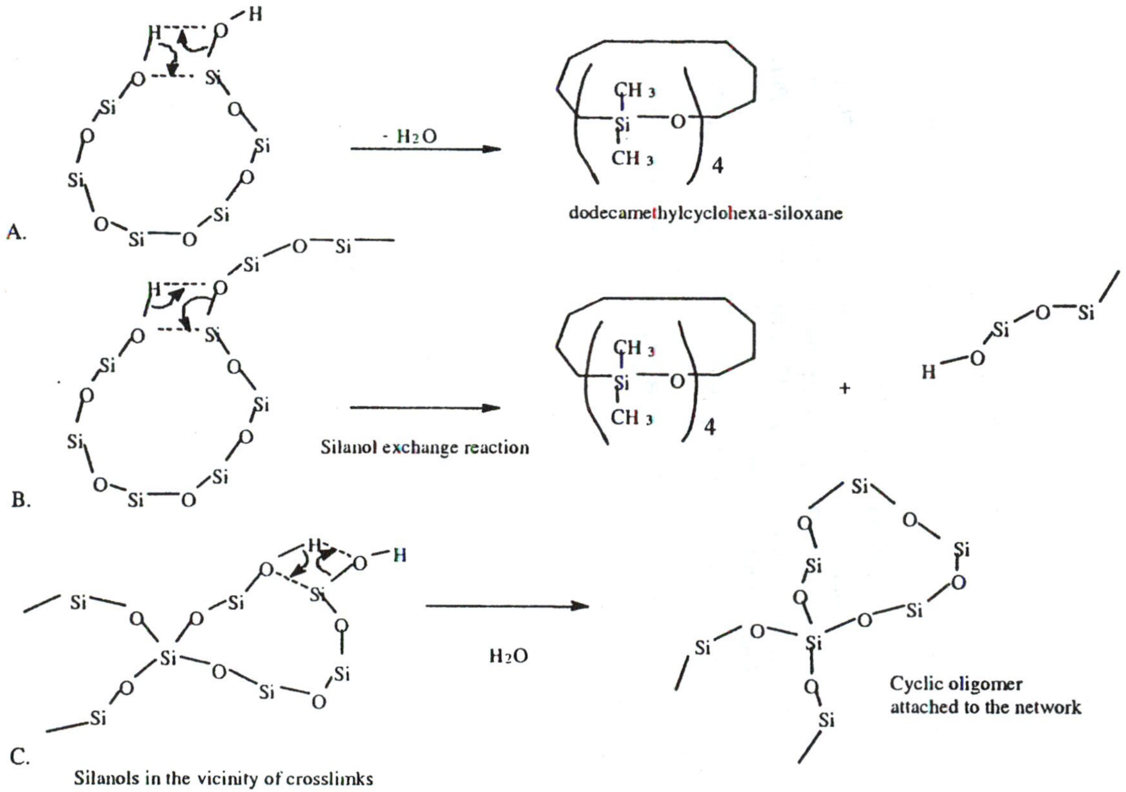

149

Figure 10: Typical degradation reactions at chain ends

Acknowledgements

The authors are grateful to Mr. R. Forster for the GC-MS experiments and Mrs. J.M York for the toluene extractable matter measurements. Mr. P.R Morrell is thanked for useful discussions on the subject.

References

1. Zeldin, M; Qian, B.; Choi, S. J. *J. Polym. Sci. Polym. Chem. Ed.* **1983**, *21*, 1361.
2. Yang, A.C.M. *Polym.* **1994**, 35(*15*), 3206.
3. Zhang, H.; Hackam, R.; Lazarescu, V. *Conference on Electrical Insulation and Dielectric Phenomena*, **1999**, ref: 0-7803-5414-1/99.
4. Mukundan, A. L.; Balasubramanian, K.; Srinivasan, K. S. V., *Polym. Communications.* **1988**, *29*, 310.
5. Ostthoff, R. C., Bueche, A. M.; Grubb, W. T. *J. Chem. Phys.* **1954**, 22, 4659.
6. Delman, A. D.; Stein, A. A.; Simms, B. B.; Katzenstein, R. J. *J. Polym. Sci.* **1966**, *Part A-1, 4,* 2307.

Chapter 14

Prediction of the Elastomeric Modulus of Poly(dimethylsiloxane) Networks Formed by Endlinking

R. F. T. Stepto, J. I. Cail, and D. J. R. Taylor

Polymer Science and Technology Group, Manchester Materials Science Centre, UMIST and University of Manchester, Grosvenor Street, Manchester M1 7HS, United Kingdom

The application of a Monte-Carlo (MC) algorithm to account fully for loop formation in RA_2 + $R'B_3$ and RA_2 + $R'B_4$ polymerisations is described. The resulting interpretation of experimental elastic moduli of polyurethane networks prepared at different dilutions shows it is essential to account for elastic losses in loop structures of all sizes. An important parameter, x, is introduced, namely the average fractional loss of elasticity per larger loop structure relative to the loss per smallest loop structure. Values of x vary between 0.50 and 0.60, depending on junction point functionality, reactant or network chain stiffness and number of skeletal bonds per smallest loop structure. Application of the MC calculations to the formation and resulting structure of poly(dimethyl siloxane) networks again predicts significant reductions in modulus due to loop structures. However, comparison with experimental modulus data shows that the reductions in modulus due to loops are outweighed by increases due to chain entanglements.

Introduction

Endlinked poly(dimethyl siloxane) (PDMS) networks formed from stoichiometric reaction mixtures are normally assumed to be model networks. This is the first quantitative investigation of defects in their structures. Networks formed at complete reaction in bulk and at various dilutions in solvent are considered. Losses in elasticity due to loop structures of all sizes occur and significant reductions in modulus are predicted. In practice, the reductions can be more than counteracted by increases in modulus due to chain entanglements.

The molecular structures and macroscopic properties of network polymers depend more closely on reactant structures (molar masses, functionalities, chain flexibilities) and reaction conditions (dilution, proportions of different reactants) than do those of linear polymers. The classical Flory-Stockmayer (F-S) treatment of the gel point and the accompanying changes in distributions of molecular species give a basic explanation of the phenomena to which the behaviour and changes in actual polymerisations may be related. However, the infinite species that occur from the gel point to complete reaction cannot be enumerated as individual molecules. In addition, F-S theory says nothing concerning the detailed topology of the network, which grows and defines its structure through the random reaction of its reactive groups with other groups on the gel and with groups on sol species. To obtain a perfect network, all reactions (sol-sol, sol-gel and gel-gel) are assumed to yield elastically active chains between junction points in the final network. Relationships between concentrations of chains and junction points that assume perfect network structures are often used when interpreting elastic properties of endlinked networks (1). The assumption is rarely true and will be examined in detail in this paper with particular application to PDMS networks.

Generally, deviations from perfect network structures may be due to topological entanglements and chain interactions (2-4), to side reactions, incomplete reaction in endlinking polymerisations (giving loose ends) (5,6) and, more fundamentally and generally, inelastic chain or loop formation due to the intramolecular reaction of pairs of groups (4,7). Hence, to understand and predict elastomeric properties, it is important to be able to model, statistically, the molecular growth leading to network formation. By considering $RA_{fa} + R'B_{fb}$ polymerisations at various dilutions of reactive groups, it is possible to evaluate the effects of loop formation resulting from intramolecular reaction.

Intramolecular Reaction

A useful measure of the propensity of a system at a given ratio of reactants for intramolecular reaction is λ_{a0}, where

$$\lambda_{a0} = P_{ab} / c_{a0} \tag{1}$$

c_{a0} is the initial concentration of A-groups, representing the concentration of groups for intermolecular reaction, and P_{ab} characterises intramolecular reaction (7).

$$P_{ab} = (3/2\pi < r^2 >)^{3/2} / N_{Av}. \tag{2}$$

It is the mutual concentration of A- and B-groups at the ends of the shortest sub-chain, of root-mean-square end-to-end distance $<r^2>^{1/2}$, that can react intramolecularly, assuming that the end-to-end distance distribution can be represented by a Gaussian function. The units of P_{ab} are moles per unit volume. It can be seen that, for a given ratio of reactive groups, λ_{a0} captures the combined effects of reactant structures and reactive-group concentrations on intramolecular reaction.

Monte-Carlo Polymerisation Algorithm

Theories to predict the modulus of a polymer network material must begin by constructing a realistic model of the network structure, including defects. Detailed characterisation of the connectivity, or topology, by conventional, experimental means is impossible. In order to investigate the effects of network topology on elastomeric properties one must use numerical simulations of the network-forming nonlinear polymerisations. These have the potential to provide the necessary detailed structural information. In such simulations, it is important to account correctly for the formation of loop-structures of various sizes resulting from intramolecular reactions, correctly weighted according to their probabilities of formation. To these ends, a Monte-Carlo (MC) nonlinear polymerisation algorithm has been developed (8-10) to simulate self-polymerisations (RA_f), and two-monomer polymerisations of the general type $RA_{fa} + R'B_{fb}$. During the course of a simulated polymerisation, populations of reactive groups on monomer units are connected together according to the relative probabilities for intramolecular and intermolecular reaction, using λ_{a0} and taking account of all possible loop sizes and the decreasing external concentration of reactive groups as a polymerisation proceeds. All the connections are recorded as a function of extent of reaction of A- or B-groups, along with the calculated sol and gel fractions, and average degrees of polymerisation. Ring-size distributions have been evaluated and have been found to be very broad (8-10).

For the purposes of evaluating elastic modulus, the loop-size distribution in the *gel molecule* at complete reaction with increasing λ_{a0} can be simply characterised by calculating extents of intramolecular reaction on that molecule resulting in smallest loops by the end (*e*) of a polymerisation, $p_{re,1}$, as a function of λ_{a0}. Such a characterisation is useful because the exact effects of smallest loops on network elasticity can be deduced (*7,8*). In addition, $p_{re,i>1}$, the extent of reaction resulting in the formation of *larger* loops may also be calculated, so that the total, final extent of intramolecular reaction may be written

$$p_{re,total} = p_{re,1} + p_{re,i>1}. \tag{3}$$

For illustration, the values of $p_{re,total}$, $p_{re,1}$ and $p_{re,i>1}$ for a series of RA$_2$ + R'B$_3$ simulations are shown in Figure 1. Notice that $p_{re,total}$ = 1/6, always. This is related to the cycle rank of the completed network. For a stoichiometric RA$_2$ + R'B$_f$ polymerisation at complete reaction (*11*), $p_{re,total} = (1/2)(1 - 2/f)$.

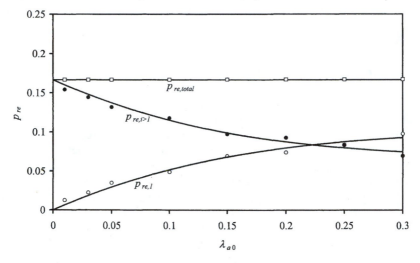

Figure 1. $p_{re,total}$, $p_{re,1}$ and $p_{re,i>1}$ versus λ_{a0} for stoichiometric RA$_2$ + R'B$_3$ simulations with 1000 branch units.

Figure 1 shows that the incidence of smallest loops ($p_{re,1}$) is negligible when λ_{a0} = 0. Essentially, in agreement with F-S statistics, all intramolecular reaction then occurs *after* the gel point, resulting in larger loop structures. However, as λ_{a0} increases, $p_{re,1}$ increases, at the expense of the proportion of larger loop structures ($p_{re,i>1}$). The distribution of ring sizes is generally broad, as illustrated in Figure 2 for the case λ_{a0} = 0.1.

For a given value of λ_{a0}, it is found that tetrafunctional systems give rise to more loop structures than trifunctional systems, due simply to the greater number of opportunities for loop-formation in the former case.

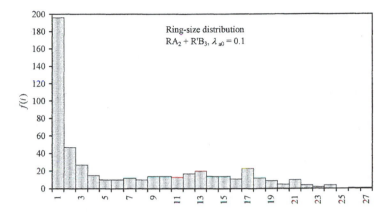

Figure 2. Frequency of ring structures, f(i), versus size, i,from simulations of stoichiometric $RA_2 + R'B_3$ polymerisations with $\lambda_{a0} = 0.1$ with 6000 A and B groups, initially.

Correlation of Model Network Topologies with Measured Network Moduli (8-10)

For a trifunctional network, each smallest loop structure renders three chains inelastic and larger loops are subject only to partial losses in elasticity. The phenomenon is illustrated in Figure 3 using the smallest ($i = 1$) and next smallest ($i = 2$) ring structures in $RA_2 + R'B_3$ networks. In general, for the completely reacted network, there are $3/2$ ($= f/2$) chains emanating from each junction. The smallest loop means that 2 junction points or 3 chains are rendered *totally* inelastic. For the next smallest ring structure, no chains are rendered totally inelastic but the chains emanating from the two junctions it contains have configurational entropies that are reduced below those of chains in a perfect (treelike) network structure. Different entropy reductions of will occur for ring structures of different sizes. In the case illustrated, $i = 2$, the entropy of the ring structure and its adjoining subchains is equivalent to a *single* elastic chain between the two junction points (●) plus the entropy of the ring, rather than the entropy of *four* chains.

These considerations lead to the expression for the modulus, relative to that of the perfect network structure,

$$G^o/G = M_c/M_c^o = 1/(1 - 6p_{re,1} - x \cdot 6p_{re,i>1}), \qquad (4)$$

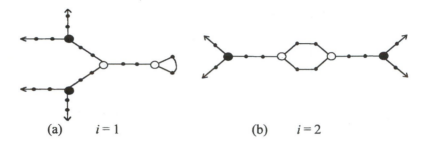

*Figure 3. Ring structures and decreases in chain elasticity for RA_2 + $R'B_3$
networks (a) smallest ring structure (i = 1); (b) second smallest ring structure
(i = 2). ● - junction points with 3 fully elastic chains emanating from them;
○ - junction points with (a) inelastic chains, and (b) chains of reduced elasticity
emanating from them.*

where G^o is the modulus of the (unswollen) perfect network, G is that of the
actual network, M_c is the average molar mass of elastically active chains
connecting pairs of junction points in the actual network and M_c^o is that in the
perfect network. x is the fractional loss of elasticity for chains in loop structures
larger than the smallest. A similar expression can be derived for tetrafunctional
network structures. They lose only 2 elastic chains per smallest loop and

$$G^o/G = M_c/M_c^o = 1/(1 - 4p_{re,1} - x \cdot 4p_{re,i>1}).$$ (5)

In the absence of theoretical evaluations of entropy reductions in loop
structures, estimates of x can be made using experimentally determined values of
M_c/M_c^o This has been done for polyurethane (PU) networks. The results, for six
series of networks prepared in bulk and at various dilutions in solvent, are shown
in Figure 4. The theoretical curves have been fitted by choosing least-squares
values of P_{ab} and x in conjunction with eqs 4 and 5. In this respect, it should be
noted that, because $p_{re,1}$ and $p_{re>1}$ depend on λ_{a0}, which is directly proportional to
P_{ab} (eq 1), M_c/M_c^o depends on P_{ab} as well as on x.

The effect of assuming $x = 0$, no entropy reduction or elastic loss in ring
structures larger than the smallest, is illustrated in Figure 5 using the results and
calculations from system 1 of Figure 4. The position of the calculated line is
very sensitive to the values of x and it is impossible to account for experimental
reductions in modulus with $x = 0$. Joining the ends of the elastic subchains of
networks together into ring structures always causes a reduction in entropy
compared with that of the independent chains, whatever the size of the structure.
It is expected that x will decrease as overall size of the ring structure increases.
The values of x used here are average values over all sizes of ring structure.

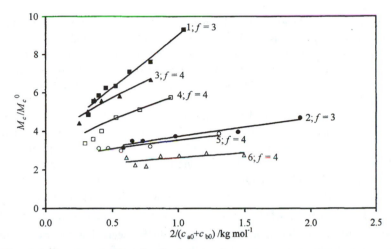

Figure 4. Experimental and calculated values of M_c / M_c^o at complete reaction as functions of the average initial dilution of reactive groups, $2/(c_{a0}+c_{b0})$, for six series of PU networks from stoichiometric reaction mixtures. 1,2: trifunctional networks from polyoxpropylene (POP) triols of different molar masses and hexamethylene diisocyanate (HDI). 3-6: tetrafunctional networks from POP tetrols and HDI.

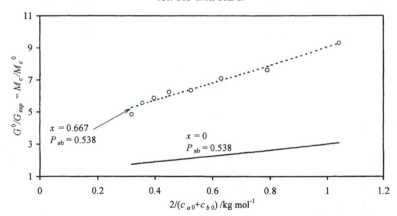

Figure 5. Illustrating the effect of assuming $x = 0$ using the results and calculations for system 1 of Figure 4. If $x = 0$ the reductions in modulus are greatly underestimated.

Predictions of Modulus for Other Networks

The values of x deduced for the six reaction systems of Figure 4 are plotted in Figure 6 versus P_{ab}. Because both chain entropy and P_{ab} are related directly to $<r^2>$, it is assumed that x versus P_{ab} is a universal function for given functionalities of reactants. Thus, the plots in Figure 6 can be used to predict reductions in modulus for networks formed from other stoichiometric $RA_2 + R'B_3$ and $RA_2 + R'B_4$ systems.

The present work now considers networks formed from linear PDMS chains of various molar masses using a trifunctional and a tetrafunctional endlinker. The systems chosen first have been used previously for a gel-point investigation (12). From the known molar masses, endlinker functionality and chain statistics of a pair of reactants, P_{ab} can be calculated and the appropriate value of x found from Figure 6. Values of $p_{re,i}$ and $p_{re,i>1}$ are known as universal functions of λ_{a0} (see Figure 1). Hence, curves of $G^o/G = M_c/M_c^o$ versus λ_{a0} can be constructed using eqs 4 and 5, as shown in Figure 7. It can be seen that the predicted reductions in modulus are large and must be accounted for when interpreting measured moduli. Significantly, the trifunctional and tetrafunctional systems give distinct curves. Amongst the tetrafunctional systems, there is little sensitivity to reactant molar masses or P_{ab}, showing that x does not vary much over the systems chosen.

Through eq 1 and knowledge of c_{a0}, the axis in λ_{a0} may be transformed to one in reactant molar mass, for polymerisations in bulk, or to one in dilution (c_{a0}^{-1}) for reactions diluted with solvent. For linear reactant RA_2 reacting with endlinker $R'B_f$ the equation is

$$\lambda_{a0} = \left(\frac{M_{A2}}{2} + \frac{M_{Bf}}{f}\right)\frac{P_{ab}}{\rho} \cdot \frac{1}{1-D} \qquad (6)$$

where ρ is the assumed uniform density of the reactants and the diluent of volume fraction D. Provided M_{A2}, is large compared with M_{Bf}, P_{ab} is proportional to $M_{A2}^{-3/2}$ and λ_{a0} becomes proportional to $M_{A2}^{-1/2}$ for bulk polymerisations. Using the curves in Figure 7 to represent bulk polymerisations, $\lambda_{a0} = 0.01$ corresponds to $M_{A2} = 10,000$ g mol^{-1}, $\lambda_{a0} = 0.02$ corresponds to $M_{A2} = 2,500$ g mol^{-1} and $\lambda_{a0} = 0.036$ corresponds to $M_{A2} = 400$ g mol^{-1}.

The present calculations can also be used to reconcile moduli of stoichiometric PDMS networks prepared at different dilutions as measured by Llorente and Mark (13) with the occurrence of entanglements in PDMS melts as characterised, for example, by Fetters et al. (14). Figures 8(a) and 8(b) show the results of Llorente and Mark. The essentially constant value of M_c/M_c^o over a range of dilutions is consistent with the predicted behaviour. The form of the relationship between λ_{a0} and D given by eq 6 means that the plots in Figure 7 transform into nearly horizontal plots with D as abscissa, as illustrated in Figure

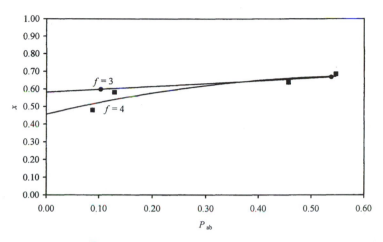

Figure 6. x versus P_{ab}, corresponding to the theoretical curves in Figure 4.

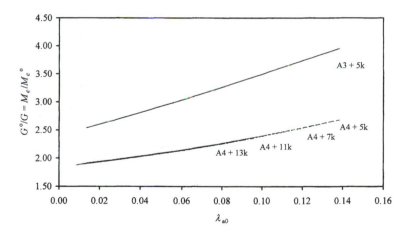

Figure 7. Predicted reductions in modulus as a function of ring-forming parameter λ_{a0}. Networks prepared from linear PDMS fractions 5k, 7k, 11k, 13k (of nominal molar masses 5000 g mol^{-1}, 7000 g mol^{-1}, 11000 g mol^{-1} and 13000 g mol^{-1}) reacting with a trifunctional endlinker (A3) and a tetrafunctional endlinker (A4). Details of the reactants are given in reference (12).

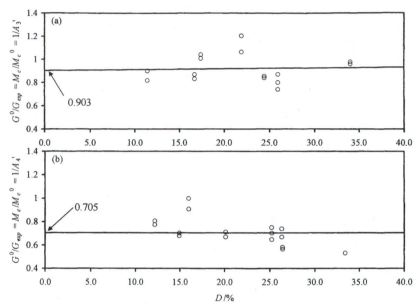

Figure 8. $G^o/G = M_c/M_c^o = 1/A'_f$ *versus dilution of preparation (D) for*
PDMS networks from stoichiometric polymerisation using a siloxane diluent.
Data of Llorente and Mark (13). (a) $RA_2 + R'B_3$ and (b) $RA_2 + R'B_4$
polymerisations.

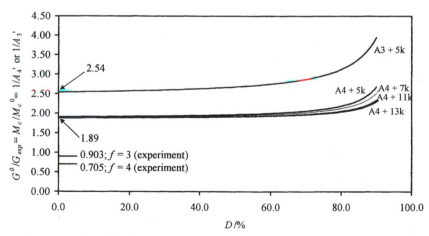

Figure 9. Predicted and experimental modulus reductions for PDMS networks.

9. Because of entanglements, the experimental values of M_c/M_c^o of 0.903 for f = 3 and 0.705 for f = 4 are less than the predicted values that account for loops.

The reconciliation of experimental and predicted values of M_c/M_c^o is straightforward. M_c^o = 18500 g mol^{-1} for the systems studied by Llorente and Mark. Hence, $M_{c,experiment}$/g mol^{-1} = 0.903 × 18500 = 16706 for f = 3 networks and $M_{c,experiment}$/g mol^{-1} = 0.705 × 18500 = 13043 for f = 4 networks. Both of these values are somewhat greater than the value of 12000 g mol^{-1} for the entanglement molar mass (M_e) in linear PDMS. However, accounting for loop structures, without entanglements one would expect $M_{c,chemical}$ = 2.54 × 18500 = 46990 for f = 3 networks and $M_{c,chemical}$ = 1.89 × 18500 = 34965 for f = 4 networks. Thus, $M_{c,chemical}$ / $M_{c,experiment}$ = 46990/16706 = 2.54/0.903 = 2.81 for f = 3 networks defines the factor by which $M_{c,chemical}$ is reduced due to entanglements. For f = 4 networks a similar reduction factor results with $M_{c,chemical}$/ $M_{c,experiment}$ = 34965/13043 = 1.89/0.705 = 2.68. Alternatively, as illustrated in Figure 10, one may say that the average number of entanglements per equivalent linear chain between junctions of the actual networks is 2.81 - 1 = 1.81 for f = 3 networks and 2.68 - 1 = 1.68 for f = 4 networks.

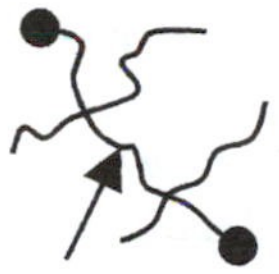

n entanglements

$$M_c \rightarrow \frac{M_c}{n+1}$$

Figure10. Illustrating the relationship between M_c and the average number of entanglements per equivalent linear chain in PDMS networks.

Conclusions

The present paper shows the importance of accounting fully for loop formation when interpreting the absolute values of the moduli of networks. Even for reactants of high molar mass, the effects of loops on elastic modulus are significant. Model networks are *not* perfect networks and the effects of loops have to be considered, alongside those of loose ends and entanglements, when interpreting modulus. For PU networks based on HDI and POP polyols, the effects of loops are dominant. Other work has shown that PU networks based on diphenylmethane diisocyanate (MDI) and POP polyols show modulus enhancement due to interchain interactions (4). For PDMS networks, using a linear reactant of molar mass greater than M_e, the reductions in modulus due to loops are significant, but are outweighed by those due to entanglements.

162

Acknowledgements

Support of the EPSRC for grant GR/L/66649 and MSI for provision of their Polymer Software is gratefully acknowledged.

References

1. Mark, J.E.; Erman, B. in *Polymer Networks – Principles of Their Formation, Structure and Properties*, Stepto, R.F.T., ed., Blackie Academic & Professional, London,1998, Chap. 7.
2. Graessley, W.W. *Adv. Polymer Sci.*, **1974**, *16*.
3. Kramer, O. in *Elastomeric Polymer Networks*, Mark, J.E.; Erman, B., eds., Prentice Hall, Englewood Cliffs, NJ, 1992, Chap. 17.
4. Stepto, R.F.T.; Eichinger, B.E.; in *Elastomeric Polymer Networks*, Mark, J.E.; Erman, B., eds., Prentice Hall, Englewood Cliffs, New Jersey, 1992, chap. 18.
5. Gottlieb, M.; Macosko, C.W.; Benjamin, G.S.; Meyers, K.O.; Merrill, E.W. *Macromolecules*, **1981**, *14*, 1039
6. Macosko, C.W.; Saam, J.C. *Polymer. Prepr., Div. Polymer Chem., Amer. Chem. Soc.*,**1985**, *26*, 48.
7. Stepto, R.F.T. in *Polymer Networks – Principles of Their Formation, Structure and Properties*, Stepto, R.F.T., ed., Blackie Academic & Professional, London, 1998, Chap. 2.
8. Stepto, R.F.T.; Taylor, D.J.R. in *Cyclic Polymers 2 nd edition*, Semlyen, J.A., ed., Kluwer Acadenic Publishers, Dordrecht, 2000, Chap.15.
9. Dutton, S.; Stepto, R.F.T.; Taylor, D.J.R. *Angew. Makromol. Chem.* **1996**, *240*, 39.
10. Stepto, R.F.T.; Cail, J.I.; Taylor, D.J.R. *Polimery* **2000**, *XLV*, 455.
11. Stepto, R.F.T. *Polym. Bull. (Berlin)*, **1990**, *24*, 53.
12. Stepto, R.F.T.; Taylor, D.J.R.; Partchuk, T.; Gottlieb, M. in *ACS Symposium Series 729, Silicones and Silicone-Modified Materials*, Clarson, S.J.; Fitzgerald J.J.; Owen, M.J.; Smith, M.D., eds., Amer. Chem. Soc., Washington DC 2000, Chap. 12.
13. Llorente, M.A.; Mark, J.E. *J. Chem. Phys.* **1979**, *71*, 682
14. Fetters, L.J.; Lohse, D.J.; Milner, S.T.; Graessley, W.W. *Macromolecules* **1999**, *32*, 6847

Chapter 15

Structure–Property Relationships for Poly(dimethylsiloxane) Networks In Situ Filled Using Titanium 2-Ethylhexoxide and Zirconium *n*-Butoxide

S. Murugesan[1], J. E. Mark[1], and G. Beaucage[2]

Departments of [1]Chemistry and [2]Materials Science, University of Cincinnati, Cincinnati, OH 45221

Abstract

In situ filling of elastomeric networks has been used as an alternative to the conventional *ex situ* filling of such materials. The *in situ* process has advantages over the *ex situ* process, for example, it is a low-temperature process, giving uniform distributions of particles compared to the energy demanding *ex situ* process. The sizes and shapes of the particles can be controlled by adjusting the reaction parameters during the synthesis. In this present work titania and zirconia filler particles were synthesized *in situ* in poly(dimethylsiloxane) networks from titanium 2-ethylhexoxide and zirconium *n*-butoxide. The advantage of the former is its long backbone chain (compared to that of titanium *n*-butoxide). Also the ethyl side group attached to the back bone is expected to reduce the initial growth of the titania particles compared to those from titanium *n*-butoxide. The sizes of the particles were obtained from SAXS (small angle X-ray scattering), and the mechanical properties were determined from stress-strain measurements.

Introduction

For the past decade or so, an alternative to the energy demanding ex-situ filling of elastomers has been *in situ* filling. The *in situ* technique is a solvent-free synthesis at room temperature that can be used to produce (in a continuous process) significant quantities of colloidal-size particles *(1-5)*. These particles can be produced from various alkoxides using either base or acid catalysts. There are many advantages to this *in situ* process, for example particle sizes and distributions *(6,7)* can be controlled by reaction conditions, low temperatures are involved, and solvent is unnecessary. SAXS (small angle x-ray scattering) is one of the important tools used to characterize such filler particles *(8-10)* with their different shapes and sizes. Using Porod's law and Guinier's law it is possible from the SAXS data to obtain values of the R_g (radius of gyration), and the slope of the Porod regime, gives an indication of the shapes of the particle. In this study an effort has been made to synthesize *in situ* zirconia and titania particles in poly(dimethylsiloxane) (PDMS) networks. Zirconium *n*-butoxide and titanium 2-ethylhexoide (TEH) were used in this study to generate the particles. The advantages of using TEH are that it has a longer chain and a side group, ethyl, which will reduce the speed at which these particles are formed (compared to those in an earlier study with titanium *n*-propoxide *(8)*).

Experimental Procedures

PDMS used in this study was obtained from Gelest and has a molecular weight (Mn) of 18000 gms/mol. Equimolar quantities of the PDMS and the crosslinking agent tetraethoxysilane (TEOS) were mixed well with a small quantity (~1%) of the catalyst stannous oleate, and left undisturbed for 3 days. Once a film was formed, it was removed and extracted first with toluene and then with a mixture of toluene and methanol with decreasing portions of toluene until it was 100% methanol. The film was then dried for 24 hrs under vacuum, and then soaked in the alkoxides for up to 24 hours. In order to monitor the speed at which these particles were formed, several strips were cut and then soaked in the alkoxide for various times. The soaked strips were then immersed in a dilute base solution of 2% w/v diethyl amine in distilled water for 24 hours. The base acted as a catalyst for the hydrolysis and condensation of the alkoxides absorbed into the PDMS network.

Some mechanical properties of the filled samples were studied using an Instron machine. Micrographs were obtained from a Hitachi scanning electron microscope, and SAXS data were obtained using pinhole 2D SAXS apparatus.

Results and Discussion

The total amount of filler *in situ* precipitated into the PDMS networks was 15wt.% for titania and 17wt.% for zirconia. Micrographs revealed that the zirconia particles occurred as strands with good interlocking (Figure 1). In the case of titania there were no fibrillar particles. Also, the reinforcement of the PDMS networks by the zirconia filled was larger than that from the titania. The SAXS data (Figure 2) showed that particle size increased up to 10-12 hours of reaction time, and after that the particle size decreased (Table I). This may be attributed to the fact that zirconia has strong tendency to form coordination compounds with the alcohol byproduct (*n*-butanol). This could make the particles bend around and thus show a decrease in size after 12 hours. Also, the particles have a tendency to form aggregates.

Titania samples also showed a similar trend, specifically a decrease in size after 10-12 hours. This may be due to fact that the 2-ethyl hexanoic acid, which is increased in size by its side group, may find it difficult to diffuse out of the PDMS network, thus enabling the particles to bend and become smaller.

The mechanical properties showed good improvement (Figure 3), especially in toughness (Table II), with zirconia giving the larger improvements. The sizes along with the stranded nature of the zirconia particles could explain these differences.

Table I Values of Rg of the filler particles

Hours in alkoxides	R_g of particles (Å)	
	Zirconia	Titania
1	50	60
6	65	79
11	76	105
24	68	75

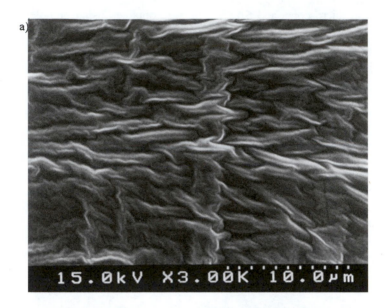

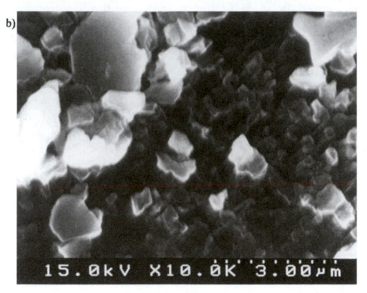

Figure 1. Micrographs of PDMS networks filled with a) zirconia
b) titania.

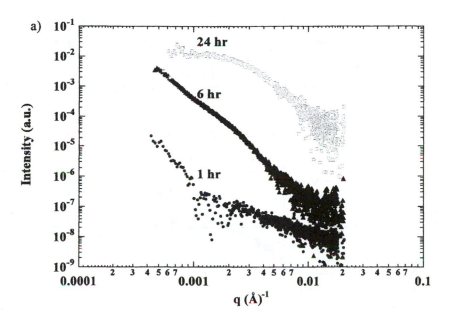

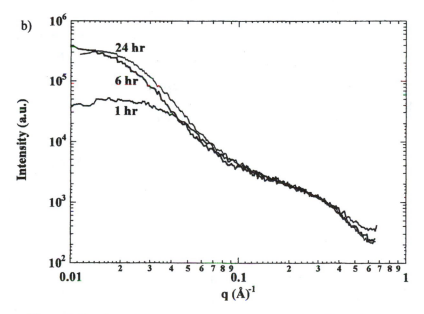

Figure 2. Small angle X-ray scattering of a) Titania b) Zirconia filled
PDMS   networks

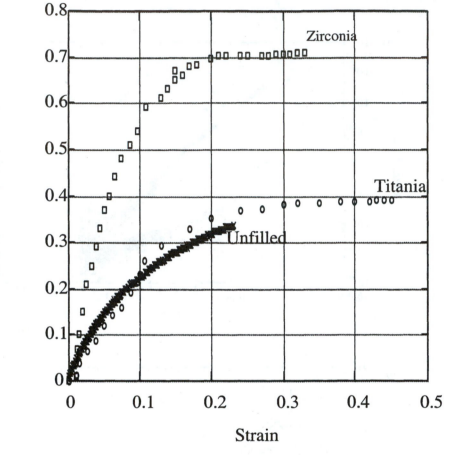

Figure 3. Plot of stress vs. strain for filled and unfilled PDMS samples

Table II Mechanical properties of TiO_2 and ZrO_2 filled PDMS

Type of filler	Toughness (MPa)
None	0.17
Titania filled	0.35
Zirconia filled	0.67

Conclusions

PDMS networks in situ filled with zirconia and titania had mechanical properties that were much improved compared to the unfilled networks. Also the nature and size of the zirconia particles were apparently the origin of their better reinforcing ability, relative to titania.

Acknowledgement

It is a pleasure to acknowledge the financial support provided by the National Science Foundation through Grant DMR-0075198 (Polymers Program, Division of Materials Research).

References

1) E. Matijevic, Chem. Mater., *5*, 412, 1993.
2) A. Garg, E. Matijevic, Langmuir, *4*, 38, 1988.
3) P. Gherardi, E. Matijevic, J. Colloid Interface Sci., *109*, 57, 1986.
4) M. Ozaki, S. Kratohvil, E. Matijevic, J. Colloid Interface Sci., *102*, 146, 1984.
5) E. Matijevic, A. D. Lindsay, S. Kratohvil, J. Colloid Interface Sci., *36*, 273, 1971.
6) S. Kan, X. Zhang, S. Yu, D. Li, L. Xiao, G. Zou, T. Li, W. Dong, Y. Lu, J. Chem. Soc., Faraday Trans, 615, 1972.
7) R. Demchak, E. Matijevic, J. Colloid Interface Sci., 31, 257, 1969.
8) J. M. Breiner, J. E. Mark, Polymer, *39*, 5483, 1998.
9) D. W. McCarthy, J. E. Mark, D. W. Schaefer, J. Polym. Sci., Part B-Polym. Phys., *36*, 1167, 1998.
10) D. W. McCarthy, J. E. Mark, S. J. Clarson, D. W. Schaefer, J. Polym. Sci., Part B-Polym. Phys., *36*, 1191, 1998.

Chapter 16

Modulus Reduction Mechanism of Trimethylsiloxy Silicates in a Polyorganosiloxane

Randall G. Schmidt, Linda R. Badour, and Glenn V. Gordon

Dow Corning Corporation, 2200 West Salzburg Road,
Midland, MI 48686–0994

Trimethylsiloxy silicates can be added to network forming linear polyorganosiloxane-based formulations to reduce the elastic modulus of the resulting cross-linked network without the loss of ultimate strength or excessive bleeding that typically occurs when either linear oligomers or polymer additives are used. The linear viscoelastic properties of uncured blends of a polyorganosiloxane with different loadings of silicates were characterized using time–temperature superposition to investigate the physical factors responsible for the modulus drop. The reduction in the plateau modulus of the polymer was found to be consistent with the power-law relation observed for nonentanglement-forming diluents, and was accompanied by a decrease in the frequency range of the plateau regime with increasing silicate content. Swelling and extraction studies on the corresponding cross-linked networks were conducted to help postulate the primary physical factors responsible for the modulus reduction. The modulus reduction enables low-durometer, high-elongation elastomeric materials useful in silicone rubbers, high-movement joint sealants and pressure-sensitive adhesives.

Introduction

The interactions resulting from the addition of trimethylsiloxy silicates to linear polyorganosiloxanes has been exploited successfully in material technologies including silicone pressure-sensitive adhesives (PSA) (*1, 2*) and paper release coatings (*3, 4*). In the former, the appearance of a mechanical transition that is manifested by a maximum in the loss tangent can be controlled to tune tack and adhesive properties; in the latter, the increase in the viscous character of a coating of a cross-linked silicone network is used to control peel forces from organic PSAs. This paper focuses on the molecular mechanisms responsible for the reduction of the plateau modulus of a high molecular weight polyorgano-siloxane gum that, when cross-linked, enables low-durometer, high-elongation elastomers useful as rubbers, mold-making materials, and high-movement joint sealants without loss of ultimate strength or excessive bleeding (*5*).

Experimental

Materials

Table I lists the materials in this study and some of their physical properties. A poly(dimethylsiloxane-*co*-phenylmethylsiloxane) ($PDPMS_1$) was used to represent a linear polyorganosiloxane exhibiting rubbery behavior. The presence of 7.5–8.1 mol% phenylmethylsiloxane suppressed the crystalline phase observed in poly(dimethylsiloxane), which enabled the use of the time–temperature superposition technique (*6*) to investigate the different viscoelastic regimes of the polymer. $PDPMS_1$ also contained 0.15 mol% methylvinylsiloxane and was endcapped with a dimethylvinylsiloxy group.

Three trimethylsiloxy silicates were investigated. A low molecular weight silicate (LS) was prepared from the acid-catalyzed hydrolysis and condensation of tetraethoxysilane and endcapped with hexamethyldisloxane. A medium (MS) and a high (HS) molecular weight silicate were obtained from the acid-catalyzed polymerization of sodium silicate followed by a reaction with trimethylchloro-silane in a process described elsewhere (*7*). For comparison, a 50-mPa·s low molecular weight poly(dimethylsiloxane-*co*-phenylmethylsiloxane) ($PDPMS_2$), with a phenylmethylsiloxane content of 9.7 mol% was used to function as a comparative oligomeric diluent for $PDPMS_1$.

The polymer was solvated in toluene to aid in mechanically mixing either a trimethylsiloxy silicate or the $PDPMS_2$ at desired concentration levels of up to 50 wt%. The solvent was subsequently removed by placing the blend in a convection oven at 60 °C overnight followed by a 130 °C exposure for an hour.

Table I. Physical Properties of Materials

Material	$\overline{M}_w$, $kg \cdot mol^{-1}$	ρ, $kg \cdot m^{-3}$	T_g, K
Dimethylsiloxane-co-phenylmethylsiloxane			
PDPMS$_1$[a]	600	980	158
PDPMS$_2$[b]	4	980	156
Trimethylsiloxy silicate			
Low molecular weight (LS)	2	1060	220
Medium molecular weight (MS)	5	1120	320
High molecular weight (HS)	14	1180	>500

[a] 7.5–8.1 mol% phenylmethylsiloxane, 0.15% methylvinylsiloxane

[b] 9.7 mol% phenylmethylsiloxane

Cross-linked siloxane networks from blends containing PDPMS$_1$ and the nonreactive LS were prepared via a hydrosilylation reaction using a platinum complex catalyst, a diethylfumerate inhibitor, and a dimethylsiloxane-co-methyl-hydrogensiloxane copolymer as a cross-linker. Molds were fashioned out of fluorosilicone-coated release liners to obtain approximately 1-cm-thick slabs following a cure cycle (5 h at 22 °C; 16 h at 70 °C; and, 1 h at 130 °C) designed to remove the solvent before any cross-linking reaction occurred.

Characterization

Linear viscoelastic properties were measured using a Rheometric Scientific™ RDAII. For uncross-linked blends, shear data from 0.1 to 500 rad·s^{-1} were collected isothermally at several temperatures with 25-mm-diameter parallel plates used to collect data above −50 °C while 8-mm-diameter plates were required for colder temperatures. The frequency range was extended to 0.01 rad·s^{-1} at 25 °C, which represented the reference temperature. Strain sweeps were conducted at each isotherm to ensure that data from the frequency sweep were obtained in the linear viscoelastic region. Thermal profiles of the cross-linked siloxane networks were obtained in rectangular torsion at a frequency of 1 rad·s^{-1} to compare the effect of either extracting the nonreactive LS silicate from the polymer network or swelling the network with LS on the shear storage modulus, G'.

Extraction of the silicate from the siloxane network, with the requisite gravimetric analysis, was carried out using reagent-grade toluene; 30-mL vials of immersed specimens were placed on a rotary shaker for 24 h before being dried thoroughly. Gravimetric analysis was performed periodically while swelling the siloxane network with LS until the desired extent of swelling was attained.

Results and Discussion

Uncross-linked Blends

Master curves for the linear viscoelastic properties of PDPMS$_1$ and its corresponding blends with either a silicate or PDPMS$_2$ were generated using time–temperature superposition (6). The temperature dependence of the modulus was taken into account from the kinetic theory of rubber elasticity. Figure 1 shows that the master curves for PDPMS$_1$ at a reference temperature of 25 °C illustrate the characteristic viscoelastic behaviors of linear polymers at different frequency regimes: glassy, transition, plateau, and terminal. The presence of a rubbery plateau is indicative of strong topological interactions that resemble the effects of covalent cross-links over a certain frequency interval, and is commonly referred to as entanglement effects. The plateau modulus, G_N^0, is used to characterize the rubbery plateau and define an average molecular weight between entanglements, M_e

$$G_N^0 = \frac{\rho RT}{M_e} \tag{1}$$

where ρ is the density, R the gas constant, and T the absolute temperature.

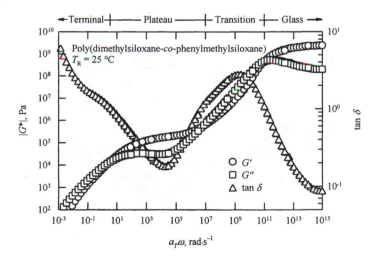

Figure 1. *Master curves for PDPMS$_1$ at a reference temperature of 25 °C.*

Figures 2 and 3 show the master curves for G' and the loss tangent, tan δ, at a reference temperature of 25 °C for blends of LS and HS in PDPMS$_1$, respectively. Using the approximation proposed by Wu (8), G_N^0 was determined from G' at the tan δ minimum in the plateau region.

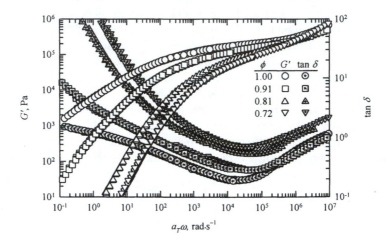

Figure 2. *Effect of volume fraction of a low molecular weight silicate, (1–ϕ), on the viscoelastic properties of PDPMS$_1$ at a reference temperature of 25 °C.*

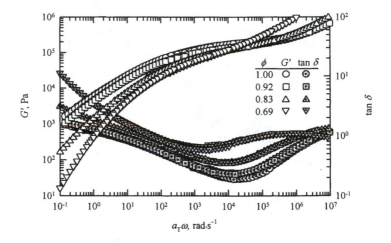

Figure 3. *Effect of volume fraction of a high molecular weight silicate, (1–ϕ), on the viscoelastic properties of PDPMS$_1$ at a reference temperature of 25 °C.*

Table II summarizes the characteristics of the rubbery plateau region for the blends as a function of the polymer volume fraction, ϕ. In this concentration range, the introduction of a second component lowered the plateau modulus relative to the base polymer. If the blend density can be described by a linear mixing rule, the primary effect of the second component from eq 1 was to dilute the concentration of polymer chains per unit volume resulting in a reduced number of pair-wise contacts and, hence, a reduction in the number of entanglements in the base polymer. This effect is generic for the addition of any non-entanglement-forming diluent to a polymer (9). The reduction in G_N^0 scales as

$$G_N^0\left(\phi\right) = G_N^0\big|_{\phi=1}\,\phi^a \tag{2}$$

where a is the concentration exponent and has a value of 2.0–2.3 for typical diluents (10). Theoretically, $a \sim 2$ if every contact between two chains has a constant

Table II. Blend Properties in the Viscoelastic Plateau Region

Blend Component	ϕ^a	G_N^0, MPa	M_e, kg mol^{-1}	Plateau region, decades of ω
PDPMS$_2$	1.00	0.18	13.7	6.2
	0.90	0.14	17.0	5.7
	0.80	0.11	22.1	5.3
	0.70	0.080	30.3	4.9
	0.60	0.059	41.1	4.8
	0.50	0.037	65.8	3.8
LS	0.91	0.15	16.7	5.2
	0.81	0.10	24.1	2.9
	0.72	0.076	32.7	2.1
	0.62	0.056	44.7	2.3
MS	0.91	0.15	16.9	5.7
	0.82	0.12	21.4	4.6
	0.68	0.069	36.7	3.0
	0.53	0.032	80.9	2.1
HS	0.92	0.14	17.9	6.0
	0.83	0.10	24.1	4.8
	0.69	0.065	39.6	3.3
	0.64	0.049	53.0	3.3

^{a} Polymer (PDPMS$_1$) volume fraction.

probability of entanglement (9) whereas scaling laws applied to conditions for the overlap of random coils predict $a = 9/4$ (11).

Figure 4 plots the effect of the low molecular weight blend component on the plateau modulus of the siloxane polymer. The concentration exponent a in eq 2 using $PDPMS_2$ was 2.2 ± 0.07 as may be expected for a good solvent. However, blends based on the trimethylsiloxy silicate exhibited a values in excess of the typical range of 2.0 to 2.3, and increased from 2.5 ± 0.16 to 2.8 ± 0.13 with increasing silicate molecular weight. Values for a as high as 3.6 have been reported previously for poly(n-butylmethacrylate)/diethylphthalate systems (12). It was hypothesized that the silicate particles not only reduced the number of polymer pair-wise contacts but also induced a stiffening effect, which reduced the convolution of the polymer chains and resulted in enhanced a values.

Further evidence for this stiffening effect, or restricted mobility of polymer chains, was from the observation that, unlike $PDPMS_2$, the silicates raised the temperature of the primary mechanical transition of the blend relative to the glass transition of $PDPMS_1$ ($T_g = 158$ K) as shown by the temperature dependence at which tan δ exhibited a maximum for the series of $MS/PDPMS_1$ (top) and $HS/PDPMS_1$ (bottom) blends in Figure 5, for example. At a given silicate volume fraction, the higher its molecular weight, the more efficient it was in increasing the transition temperature and in reducing the plateau modulus of the base polymer.

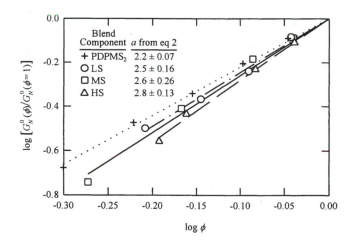

Figure 4. *Effect of the low molecular weight blend component on the plateau modulus of $PDPMS_1$, G_N^0 ($\phi=1$), as a function of $PDPMS_1$ volume fraction, ϕ. Also tabulated are the estimate and 95% confidence interval for the concentration exponent a in eq 2 obtained by the linear regression analysis of the slope.*

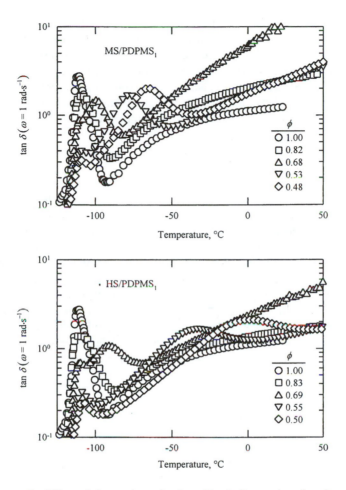

Figure 5. *Effect of the medium (top) and high (bottom) molecular weight silicate on the primary mechanical transition of PDPMS₁ as a function of polymer volume fraction, φ.*

Table II also shows that a decrease in the extent of the frequency range of the rubbery plateau region accompanied the reduction in G_N^0 as the fraction of the low molecular weight component increased, where a more significant effect again was obtained using the silicates. This was consistent with the hypothesis that the addition of the trimethylsiloxy silicate reduced the number of effective polymer chain entanglements, thereby enabling the onset of long range translational flow at frequencies higher than that exhibited by the base polymer alone.

Organic tackifier resins used in organic PSA formulations are blended with a base polymer and, by definition, raise the T_g and lower the plateau modulus of the system (13). Tse (14) reported that the addition of a resin tackifier to styrene–isoprene block copolymers decreased the frequency range of the plateau region and lowered G_N^0 with a concentration exponent a of approximately 2. The trimethylsiloxy silicate materials in this study have been shown previously to be effective tackifier resins for linear polyorganosiloxanes.

Silicates in a Cross-linked Siloxane Network

The formation of a covalently cross-linked network effectively locks into place polymer chain entanglements, and provides a scheme in which to further elucidate the effect of the silicates on a highly entangled linear polyorganosiloxane. Figure 6 shows the temperature dependence of G' at 1 rad·s^{-1} for the PDPMS$_1$-based networks. The incorporation of 0.38 volume fraction nonreactive LS prior to cross-linking reduced considerably the network modulus (■) relative to the unfilled siloxane network (●). After extracting more than 98 wt% LS from the network, the network modulus (□) remained essentially unchanged. It was hypothesized that the silicate effectively reduced the number of entanglements in PDPMS$_1$ prior to cross-linking—thus, reducing the plateau modulus—and that this effect was maintained even after the silicate was extracted from the cross-linked network.

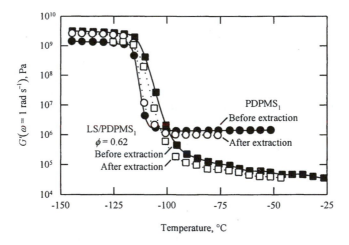

Figure 6. *Effect of the non-reactive low molecular weight silicate (LS) on a cross-linked PDPMS$_1$ network before and after extraction.*

Another experiment to obtain additional insight into the modulus reduction mechanism involved swelling a cross-linked PDPMS₁ network with the silicate. Figure 7 shows that although a slight reduction was evident in the network modulus of PDPMS₁ when swollen with 0.42 volume fraction LS, the decrease was significantly less than if the non-reactive silicate was blended into the polymer prior to being cross-linked. Hence, the silicate was unable to alter the distribution of physical entanglements, which is the hypothesized mechanism for modulus reduction, if it is subsequently incorporated into a cross-linked network. This was consistent with the prediction that the reduction of the elastic modulus of a swollen network scales as $\phi^{1/3}$ in a θ solvent (*15, 16*) or the experimental observation $\phi^{7/12}$ in a good solvent (*17*), unlike the $\phi^{2.5}$ behavior when LS was blended into the precursor polymer.

The utility of the modulus reduction mechanism can be exploited to enable low-durometer, high-elongation elastomers useful as rubbers, mold-making materials, and high-movement joint sealants. For example, trimethylsiloxy silicates can be added to standard silicone sealant formulations at loadings of 5–20 wt% to significantly reduce the modulus (> 30%) and enhance the elongation (> 30%) of the fully cured sealant without sacrificing strength. When a linear oligomer or polymer is utilized to reduce modulus, excessive bleeding of the diluent usually occurs that results in undesirable surface wetness and staining. The particulate nature of the silicates minimizes bleed and there is no surface wetness if solid silicates are employed.

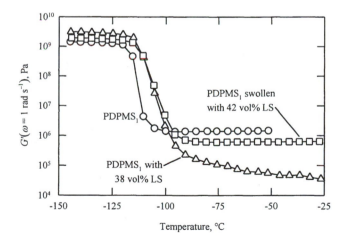

Figure 7. *Effect of the non-reactive low molecular weight silicate on a PDPMS₁ network: PDPMS₁ cross-linked with 0.38 volume fraction LS (Δ); and, PDPMS₁ network swollen with with 0.42 volume fraction LS (□).*

Conclusions

Trimethylsiloxy silicates were shown to reduce the rubbery plateau modulus of a linear polyorganosiloxane and significantly reduce the extent of the plateau region in simple blends. The extent in the reduction of the polymer plateau modulus, as a function of the silicate concentration in the blend, exceeded that predicted and typically observed when using simple diluents. It was hypothesized that the silicate particles not only reduced the number of polymer pair-wise contacts but also induced a stiffening effect, which reduced the convolution of the polymer chains and enhanced the modulus reduction. The ability of these trimethylsiloxy silicates to both increase the primary transition of the blend and lower the plateau modulus relative to the base polymer makes them well suited to function as tackifier resins in polyorganosiloxane systems.

References

1. Wengorvius, J. H.; Burnell, T. B.; Zumbrum, M. A.; Krenceski, M. A. *Polym. Preprints* **1998**, *39(1)*, 512.
2. Lin, S. B.; Krenceski, M. A. *Proc. Annu. Meet. Adhes. Soc.* **1996**, 363.
3. Gordon, G. V., Tabler, R. L.; Perz, S. V.; Stasser, J. L.; Owen, M. J.; Tonge, J. S. *Polym. Preprints* **1998**, *39(1)*, 537.
4. Gordon, G. V.; Schmidt, R. G. *J. Adhesion* **2000**, *72*, 133.
5. Juen, D. R.; Rapson, L. J.; Schmidt, R. G.; U. S. Patent 5,373,078, 1994 Dow Corning Corp.
6. Williams, M.L.; Landel, R. F.; Ferry, J. D. *J. Amer. Chem. Soc.* **1955**, *77*, 3701.
7. Daudt, W.; Tyler, L. U. S. Patent 2,676,182, 1954 Dow Corning Corp.
8. Wu, S. *J. Polym. Sci., Polym. Phys.* **1987**, *25*, 557; 2511.
9. Ferry, J. D. *Viscoelastic Properties of Polymers*; Wiley: New York, 1980.
10. Graessley, W. W. *Adv. Polym. Sci.* **1974**, *16*, 1.
11. DeGennes, P. G. *Macromolecules* **1976**, *9*, 587; 594.
12. Stern, D. M.; Berge, J. W.; Kurath, S. F.; Sakoonkim, C.; Ferry, J. D. *J. Colloid Sci.* **1962**, *17*, 409.
13. Chu, S. G. In *Handbook of Pressure Sensitive Adhesive Technology*; 2nd ed.; Satas, D., Ed.; Van Nostrand Reinhold: New York, 1989; p 176.
14. Tse, M. F. *J. Adhesion Sci. Technol.* **1989**, *3*, 551.
15. James, H.; Guth, E. *J. Polym. Sci.* **1949**, *4*, 153.
16. Flory, P. J. *Principles of Polymer Chemistry*; Cornell University Press: Ithaca, NY, 1953.
17. Obukhov, S. P.; Rubenstein, M.; Colby, R. H. *Macromolecules* **1994**, *27*, 3191.

Chapter 17

NMR Spin Relaxation, Self-Diffusion and Viscosity Studies of Poly(dimethylsiloxane)s Blended with Silicate Nanoparticles

Claire Roberts[1], Terence Cosgrove[1,*], Randall G. Schmidt[1], Glenn V. Gordon[2], Andrew J. Goodwin[3], and Axel Kretschmer[3]

[1]School of Chemistry, University of Bristol, Cantock's Close, Bristol BS8 1TS, United Kingdom
[2]Dow Corning Corporation, 2200 West Salzburg Road, Midland, MI 48686–0994
[3]Dow Corning Ltd., Cardiff Road, Barry, S. Glamorgan CF63 2YL, United Kingdom

Nuclear magnetic resonance spin–spin relaxation and self-diffusion measurements were used to investigate the mobility of poly(dimethylsiloxane) (PDMS) polymers when mixed with trimethylsilyl-treated silicate nanoparticles. The silicate can either solvate or reinforce the composite depending on the size of the particle. The smaller nanoparticle solvated all the polymer samples studied. The larger nanoparticle reinforced the two lowest molecular weight polymers at any concentration. When mixed with the higher molecular weight PDMS, the larger polysilicate caused a dramatic reduction in the mobility of the polymer chains above a critical concentration; however, an increase in mobility was observed below this critical concentration. This apparent increase in the mobility of the polymer chains was also manifested by a decrease in the zero-shear-rate viscosity of the mixture.

Introduction

Fillers are frequently added to commercial polymers to enhance their mechanical properties without great expense. Despite widespread use of such systems, there remains a need to provide a fundamental understanding of how the mechanical properties of these systems are altered, and few techniques have proved suitable for studying such systems at a molecular level. Methods that have been used include nuclear magnetic resonance (NMR) relaxation (1–4), neutron scattering and reflection (5, 6), and other atomic/molecular scattering techniques (7). Quasielastic neutron scattering was also used to detect adsorption (8). In this paper, we review our recent work on composite materials formed by blending trimethylsilyl-treated silicates with poly(dimethylsiloxane)s (PDMS).

Silica is frequently used as a reinforcing agent in polymers such as PDMS and strong interactions between the two components have been observed (9). Surface modification of silica, such as trimethylsilylation, can be used to weaken the interaction with the PDMS to avoid the complications of agglomeration when studying these interactions. It was envisaged that the understanding gained from this approach could be extended to active fillers such as fumed silica where stronger interactions play an important role.

The mobility of polymer melts has been widely discussed in the literature, both from theoretical (9) and experimental (10) points of view. In particular, NMR relaxation data are useful in determining the segmental mobility of chains, especially in different physical environments (11, 12). Upon adsorption at an interface, polymer segments can no longer orient isotropically; consequently, the spin–spin relaxation times will be reduced and may exhibit some degree of non-exponential behaviour. The transverse relaxation time, T_2, decreases with decreasing segmental mobility until the sample is solid on the NMR timescale. For polymers adsorbed as trains—in direct contact with the surface—very short values of T_2 can be found (13). For segments that retain some mobility, T_2 will depend on the local monomer concentration and the segmental mobility. In a more recent paper (14), the measured T_2 values for silica-filled PDMS mixtures revealed that the presence of the filler particles dramatically influenced the mobility of the polymer chains. For silica concentrations greater than 25% by weight, most of the polymer segments in the dispersion experienced some degree of restricted mobility. This effect was molecular weight dependent with the high molecular weight chains being more influenced at a given silica concentration.

In the present study, NMR spin–spin relaxation and diffusion measurements were used to investigate the mobility of PDMS mixed with a trimethylsilyl-treated silicate. Along with supplementary viscosity data, this paper will focus on NMR results of four different molecular weight polymers and two different sized silicate particles.

Experimental

Materials

The linear PDMS and the polysilicate materials were supplied by Dow Corning Corporation, where some of their physical properties are listed in Table I. An anionic polymerization process using hexamethylcyclotrisiloxane was applied to obtain the two relatively monodisperse low molecular weight PDMS samples, designated as 5K and 12K. Solvent fractionation was applied to acquire higher molecular weight polymers designated as 120K and 633K, which were above the critical molecular weight—marking the onset of entanglement effects—for PDMS ($M_c \sim 30$ kg mol^{-1}). The lower molecular weight silicate designated R1 was made from the acid-catalyzed hydrolysis and condensation of tetraethoxysilane and end-capped with hexamethyldisiloxane. The silicate designated R3 was obtained from an acid-catalyzed polymerization of sodium silicate, followed by a reaction with trimethylchlorosilane in a process described elsewhere (15).

Information on molar dimensions was obtained by size exclusion chromatography (SEC) techniques. Fluid density was measured at 25 °C using an Anton PAAR DMA48 density meter, whereas the density of R3 was obtained using helium pycnometry (Quantachrome Micropycnometer MPY–2).

The polymer and silicate components were blended together at different weight ratios and measurements were conducted approximately one month after preparation.

Table I. Physical Properties of PDMS and Silicate Materials[a]

Material	$\overline{M}_w$, kg·mol^{-1}	$\overline{M}_w/\overline{M}_n$	R_g, nm	ρ, kg·m^{-3}	η_0, Pa·s
PDMS					
5K	5.2	1.07	2.0	960	$10^{-1.33}$
12K	12.2	1.03	3.1	965	$10^{-0.92}$
120K	120.3	1.53	9.2^b	970	$10^{2.06}$
633K	633.7	1.25	21.1^c	978	$10^{4.75}$
Silicate					
R1	0.5	1.15	0.35	880	$10^{-2.44}$
R3	14.1	2.99	2.2	1170	$> 10^{26\ c}$

[a] $\overline{M}_w$ is the weight-average molecular weight, $\overline{M}_w/\overline{M}_n$ the polydispersity, R_g the radius of gyration, ρ the density, and η_0 the zero-shear-rate viscosity.

[b] Estimated from reference 16.

[c] Extrapolated from data of blends with PDMS.

NMR Measurements

Liquid-state NMR relaxation data were collected using a JEOL FX200 modified with a Surrey Medical Image Systems vector processor and a digital radio frequency console. Results were obtained using the Carr–Purcell–Meiboom–Gill (CPMG) pulse sequence (17) with a 180° pulse spacing of 200 – 500 μs. Up to 8192 data points were collected by sampling the echo maxima. For the solid R3 silicate, T_2 was obtained from a solid-state experiment performed on a Bruker MSL 300 spectrometer using a single 90° pulse with a length of 2.7 μs. Pulsed field gradient (PFG) NMR experiments were performed on a JEOL FX-100 spectrometer at 25°C using an external ^2D lock. Before performing the experiments, the applied gradient was calibrated using water, which has a diffusion coefficient of 2.3×10^{-9} $m^2\,s^{-1}$, and the sample temperature was controlled at 22 ± 1 °C. Results were analysed using the basic equation for unrestricted diffusion.

Viscosity Measurements

The zero-shear-rate viscosity, η_0, at 25 °C was obtained using a Rheometric Scientific RDAII equipped with cone-and-plate fixtures. For the 633K PDMS and blends with a high concentration of the R3 silicate, η_0 was approximated from dynamic measurements using 8-mm-diameter parallel plates.

Results and Discussion

Pure Components

The first set of experiments was made on the pure components given in Table 1. Figure 1 shows the spin–spin relaxation decays obtained using the CPMG pulse sequence. The silicate nanoparticle R1 clearly was the most mobile species (see also η_0 in Table I) and was effectively a molecular solvent. The 12K PDMS, which is to the right of the decay for R1, was evidently less mobile and its relaxation can be described by the Brereton model (11). The higher molecular weight polymers had substantially shorter relaxation times and this can be explained in part by the formation of molecular entanglements, which exponentially increases the bulk viscosity and reduces translational diffusion. Despite a rather modest increase in size and molecular weight, the R3 silicate was essentially a solid as perceived by NMR and the CPMG sequence, which cannot restore the transverse magnetisation lost by strong dipolar coupling and only provided a qualitative estimate of the relaxation time.

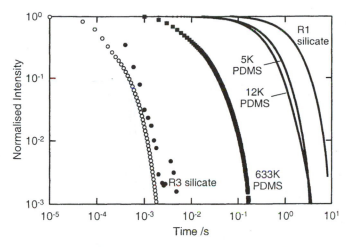

Figure 1. Relaxation decays for select components using CPMG pulse sequence except for o *calculated from T_2 values obtained from fitting solid-state spectrum.*

Figure 2 shows the solid-state spectrum obtained for the R3 silicate, and it clearly has a complex structure due to methyl groups in different environments as revealed by the deconvolution of the spectrum into Gaussian and Lorentzian functions. Given the pulse delays used in subsequent experiments, the R3 nanoparticle was effectively invisible in the CPMG relaxation decays.

Figure 3 shows the relaxation decays found by mixing the 12K PDMS with the smaller R1 silicate nanoparticle. The relaxation decays essentially became longer as R1 simply functioned as a solvation agent. This was consistent with the viscosity data shown inset for the blends based on the 5K and 12K polymers.

However, the opposite trend occurred when these polymers were mixed with the larger R3 silicate. The blend viscosity was greater relative to the viscosity of the base polymer at any concentration of R3. Figure 4 shows that the relaxation decays based on the 12K PDMS become faster indicating a large reduction in mobility on mixing the two components, which typifies reinforcing behaviour. The intensity of the initial decays have not been normalised to indicate that a proportion of the signal had been lost. The decay can be fitted to a multiple exponential function and extrapolated to zero time. However, after making this correction, the initial heights were still not linearly proportional to the fraction of PDMS protons in the mixture. Recalling that the R3 silicate was invisible under these experimental conditions, this indicated that a proportion of the PDMS was severely restricted in mobility. It was likely that this fraction corresponded to molecular segments in direct contact with the nanoparticle surface (i.e., train segments) and accounted in part for the increased rigidity of the dispersion.

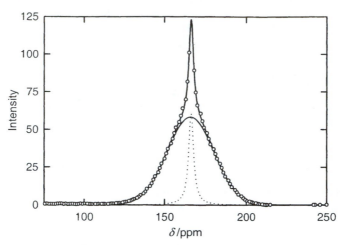

Figure 2. Solid-state ^1H spectrum for R3 silicate at 300 MHz showing fits to a Gaussian, Lorentzian (⋯⋯), and sum of a Gaussian and Lorentzian function.

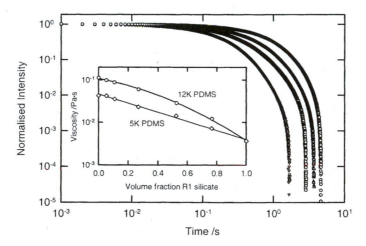

Figure 3. Relaxation decays for blends of the R1 silicate with the 12K PDMS (all normalised to 1): 100 (Ⓞ), 52 (Ⓒ), 27 (Ⓤ), and 0 (Ⓢ) vol% R1. Shown inset is the effect of R1 content on the viscosity of the 5K and 12K polymers.

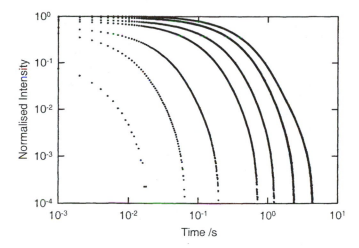

Figure 4. T_2 relaxation decays for blends of 12K PDMS and the R3 silicate normalised to the pure polymer. PDMS concentration increases going from left to right and correspond to 34, 45, 55, 69, 78, 92, and 100 vol% PDMS.

The same sets of blends were also studied using PFG NMR, which measures translational diffusion. Figure 5 shows the attenuation plots as function of the field gradient width, δ, for different concentrations of R1 silicate in 12K PDMS. The self-diffusion coefficient, D, depends on the attenuation approximately as $\sim exp(-D\delta^2)$. Hence, the faster the decay, the larger D. As both the silicate and PDMS contribute to the decay, a double-diffusion attenuation was seen and two independent diffusion coefficients can be obtained. The data can be analysed using the Stokes–Einstein equation and the measured viscosity of R1 (Table I) to give a hydrodynamic radius, R_H, for both components. The average value for R1 was 0.3 ± 0.1, and decreased slightly with increasing polymer volume fraction. However, the R_H values for the polymer decreased continuously with increasing concentration suggesting that the coil collapses from a good solvent environment to that of a theta solvent (pure melt).

A rather different analysis was required when the same polymer was mixed with the larger R3 silicate. The decay was now single diffusion but with a non-zero baseline noting that the signal from R3 was invisible. These decays can be used in two ways (*18*) to estimate the adsorbed amount and the polymer layer thickness at the interface. The first parameter was found by assuming that the baseline decay was due to locally mobile polymer segments that were attached to the particle and hence translate too slowly for their diffusion to be observed directly with the current experimental conditions of gradient strength and pulse

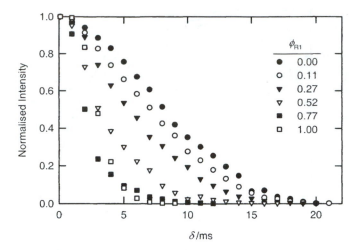

Figure 5. Attenuation decays for blends of the 12K PDMS and R1 silicate.

duration times. The layer thickness can be found by using an obstruction model (*19*) and the viscosity data (Table I). This gave a larger hydrodynamic size than that expected from the nanoparticle with no attached polymer layer. The results for the 12K polymer were unfortunately inconsistent due to sensitivity; however, for the smaller (5K) polymer, an adsorbed amount of 0.6 mg·m^{-2} was obtained, which was consistent with other data and set a lower limit for the adsorption of the 12K PDMS. The calculated hydrodynamic layer thickness as a function of the volume fraction of R3 silicate is shown in Figure 6 for the 12K PDMS. An average layer thickness of 0.35 ± 0.04 and 0.52 ± 0.2 nm was obtained for the 5K and 12K samples, respectively. These two values can be used to obtain an approximate scaling exponent α, $\delta \sim M^{\alpha}$, of 0.35, which is somewhat less than the scaling prediction of 0.5 for a bulk polymer.

Mixtures of High Molecular Weight PDMS with the R3 Silicate

The behaviour of mixtures formed using the larger R3 silicate with the higher molecular weight polymers was substantially different. Figure 7 shows the normalised relaxation curves at 22 °C for blends of the 633K PDMS with the R3 silicate. At R3 concentrations of up to 17 vol%, the inclusion of the silicate nanoparticles was observed to increase the mobility of the polymer as shown by

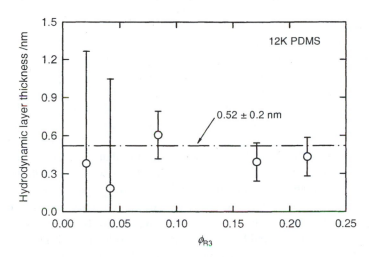

Figure 6. Hydrodynamic layer thickness as a function of the volume fraction of R3 silicate in the 12K polymer. The horizontal line represents the average value.

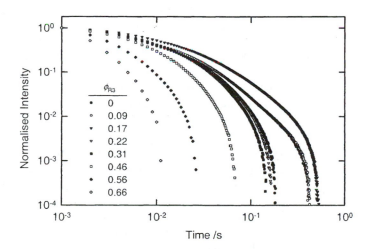

Figure 7. Normalised NMR spin–spin relaxation data for blends of the 633K PDMS and R3 silicate at 22 °C as a function of the volume fraction of R3.

the shift of the relaxation curves to the right (longer T_2) but beyond this concentration, a reinforcement effect was seen. However, the relaxation decay does not become longer than the pure melt until 22% incorporation has been reached. Although the decays cannot be fitted to a single exponential, it is revealing to plot the 1/e time for these data and this is shown in Figure 8 where a maximum in T_2 was evident.

The rheological properties of the blends were also measured as a function R3 content. Figure 9 shows that at high R3 inclusions ($\phi_{R3} > 0.35$) $\eta_{0, \text{ blend}}$ increased exponentially. However, there was a distinct decrease in viscosity up to a concentration range of between 20 to 25 vol% R3. This bulk plasticisation effect at low silicate concentrations is also manifested by a decrease in the rubbery plateau modulus (20). Above this silicate concentration, the mobility of the polymer began to decrease. At 22 vol% R3, the mobility of the blend was less than that of the 633K PDMS, thereby indicating that the silicate nanoparticles were starting to reinforce the polymer. This trend continued with increasing particle concentration resulting in an enhanced reinforcement of the polymer.

A lower molecular weight (120K) polymer was also studied with its 1/e data also plotted in Figure 8. The trends in these data were similar to the 633K polymer except that reinforced behaviour was observed for R3 concentrations of 5 vol% and above, and the maximum was much weaker. In contrast, the data in Figure 9 revealed a continuous reduction in viscosity up to 22 vol% R3, noting that the time scale associated with terminal flow behavior is significantly longer relative to that in NMR experiments.

The plasticisation effect of the trimethylsilyl-treated silicate on topologically constrained polymers is quite atypical and three possible scenarios exist as to its origins. The first is that on adsorption on the filler surface the polymer chains shrink in size compared to their bulk dimension. However, experimental evidence and Monte Carlo calculations show that when the polymer chains are larger than the filler particle—as was the case for the R3 silicate with either the 120K or 633K PDMS—there is a relative increase in chain dimensions at low filler concentrations (21). Hence, another possibility is that the particles effectively introduce free volume into the system and indeed these effects are only apparent when the polymer is substantially larger than the particle. A final scenario is that by dispersing amongst the polymer chains at a molecular level, the silicate particles dislocate a fraction of the physical entanglements in the melt, which presupposes a very weak interaction between the polymer and the particle surface. Evidence to this effect in related siloxane-based systems is described in detail elsewhere in this volume (20). Work is continuing to explore this phenomenon in greater detail and this will be published subsequently.

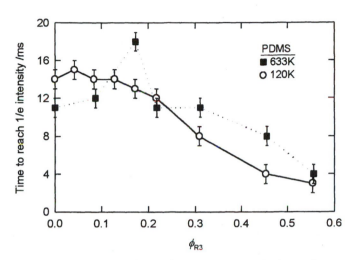

Figure 8. Time at which 1/e of initial intensity is reached as a function of R3 silicate volume fraction in high molecular weight PDMS polymers.

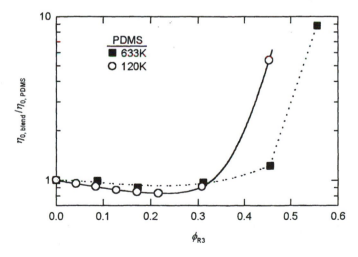

Figure 9. Effect of R3 content on blend viscosity relative to the base polymer.

Conclusions

NMR and rheology measurements were used to investigate interactions between trimethylsilyl-treated silicate nanoparticles and PDMS melts. Evidence of an immobile adsorbed layer was found with corresponding thickness somewhat less than the bulk radius of gyration. For the high molecular weight polymers, an atypical behaviour was found in which the particles first plasticise then reinforce the melts as a function of silicate loading.

Acknowledgments

The authors would like to like to acknowledge EPSRC (CR) and Dow Corning Ltd. for the provision of a CASE award.

References

1. Cosgrove, T.; Griffiths, P. C. *Adv. Colloid Interface Sci.* **1992**, *42*, 175.
2. Addad, J. P. C.; Touzet, S. *Polymer* **1993**, *34*, 3490.
3. Addad, J. P. C.; Morel, N. *J. Phys. III* **1996**, *6*, 267.
4. Kirst, K. U.; Kremer, F.; Litvinov, V. M. *Macromolecules* **1993**, *26*, 975.
5. Auvray, L.; Auroy, P.; Cruz, M. *J. Phys. I* **1992**, *2*, 943.
6. Field, J. B.; Toprakcioglu, C.; Dai, L.; Hadziioannou, G.; Smith, G.; Hamilton, W. *J. Phys. II* **1992**, *2*, 2221.
7. Steiner, U.; Chaturvedi, U. K.; Zak, O.; Krausch, G.; Schatz, G.; Klein, J. *Makromol. Chem., Macromol. Symp.* **1991**, *45*, 283.
8. Arrighi, V.; Higgins, J. S.; Burgess, A. N.; Floudas, G. *Polymer* **1998**, *39*, 6369.
9. Chahal, R. S.; St. Pierre, L. E. *Macromolecules* **1969**, *2*, 193.
10. Cosgrove, T.; Griffiths, P. C.; Hollingshurst, J.; Richards, R. D. C.; Semlyen, J. A. *Macromolecules* **1992**, *25*, 6761.
11. Brereton, M. G.; Ward, I. M.; Boden, N.; Wright, P. *Macromolecules* **1991**, *24*, 2068.
12. McCall, D. W.; Douglas, D. C.; Anderson, E. W. *J. Polym. Sci.* **1962**, *59*, 301.
13. Barnett, K. G.; Cosgrove, T.; Vincent, B.; Sissons, D. S.; Cohen Stuart, M. *Macromolecules* **1981**, *14*, 1018.
14. Cosgrove, T.; Weatherhead, I.; Turner, M. J.; Schmidt, R. G.; Gordon, G. V.; Hannington, J. P. *Polym. Preprints* **1998**, *39(1)*, 545.

15. Daudt, W.; Tyler, L. U.S. Patent 2,676.182, 1954 Dow Corning Corp.
16. Fetters, L. J.; Lohse, D. J.; Richter, D.; Witten, T. A.; Zirkel, A. *Macromolecules* **1994**, *27*, 4639.
17. Meiboom, S.; Gill, G. *Rev. Sci.Instrum.* **1958**, *23*, 68.
18. Roberts, C.; Cosgrove, T.; Schmidt, R. G.; Gordon, G. V. *Macromolecules* **2001**, *34*, 538.
19. Batchelor, G.K. J. Fluid Mech. **1971**, *74*, 1.
20. Schmidt, R. G.; Badour, L. R.; Gordon, G. V. In *Synthesis and Properties of Silicones and Silicone-Modified Materials*; Clarson, S. J.; Van Dyke M. E.; Fitzgerald, J. J.; Owen, M. J.; Smith, S. D., Eds.; American Chemical Society: Washington, DC, 2002.
21. Nakatani, A. I.; Chen, W.; Schmidt, R. G.; Gordon, G. V.; Han, C. C. *Polymer* **2001**, *42*, 3713.

Surfaces and Interfaces

Chapter 18

Investigating the Energetics of Bioadhesion on Microengineered Siloxane Elastomers

Characterizing the Topography, Mechanical Properties, and Surface Energy and Their Effect on Cell Contact Guidance

Adam W. Feinberg[1,2], Amy L. Gibson[2], Wade R. Wilkerson[1,2], Charles A. Seegert[1,2], Leslie H. Wilson[2], Lee C. Zhao[1,2], Ronald H. Baney[2], James A. Callow[3], Maureen E. Callow[3], and Anthony B. Brennan[1,2,*]

[1]Biomedical Engineering Program and [2]Department of Materials Science and Engineering, University of Florida, Gainesville, FL 32611
[3]University of Birmingham, School of Biosciences, Birmingham B15 2TT, United Kingdom

The energetics of a polydimethylsiloxane (PDMS) elastomer biointerface were micro-engineered through topographical and chemical modification to elicit controlled cellular responses. The PDMS elastomer surfaces were engineered with micrometer scale pillars and ridges on the surface and variable mechanical properties intended to effect directed cell behavior. The topographical features were created by casting the elastomer against epoxy replicas of micropatterned silicon wafers. Using UV photolithography and a reactive ion etching process, highly controlled and repeatable surface microtextures were produced on these wafers. AFM, SEM and white light interference profilometry (WLIP) confirmed the high fidelity of the pattern transfer process from wafer to elastomer. Ridges and pillars 5 μm wide and 1.5 μm or 5 μm tall separated by valleys at 5 μm, 10 μm, or 20 μm widths were examined. Mechanical properties were modulated by addition of linear and branched nonfunctional trimethylsiloxy terminated silicone oils. The modulus of the siloxane elastomer decreased from 1.43 MPa for the unmodified formulation to as low as 0.81 MPa with additives. The oils had no significant effect on the surface energy of the siloxane elastomer as measured by goniometry. Two main biological systems were studied: spores of the green alga *Enteromorpha* and porcine vascular endothelial cells (PVECs). The density of *Enteromorpha* spores that settled increased as the valley width decreased. The surface properties of the elastomer were altered by Argon plasma, radio frequency glow discharge (RFGD) treatment, to increase the hydrophilicity for PVEC culture. The endothelial cells formed a confluent layer on the RFGD treated smooth siloxane surface that was interrupted when micro-topography was introduced.

Introduction

Understanding the role of surface topography and chemical functionality on adhesion and proliferation of cells to biomaterials is critical in developing and improving biointerfaces. Many of the current limitations of existing devices and problems associated with developing technologies is the inability to start, stop and otherwise control biological growth, i.e., biofouling. Beginning with the deposition of proteins and complex carbohydrates on a surface, the resulting biological cascade can induce unwanted and irreversible effects[1]. Therefore, it is crucial to develop engineered biosurfaces that can elicit a desired cellular response.

While many researchers have evaluated cellular response to topographically and chemically modified polymer surfaces and have observed qualitative effects, few have correlated these results with an in depth characterization of the bulk and surface properties of the material[2-6]. In this paper, an analysis of the interfacial energies of the system is presented through characterization of the mechanical and morphological properties and surface energy of the biomaterial. The goal is to characterize the bioadhesive properties of the biologically active polymer surface.

Siloxane elastomers were chosen for the model polymer system because of the broad applications ranging from anti-fouling marine coatings to acute and chronic use medical devices. In particular, silicone elastomers represent a class of materials with a well understood chemistry, ease of manufacturability and extensive data on performance in biological systems. The engineered siloxane elastomer surfaces, in this study, have a combination of spatially controlled topographical and chemical modification. Features on the scale of 5 μm to 20 μm have been shown to influence cell function in the literature and these results have been replicated with patterns of ridges and pillars in these experiments[3-5,7-9]. Nonfunctional silicone oils were added to the bulk siloxane elastomer to determine the effects on modulus and surface energy. Due to a lower surface energy, silicone oils segregate to the surface, hence changing bulk properties on a macroscopic scale and imparting liquid like properties to the surface. We theorize that both topographical and chemical features define the surface energy, which in combination with elastic modulus, will dictate biological activity.

In order to understand the biological response to the surface, two types of cell were allowed to attach on the microengineered siloxane elastomers. Settlement (attachment) of motile spores of the marine fouling alga *Enteromorpha* was studied to evaluate the anti-fouling and foul release performance of the surface for marine applications. Porcine vascular endothelial cells (PVECs) were cultured on the surfaces as a model system to determine the bioactivity for potential medical applications such as vascular grafts and orthopedic implant coatings. The grooves formed on these substrates have shown significant control over growth directions of cells. This phenomenon, commonly referred to as "contact guidance," typically demonstrates cellular alignment along the grooves depending on the dimensions of the features. Current literature focuses on the mechanisms behind the alignment of these cells to the surface topography. An area of investigation is whether the actual geometries of the features were the defining factor or the change in surface free energy due to edges and disruptions in the planar surface[10]. It has been demonstrated that parameters such as surface free energy and hydrophilicity influence cell growth, but not necessarily the shape or orientation of cells[3]. With the addition of silicone oils on the molecular level and surface texture on the microscopic level, a hierarchical model was developed and examined to determine its effects on biological adhesion.

Experimental

Substrate Production

The micropatterned surface was initially etched into a silicon wafer using the Bosch process. The patterns chosen were 5 μm ridges separated by 5 μm, 10 μm and 20 μm valleys and 5 μm high cubes separated by 5 μm, 10 μm and 20 μm as seen in figure 1. These features were on the same size scale as cells and have been shown in the literature, and our own observations, to influence cell growth. Standard photolithographic microprocessing was used to fabricate the patterns, briefly: wafers were coated with photoresist (Clariant AZ1529 positive photoresist) and then exposed to UV light through a photomask. The photoresist was developed and the wafers were etched to a depth of 5 μm (Unaxis, Tampa, FL) or etched to a depth of 1.5 μm (University of Florida). Wafers were etched to a depth of 5 μm at Unaxis using the Bosch process to obtain nearly vertical sidewalls. This was in contrast to the 1.5 μm deep features produced at the University of Florida (UF) where a standard reactive ion etcher (RIE) produced sloping features due to lower anisotropy in the vertical direction. Molds were made of the etched wafers using the PDMS elastomer and then used to make nearly identical copies of the silicon wafer using a low shrinkage epoxy (Epon 828 with Jeffamine D230). These epoxy copies then became the masters for mass replication with siloxane elastomer and the original silicon wafer was stored to maintain the original pattern.

Silastic T-2 (Dow Corning) was chosen for its excellent dimensional stability, self-degassing properties and optical transparency (for laser confocal and light microscopy). It was also chosen because there were minimal unknown or non-disclosed additives, allowing for comprehensive control of additives and chemical modifications. The platinum catalyzed hydrosilylation cured siloxane elastomer was mixed in a 10:1 wt/wt base resin to curing agent ratio as specified by the manufacturer. It was then degassed under vacuum for 20 minutes and cast on the epoxy copies. Samples were then cured for 24 hours at room temperature before removing from the mold.

Silicone oils of linear structure were added to the Silastic T-2 during mixing and then cured to vary bioadhesive properties. Due to the nonfunctional trimethylsiloxy terminated polydimethylsiloxane oils used, no chemical incorporation of the oils into the vinyl functionalized elastomer occurred. Energetics dictate that the oils wet the siloxane elastomer surface, thus a controlled release of the oils to the surface occurs. The linear silicone oils chosen were 50 cSt, 500 cSt and 5000 cSt viscosities, which correlates with molar mass of 3.8, 17.3 and 49.4 kg/mol respectively, as reported by the

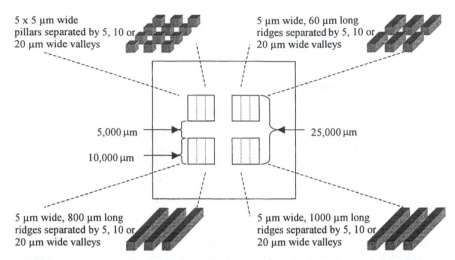

5 x 5 µm wide pillars separated by 5, 10 or 20 µm wide valleys

5 µm wide, 60 µm long ridges separated by 5, 10 or 20 µm wide valleys

5,000 µm

25,000 µm

10,000 µm

5 µm wide, 800 µm long ridges separated by 5, 10 or 20 µm wide valleys

5 µm wide, 1000 µm long ridges separated by 5, 10 or 20 µm wide valleys

Figure 1: Layout of the micro-patterned substrate etched to 1.5 µm or 5 µm depth. There were four 1 cm² main areas each with the feature type indicated in the diagram. Each main area was subdivided into three smaller 1/3 cm² areas with either 5, 10 or 20 µm spacing between features.

manufacturer (Gelest Inc.). In addition, low molar mass, non-functional, branched oils tetrakis(trimethylsiloxy)silane and tris(trimethylsiloxy)silane of 385 g/mol and 297 g/mol respectively (Gelest Inc.) were used as additives. The oils were mixed independently into the Silastic T-2 at 5%, 10% or 20% by weight prior to cure.

Morphological Analysis

Verifying the uniformity and consistency of dimensions in the micropatterning process was necessary to ensure the engineered patterns were properly transferred to the elastomer surface. This was accomplished with a combination of scanning electron microscopy (SEM), atomic force microscopy (AFM) and white light interference profilometry (WLIP). The SEM (Jeol 6400) allowed high magnification of the surface in order to identify defects and irregularities. Samples were coated with Au/Pd and examined at magnifications from 100x to 5,000x. AFM (Digital Instruments Nanoscope IIIa controller operating a Dimension 3100 scanner) operated in tapping mode in air provided the 3-dimensional shape to be observed although artifacts were potentially introduced due to the high aspect ratio of some features. The WLIP (Wyko NT 1000, Veeco Metrology) was operated in vertical scanning-interferometry (VSI) mode and used white light interference fringes to reconstruct the surface topography in a non-contacting and nondestructive manner. Images were processed using the Vision32 software package (Veeco Metrology) to examine the 3-dimensional topography and 2-dimensional height profiles of the surface.

Mechanical Properties

Tensile specimens were cut from cured 1 mm thick silicone films using an ASTM D1822-68 type L dog bone die. Tensile measurements were made according to ASTM D412-97 on an Instron model 1122 equipped with TestWorks 3.07 software for analysis. Strain measurements were based on crosshead displacement of 2 inches per minute. It is acknowledged that some slippage at the grips did occur, future studies will address this error through use of a laser extensometer. The various surface topographies were not evaluated with this method because the small feature size was insignificant to the bulk properties.

Surface Energy

Contact angles were obtained with a goniometer using the sessile bubble technique for all siloxane elastomer formulations within seconds of placement. Zisman plots were constructed from the contact angles of 2 μL droplets each of HPLC grade water, methylene iodide, 1-propanol, N,N-dimethylformamide and acetonitrile to extrapolate substrate surface energy (n=20 per liquid, per elastomer type)[11,12]. The effect of the oil additives and micro-topography on surface energy were evaluated by the Wilhelmy plate technique using a Cahn DCA. PDMS elastomer strips were used with dimensions approximately 10 mm wide, 30 mm long and 3 mm thick for the smooth, non-textured films, but varied for the textured surfaces since they were cast off an epoxy copy with no back plate. The micro-texture was only on one side and both textured and smooth samples were dipped into HPLC grade water (Fisher Scientific) at a dipping rate of 100 μm/sec.

Enteromorpha Studies

Glass cover slips coated with microtextured siloxane elastomer were evaluated at the University of Birmingham, UK, for attachment studies of the green alga *Enteromorpha* spore. The samples were soaked in sterile seawater overnight and then incubated with *Enteromorpha* spores at $\sim$2 x 10^6 spores/cc density for 1 hour[13]. The surfaces were then rinsed, fixed, and examined using a Zeiss imaging system attached to a fluorescent microscope to determine the density of settled (attached) spores on each surface.

Endothelial Cell Culture

The elastomer samples were first treated with argon RFGD plasma to increase hydrophilicity and then sterilized with ethylene oxide. The silica-like surface layer created by the plasma treatment is hydrophilic after exposure and then slowly reverts back to a hydrophobic state[14]. Gedde et al have indicated that hydrophobic recovery occurs by oligomer PDMS from the bulk recoating the surface by diffusion through the silica-like layer or through cracks in the silica-like layer caused by mechanical stress. An exposure time of 5 minutes at 50 mW and 4 cm distance from the RF coil prevented measureable hydrophobic recovery in the 5 days between plasma treatment and cell seeding. Individual feature types were separated and cultured in standard polystyrene culture 24 well plates with vascular porcine endothelial cells. For culture, RPMI 1640 medium (Life Technologies, Grand Island, NY) containing antibiotics (100 units/ml of penicillin, 100 µg/ml of streptomycin, 20 µg/ml of gentamicin, and 2 µg/ml of Fungizone) was used. There was no serum added during the culture process. Samples were incubated in 5% CO_2 atmosphere for 3 days to allow cells to reach confluence. Samples were removed, fixed in a 10% formalin solution, stained with crystal violet and then examined under light microscopy.

Results and Discussion

Morphology

SEM, AFM and WLIP images indicate the high fidelity reproduction of the silicon wafer micropattern in the PDMS elastomer. Typical SEM and WLIP images of the silicon wafer, epoxy copies and siloxane elastomer replicates reveal a surface with varying degrees of visible defects but acceptable overall fidelity. There was no observable loss in any of the micrometer scale dimensions and this was consistent for both ridge and pillar type features. As mentioned in the methods, due to differences in the type of RIE used, the 1.5 µm high features have a sloping sidewall indicative of the lower anisotropic etching process available at UF. An important observation, seen in Figure 2 a-c, was that the pillars did not have a cube-like shape but rather appear cylindrical. This phenomenon was partly a result of the exposure and development of the photoresist prior to etching. Light scattering led to features that were octagonal instead of square. These areas of the mask were further rounded with repeated levels of polymer replication so that by the end of the process they had attained this nearly cylindrical shape. This same defect also occurs at the ends of the ridges. Occasional nanometer scale defects were seen in the SEM images in figure 2b and 2e due to errors in the replication process. These small defects were unavoidable and have been ignored in terms of evaluating the effect of micrometer scale topographical features. The AFM images in figures 2a and 2d were from a 1.5 µm high pillar and 5 µm high ridges respectively and the sloping side wall was believed to be an imaging artifact due to a combination of feature aspect ratio and TappingMode tip geometry. While limited in the ability to image high aspect ratio features, the AFM indicated extremely smooth planar areas on top of the ridges and pillars and the bottom of the valleys. The WLIP images corroborate the SEM and AFM results and due to the exaggerated z axis scale make the finite radius of curvature of the feature edges visible. The nanometer scale radius of curvature was approximately three orders of magnitude smaller than the cell and therefore has not been evaluated in the scope of this paper.

The oils added to the Silastic T-2 can have an impact on the fidelity of the pattern reproduction. This has been examined thoroughly for the 1.5 µm features for the various oil types. The samples with 5000 cSt linear oils added to the Silastic T-2 had a visible oil layer on the surface. The patterns would disappear optically to both the naked eye and WLIP after 24-48 hours indicating that oils were segregating to the surface and filling in the pattern. This behavior was not noticed at the lower oil viscosities. Figure 3 shows a typical profile WLIP image of the 5 µm ridges separated by 5 µm valleys of the 5000 cSt, 20 wt% oil modified elastomer that was achieved by removing the surface oils by wiping with toluene just prior to imaging. The feature height and valley depth should be constant compared to other sample types, but was variable with a measured distance of over 2 µm for the amplitude at some points. Since the epoxy mold was verified to have a ~1.5 micrometer feature height, this change in height of the elastomer was due to surface distortions and possible material loss due to the removal of the 5000 cSt PDMS oils. From these data it was concluded that oil at the surface does change the topography with the effect dependent on oil viscosity and a time dependence on quantity. In terms of the surface seen by microorganisms, the question is whether a particular cell type, in a given biological environment, can displace the oils and contact the underlying solid surface.

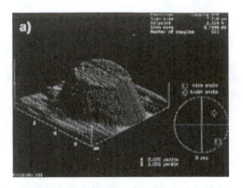

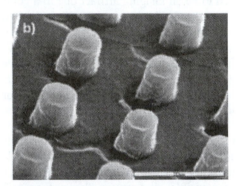

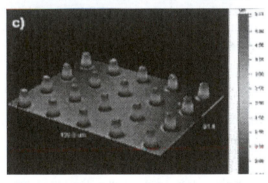

Figure 2: Atomic force micrographs (a and d), scanning electron micrographs (b and e) and white light interference profilometry (c and f) of the microtextured siloxane elastomer surface. The pillars and ridges were easily seen. a) Single 5 μm wide pillar, b) 5 μm high pillars separated by 5 μm valleys, c) 5 μm high pillars separated by 10 μm valleys (Si wafer), d) 5 μm high, 5 μm deep and 800 μm long ridges separated by 10 μm valleys, e) 5 μm high, 5 μm deep and 60 μm long ridges separated by 5 μm valleys, f) 1.5 μm high, 1000 μm long ridges separated by 10 μm valleys (Si wafer).

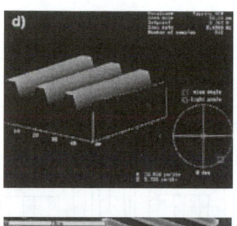

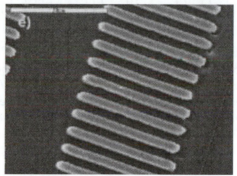

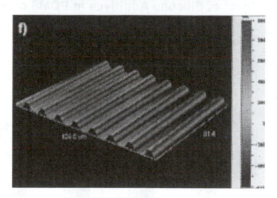

Figure 2. *Continued.*

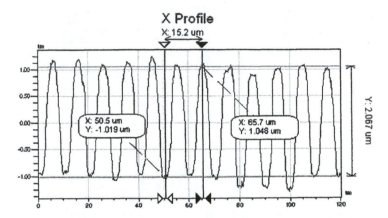

Figure 3: 5000 cSt, 20 wt% oil modified elastomer wiped with toluene immediately before imaging. A 2-dimenional profile WLIP image sliced perpendicular to the long axis of the 5 μm ridges separated by 5 μm valleys. The variation in valley and peak heights can be clearly seen as well as the variable amplitude ranging from 1.7 to 2.5 μm.

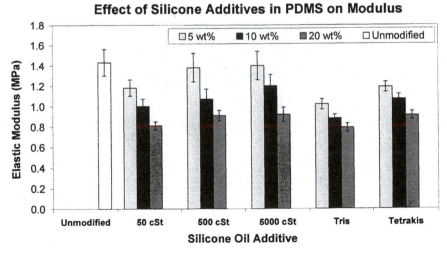

Figure 4: Graph of Young's Modulus obtained from tensile testing for siloxane elastomer modified with non-functional trimethylsiloxy terminated siloxane oils

Mechanical Properties

Instron tensile testing of various nonfunctional silicone oil incorporations yielded less than an order of magnitude change in elastic modulus and was highly dependent on concentration, as shown in Figure 4. Future studies will incorporate functionalized silicone oils that vary the crosslink density thus significantly changing the elastic modulus. Modulus has been shown to be a significant factor in controlling bioadhesion. While traditional theories related the relative adhesion of a surface primarily to surface energetics, it is now recognized that a factor of (surface energy *Elastic modulus)$^{1/2}$ is a better representation[15,16]. This indicates why silicones are found to have lower adhesive values than lower energy surfaces such as higher modulus fluoropolymers.

Surface Energy

Surface energy was changed as a function of the chemical and topographical features of the surface[17,18]. The effect of the oil additives were evaluated using goniometry and DCA. Results for nonfunctional oil additives as well as the unmodified Silastic T-2 elastomer indicated no significant change in surface energy for any formulation as seen in Table 1. The lower surface tension of the liquid PDMS as compared to the surface energy of the bulk elastomer caused oils to segregate to the surface. Since surface energy analysis was only sensitive for one molecular monolayer, the methyl side groups of the PDMS oils presents a surface that was indistinguishable from the cured elastomer formulations[12,19]. The surface energies were calculated from the contact angles using a modified Zisman plot of Cos θ vs. $1/(\gamma^{1/2})$, applicable to low surface energy materials like PDMS[20]. DCA was used to evaluate the affect oil additives had on the hysteresis of the advancing and receding contact angles as seen in Table 2. The large hysteresis for the first four samples was thought to be due to the rearrangement of the polymer backbones to express more hydrophilic moieties for the receding contact angle. Of note is the significantly smaller hysteresis for the 20% 5000 cSt sample where the visible oil layer seems to have interfered with the surface rearrangement. The effect of micrometer scale topography was evaluated using DCA. Direct measurement of the contact angle was not possible due to software limitations, however, a change in slope of the force verse position curves was observed between the smooth and textured regions. This change in slope, and hence contact angle, indicates that micropatterning causes a measureable difference between the surface energy of different topographical domains.

Table 1: Contact angles and calculated surface energy of PDMS elastomer modified with non-functional silicone oils

Samples	Water	MeI	DMF	ACN	1-Prop	γ_c (mN/m)
Unmodified	109.1 ± 3.5	67.2 ± 3.8	54.5 ± 2.1	46.7 ± 3.5	31.5 ± 2.1	23.0 ± 0.4
5% 50 cSt	110.3 ± 2.6	65.5 ± 1.4	55.7 ± 2.2	44.2 ± 2.8	30.6 ± 1.4	23.3 ± 0.3
20% 50 cSt	109.2 ± 1.4	64.0 ± 2.3	54.3 ± 2.2	45.8 ± 2.8	24.5 ± 1.4	23.7 ± 0.4
5% 5000 cSt	107.7 ± 2.0	64.6 ± 1.9	54.7 ± 1.3	47.8 ± 1.9	26.1 ± 1.5	23.3 ± 0.2
20% 5000 cSt	103.4 ± 2.4	64.4 ± 1.9	51.9 ± 2.3	48.1 ± 1.8	26.0 ± 1.9	23.1 ± 0.3
20% Tris	106.4 ± 2.3	64.3 ± 2.3	55.5 ± 1.8	46.2 ± 1.9	28.8 ± 1.2	23.1 ± 0.2

Table 2: DCA Data for PDMS elastomer modified with non-functional oils

Viscosity	Wt. %	θ_{adv}	θ_{rec}	$\Delta\theta$
No Oil	N/A	115.1 ± 3.8	68.7 ± 2.2	46.4 ± 1.7
50 cSt	5%	113.9 ± 1.8	77.5 ± 1.8	36.4 ± 0.3
50 cSt	20%	100.5 ± 1.3	65.1 ± 2.1	35.4 ± 1.6
5000 cSt	5%	106.1 ± 0.7	71.6 ± 2.2	34.5 ± 2.1
5000 cSt	20%	100.9 ± 1.1	91.8 ± 2.4	9.1 ± 1.3

Bioadhesion Studies on *Enteromorpha* spores and Porcine Vascular Endothelial Cells (PVECs)

Preliminary *in vitro* cell culture studies demonstrate that the type of contact guidance and bioadhesion to the surface was dependent on the cell type. Evaluating the changes in surface topography using the unmodified PDMS elastomer, the textured surfaces showed a substantial increase in spore accumulation compared to non-textured surfaces as seen in Figure 5. The flat elastomer surface provided the least favorable substrate in terms of spore adhesion, and the 5 μm widths between features had the greatest spore accumulation. This increase in spore density on smaller widths also corresponds to the highest density of microfeatures, which may also be related to the length of the feature. The spores appeared to attach inside the grooves in the angle between the valley floor and the side wall.

Preliminary endothelial cell (EC) studies show a decrease in confluence on textured surfaces. Figure 6 are light microscope images of stained EC's on textured siloxane substrate. The cells form a confluent layer in the smooth region, but were less dense on the ridge and pillar areas. This disruption of the normally confluent endothelial cell mono-layer suggests that these types of patterns might be appropriate for medical applications where cell adhesion is undesirable, such as surgical adhesions.

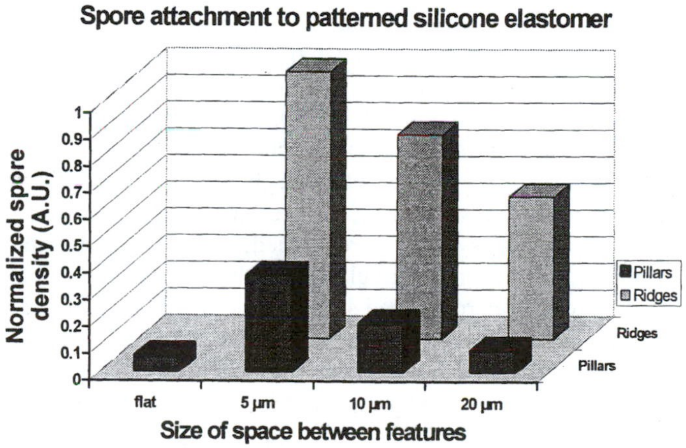

Figure 5: Spore attachment to patterned silicone elastomer. Features were 5 μm high.

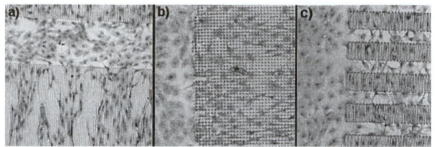

Figure 6: Optical microscopy of stained PVECs on a textured RFGD treated siloxane elastomer substrate. a) smooth and 5mm high, 1000 mm long textured ridges (10X), b) 5 mm high textured pillars (10X), c) smooth and 5mm high, 60 mm long textured ridges (10X).

Conclusions

The ability to characterize the surface and bulk properties of biomaterials is essential in understanding how they influence cellular adhesion and growth. The results of this paper demonstrate the ability to both evaluate and control the topographical and chemical features engineered into siloxane elastomers. The AFM, SEM and WLIP images are evidence that the replication of the micropatterned silicon wafer was high fidelity limiting the possible topographies to the capabilities of the micromachining process itself. The mechanical analysis indicates that surfaces can be engineered with varying moduli to moderate interactions with specific cell types. Combining these findings with qualitative evaluations of *in vitro* cell culture and *in vivo* device implants will allow modeling of the cellular response to morphological, mechanical and surface energy properties. The goal is to develop reliable models for different cell types to engineer specific surfaces for each unique biological application.

Two main biological systems were studied: spores of the marine green alga *Enteromorpha* and porcine vascular endothelial cells. Spores settled preferentially onto the microtextured surfaces and the density of attached spores increased as the space between features decreased. The endothelial cells appear to prefer the smooth siloxane surface and the confluence of cell coverage was decreased when micro-topography was introduced. From these results, it is clear that the cell type has a significant impact on the response to micro-engineered surfaces. The complexity of biological systems requires a better fundamental understanding of the substrate/cell interface in order to engineer biomaterials for specific applications.

Acknowledgements

The authors acknowledge the financial support of the Office of Naval Research (US Navy N0014-99-1-0795) to fund this research. The 5 μm deep micropatterned silicon wafers were reactive ion etched by Abdul Latif at Unaxis (Tampa, FL). We thank the research group of Dr. Edward R. Block for supplying endothelial cells, and C. Keith Ozaki, M.D. and Zaher Abouhamze, B.S. for technical assistance and advice.

References

1. *Biomaterials Science: An Introduction to Materials in Medicine*; Ratner, B.D.; Hoffman, A.S.; Schoen, F.J.; Lemons, J.E., Eds.; Academic Press: San Diego, CA, 1996.
2. Curtis, A., Wilkinson, C. *Biomaterials* **1997**, *18*, 1573-1583.
3. den Braber, E.T., de Ruijter. *Journal of Biomedical Materials Research* **1995**, *29*, 511-518.
4. Flemming R.G., Murphy C.J., Abrams G.A., Goodman S.L., Nealey P.F. *Biomaterials* **1999**, *20(6)*, 573-588.
5. Bahatia, S.N.; Chen, C.S. *Biomedical Microdevices* **1999**, *2:2*, 131-144.
6. Singhvi, R., Kumar, A., Lopez, G.P., Stephanopoulos, G.N., Wang, D.I.C., Whitesides, G.M., Ingber, D.E., *SCIENCE* **1994,** *264*, 696-698.
7. Tan, J., Shen, H., Carter, K.L., Saltzman, W.M. *J. Biomed. Mater. Res.* **2000,** *51*, 694-702.
8. Folch, A., Byong-Ho, J., Hurtado, O., Beebe, D.J., Toner, M., *J. Biomed. Mater. Res.* **2000**, *52*, 346–353.
9. *Cell Adhesion: Characterization of Adhesive Forces and Effect of Topography*, Zhao, L.C., Masters of Science Thesis, University of Florida, Gainesville, FL, 2000.
10. von Recum, A.F. and T.G. Van Kooten. *J. Biomater. Sci. Polymer Edn.* **1995**, *7(2),* 181-198.
11. Johnson, R.E. and Dettre R.H. *Surface and Colloid Science* **1969,** *2,* 85-153.
12. Good, R.J. *J.Adhesion Sci. Technol.* **1992,** *6(12),* 1269-1302.
13. Callow, M.E., Callow, J.A., Pickett-Heaps, J.D., Wetherbee, R. *Journal of Phycology* **1997**, *33*, 938-947.
14. Hillborg, H., Sandelin, M., Gedde, U.W. *Polymer* **2001,** *42(17)*, 7349-7362.
15. Kohl, J.G. and Singer I.L. *Progress in Organic Coatings* **1999**, *36*, 15-20.
16. Brady, R.F. and Singer, I.L. *Biofouling* **2000**, *15(1-3)*, 73-81.
17. Oner, D. and McCarthy, T.J. *Langmuir* **2000**, *16*, 7777-7782.
18. Schmidt, J.A. and von Recum, A.F. *Biomaterials* **1992**, *13(10),* 675-681.
19. Shafrin, E.G. and Zisman W.A. *Langmuir* **1960**, *64*, 519-524.
20. Good, R.J., *J. Ahesion Sci. Technol.*, **1992,** *6(2),* 1269-1302.

Chapter 19

Exploiting Favorable Silicone–Protein Interactions: Stabilization against Denaturation at Oil–Water Interfaces

Paul M. Zelisko, Vasiliki Bartzoka, and Michael A. Brook[1,*]

Department of Chemistry, McMaster University, 1280 Main Street West, Hamilton, Ontario L8S 4M1, Canada

Water-in-silicone oil emulsions were prepared using either silicone surfactants containing pendant polyethylene oxide (PEO) side chains (DC3225C) or terminal $Si(OEt)_3$ groups. The aqueous solutions contained either the enzymes α-chymotrypsin or alkaline phosphatase (DC3225C emulsions) or the surface active protein human serum albumin (TES-PDMS emulsions). Labelled albumin was shown to reside almost exclusively at the oil/water interface in the emulsion. The activity of the enzymes was followed over time by first breaking the emulsion and then performing standard enzyme assays on the aqueous phase. The enzymes were observed to undergo denaturation, as measured by reduced enzymatic activity, as a result of the mechanical energy used to make and break the emulsion. However, the rate of enzyme denaturation in the emulsions was lower than that observed for the aqueous control which had not been exposed to silicone. These results are consistent with favorable interactions between PEO or $Si(OEt)_3$ groups (or hydrolytic byproducts in the latter case) and the proteins that not only stabilize the interface of a water-in-oil emulsion, but also the protein, which normally would otherwise undergo efficient denaturation in the presence of silicone oil.

Introduction

Silicones are exceptionally hydrophobic materials, which is partly the source of their impressive surface activity.[2] By contrast, proteins are generally very hydrophilic, at least at their external surface, with the more hydrophobic amino acid residues typically being sequestered in the core of the folded protein. Relatively little energy is normally needed to perturb protein tertiary structure. Favorable interactions between the more hydrophobic protein core with silicone oils have been attributed to the facile denaturation that can occur if proteins are exposed to dimethylsilicone fluids (e.g., during emulsification):[3] in order for the protein and silicone to favorably interact, the protein must change its conformation such that the hydrophobic moieties of its core are available at the external surface. In such a case, the hydrophobic domains on both polymers can associate. However, this change in conformation typically results in the catastrophic denaturation of the protein.

Silicones are not, however, always deleterious to protein structure. Microparticles consisting of a starch/protein mixture, to which a surface coating of functional silicone ($[(EtO)_3Si(CH_2)_nSiMe_2(OSiMe_2)_m]_2O$, TES-PDMS) had been added, served as a vehicle to deliver the entrapped protein to mice. That is, the entrapped protein elicited an immune response when administered orally. No such response was observed when the particles were coated with Me_3Si-capped silicone oils of comparable molecular weight.[4]

It was extremely surprising to learn that such a subtle difference in structure (two $Si(OEt)_3$ groups on a polymer of MW 28000!) could have such a profound effect on the bioavailability of the protein in these microparticles. However, it was not clear whether the silicone was simply acting as macromolecular protecting group, or whether a more profound interaction was taking place between the biological and the synthetic polymer. Assessing these interactions was particularly difficult with the starch microparticles, because of the presence of many solid/liquid and liquid/liquid interfaces. Therefore, we have examined, and describe below, the rate of denaturation of various proteins, including enzymes, when contained in water-in-silicone oil emulsions that contain a single liquid/liquid interface. We also examine the affinity of the protein for the oil/water interface.

Experimental Section

Reagents. D_4 ($Me_2SiO)_4$ and DC3225C were obtained from Dow Corning and were used without further purification. Two structurally different functional polymers were utilized in this study: commercially available Dow Corning 3225C, and $(EtO)_3Si(CH_2)_3$-modified silicones, TES-PDMS. The latter polymers, with functional groups ($(EtO)_3Si$) at the termini, were prepared using methodology previously reported with molecular weights ranging from about 500-30000.[5]

α-Chymotrypsin, alkaline phosphatase, human serum albumin (HSA), *N*-glutaryl-*L*-phenylalanine-*p*-nitroanilide, and *p*-nitrophenyl phosphate (Sigma, Inc.) were used without further purification. The HCl and trishydroxymethylaminomethane (Tris) used for preparing buffers were obtained from BDH while K_2HPO_4 and absolute ethanol were supplied by Anachemia. The buffers comprising the aqueous phase of the emulsion were formulated using deionized, organic-free, distilled water containing 1.25×10^{-5} g mL^{-1} of sodium azide (Aldrich). Calcium chloride was obtained from Aldrich. 5-(4,6-Dichlorotriazinyl)aminofluorescein (5-DTAF) was obtained from Molecular Probes and used without further purification. Allyltriethoxysilane, hydride-terminated polydimethylsiloxanes (28000 MW), and platinum divinyl-tetramethyldisiloxane complex in vinyl-terminated polydimethylsiloxane (Karstedt's catalyst) were obtained from Gelest, and used without further purification.

Fluorescein-labelled HSA. The 5-DTAF (0.5 mg) was added to 20.0 mL of an HSA solution (0.5 mg mL^{-1}) in 1.0 M Tris-HCl buffer (pH 8). The mixture was protected from light and allowed to stand for 1 hour at room temperature. Dialysis was then performed using the Tris-HCl buffer as the exchange medium until such time as the exchange medium no longer displayed an absorbance peak at 492 nm.

Emulsion Formulation for Enzyme Stability Studies Using DC3225C. The silicone surfactant DC3225C (1.0 g) was dissolved in D_4 (29.2 g) (continuous phase) at a concentration of 3.54 wt% and added to the mixing vessel. α-Chymotrypsin (0.0118 g, at a concentration of 1.09×10^{-4} g mL^{-1}) was dissolved in 0.08 M Tris-HCl buffer (pH 7.8)/1.0 M $CaCl_2$ (108 mL). Alkaline phosphatase (8.7×10^{-5} g, at a concentration of 8.06×10^{-7} g mL^{-1}) was dissolved in a 0.1 M Tris-HCl buffer solution (108 mL, pH 7.8). These solutions constituted the dispersed phase of their respective emulsions. All enzyme solutions were prepared on the day of emulsification. The aqueous dispersed phase (10.0 g), containing the desired enzyme, was added drop-wise to the silicone continuous phase under dual blade, turbulent mixing conditions at 2780 rpm using a Caframo mixer over a period of 10 min. Emulsions were made in a Pyrex 180 mL beaker model number 1140. The bottom of the upper mixing blade (pitched at a 45° angle) was positioned at a height of 2/3 the liquid depth (from the bottom of the mixing vessel), while the bottom of the lower blade (at a 90° angle) was positioned at a position equal to 1/3 the diameter of the mixing vessel (from the bottom of the mixing vessel). The total height of the fluid in the mixing vessel (diameter 4.5 cm) was 3.0 cm. The emulsion was allowed to stir for an additional 20 min following addition of the dispersed phase. The enzyme solution (3.0 g) was stored in a sealed container at room temperature as a control.[6,7,8]

Emulsion Formulation using HSA and TES-PDMS. HSA (0.9 g, at a concentration of 0.03 g mL^{-1} (dispersed phase)) was dissolved in 1.0 M Tris-HCl buffer (30.0 mL, pH 7.8). A portion of the aqueous dispersed phase was stored in a sealed container at ambient temperature as a control. (Triethoxysilyl)ethyl-terminated PDMS (4.0 g) was dissolved in D_4 (16.0 g, continuous phase) in the mixing vessel. The aqueous dispersed phase (20.0 g) was added to the silicone oil phase in a continuous, drop-wise manner over a period of 2 h under dual blade, turbulent mixing conditions at 2780 rpm using a Caframo mixer. The emulsion was allowed to mix for an additional 2 h following the addition of the dispersed phase.

Enzyme Extraction. The enzymes (α-chymotrypsin or alkaline phosphatase) were extracted from the emulsion prior to the assessment of the enzyme activity. To do so, an aliquot (5.0 mL) of the emulsion was centrifuged at 2500 rpm for 60 min, and the silicone supernatant was discarded. The concentrated emulsion was transferred to the mixing vessel. The Tris-HCl buffer (2.0 mL) was added drop-wise to the concentrated emulsion at a rate of 0.1 mL min^{-1} while stirring at 3000 rpm. The mixture was then allowed to stir for an additional 30 min following the addition of the buffer. Once extracted, the aqueous phase was drawn into a 5.0 mL syringe, filtered through a 0.2 μm pore syringe filter into a vial, and stoppered.

N-Glutaryl-L-Phenylalanine-p-Nitroanilide Solution. The substrate solution for α-chymotrypsin was prepared by combining N-glutaryl-L-phenylalanine-p-nitroanilide (0.1 g) with Tris-HCl buffer (20.0 mL, pH 8.0). Fresh substrate solutions were prepared each day that an activity assay was performed.

p-Nitrophenyl Phosphate Solution. The substrate solution for alkaline phosphatase was prepared from p-nitrophenyl phosphate (0.0393 g, 3.93 mM) in Tris-HCl buffer solution (10 mL, pH 8.0). Fresh substrate solutions were prepared each day that an activity assay was performed.

Activity Assay for α-Chymotrypsin. After the emulsions were broken, enzyme activity assays were performed on days 1, 3, 5, 7, and 9 using a Cary 400 spectrophotometer. Concentrations of enzyme extracted from the emulsion were calculated at 256 nm using Beer's Law and an extinction coefficient (ε) of 26520 M^{-1} for α-chymotrypsin. The extinction coefficient for α-chymotrypsin at 256 nm was calculated using known concentrations of the enzyme. The concentrations of extracted enzymes were normalized with the control. Assays were performed by adding Tris-HCl (1.25 mL) to N-glutaryl-L-phenylalanine-p-nitroanilide solution (1.25 mL) in an absorbance cuvette and equilibrating at room temperature for 5-10 minutes. Distilled, deionized water (10.0 μL) was added to the blank, and an enzyme solution (extracted enzyme or control, 10.0 μL) was added to the sample cuvette. The increase in absorbance at 400 nm was measured for 10 min to determine enzyme activity. The slope of the initial linear portion of the curve was taken as being representative of enzymatic activity.

Activity Assay for Alkaline Phosphatase. The hydrolytic activity of alkaline phosphatase was quantified by following spectrophotometrically the formation of p-nitrophenoxide at 405 nm. The p-nitrophenol substrate solution (1.25 mL) and Tris-HCl buffer (1.25 mL) were transferred to a 4.0 mL polystyrene absorbance cuvette. An enzyme solution (10 μL), either control or extracted from the emulsion, was added to the cuvette and assayed. The molar extinction coefficient of alkaline phosphatase at 278 nm, in Tris-HCl buffer at pH 8.0, was taken as 18000 M^{-1} cm^{-1}.[9]

Results and Discussion

Water-in-silicone oil emulsions were formed by the simple expedient of slow addition of the aqueous protein-containing phase to the stirring silicone oil containing the silicone surfactant. The speed of a dual blade mixer was adjusted such that the tip of the blade had a velocity of approximately 4.6 m s^{-1}. With DC3225C (Chart 1), emulsions were readily formed with D_4 in the absence of protein. Such emulsions were extremely stable and are, in fact, relatively difficult to break. By contrast, surfactant TES-PDMS (MW 28000) (Chart 1) only led to stable emulsions if a protein, such as HSA, was present. The ease of forming the emulsion and the subsequent emulsion stabilities were also dependent on the specific protein and the concentration of the two co-surfactants.[7] Stable water-in-oil emulsions could be prepared with water in concentrations exceeding 80% using DC3225C (data not shown). Emulsion droplets ranged in diameter from 1-10 μm with the majority having a diameter between 3-5 μm (Figure 1A).

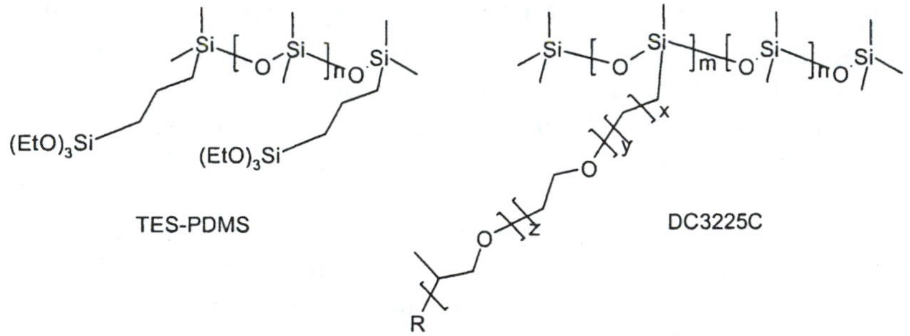

Chart 1

Proteins typically deposit on solid surfaces.[10,11,12] To address how proteins behave at silicone-water interfaces, that is, to determine if they are similarly attracted to more flexible liquid/liquid interfaces, human serum albumin was labelled with a commercially available fluorescein derivative. Stable emulsions of HSA were made in D_4 with linear TES-PDMS (MW 28000, di-terminally functional). Previous studies had demonstrated that optimum emulsion stabilization in such systems arises at HSA concentrations of 0.03wt%.[7] At this protein concentration, HSA primarily resides at the interface; confocal microscopy (observing a single 2D plane) with the fluorescently-labelled protein shows a protein corona, indicating preferential adsorption of HSA at the water-silicone oil interface over the bulk aqueous solution (Figure 1B).

It was of interest to determine, during the formation and aging of the emulsion, to what extent the proteins were undergoing thermal denaturation.[8,13] This question can be addressed, to some degree, by following the ability of the protein to process substrate. In preliminary experiments, alkaline phosphatase (AP) was entrapped in a TES-PDMS/D_4 emulsion. After two months, the protein was extracted from the emulsion by the addition of water and the substrate p-nitrophenyl phosphate was added to the aqueous phase. The solution immediately changed color due to the presence of p-nitrophenoxide. This indicated that the enzyme was still capable after two months of catalyzing the transformation of the substrate. With this result in hand, the rates of denaturation of AP and α-chymotrypsin were tested in DC3225C/D_4 emulsions. Chromogenic assays are readily available for both of these enzymes (Scheme 1).[14,15]

Approximately 1.1×10^{-2} wt% in water of the α-chymotrypsin and 8.1×10^{-5} wt% in water of the alkaline phosphatase were emulsified, as noted above, in DC3225C/D_4 at room temperature. Aqueous solutions of the protein in buffer were left at room temperature as controls. Each second day, for 9 sequential days, an aliquot of the emulsion was broken and the enzyme activity was measured using chromogenic assays (*vida supra*). The rates of substrate processing, which reflect residual active protein, are shown against time (Figure 2A,B). Protein that was never exposed to silicone is indicated by triangles (σ), and protein extracted from the emulsion is indicated by squares (ν). It is immediately apparent that there is initially much less active protein in the silicone emulsion than in the control. However, the rate of denaturation for both proteins, as shown by the respective negative slopes of the two lines, is far higher for the control.

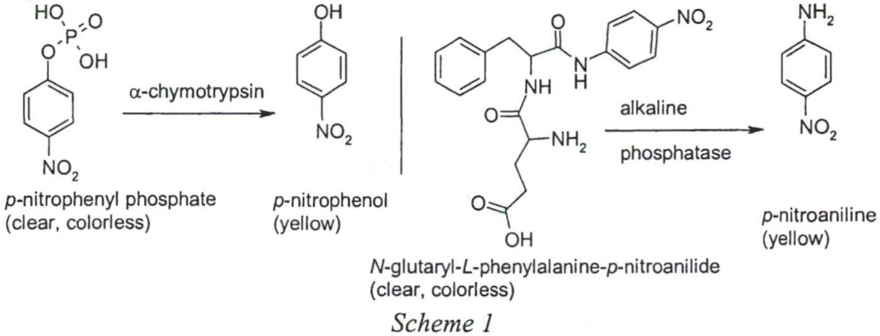

Scheme 1

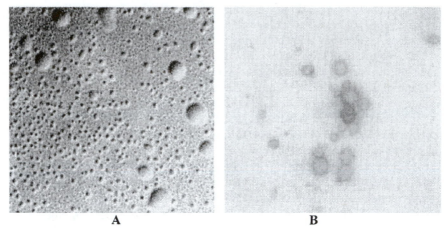

Figure 1: A water-in-oil emulsion of HSA/Tris-buffer in 28000 MW (EtO)₃Si(CH₂)₃-modified silicones/D₄. Average size of small (stable) droplets is about 3-5 microns. A: optical microscopy, B: confocal microscopy.

The mechanical energy used to form an emulsion is detrimental to protein structure. The classic example of this is denaturation of egg whites (ovalbumin) when whipped. Proteins (e.g., α-chymotrypsin) in buffer solutions that were subjected to the same mixing stresses involved in the emulsification process, but in the absence of silicone oil and silicone surfactants, were irreversibly denatured (data not shown). Similar degradation can be observed in Figure 2 although the magnitude is lower. The damage done to the proteins during emulsification and in breaking the emulsion manifests itself in an initially lower concentration of active protein. However, over time, little further change in enzyme activity is observed with the emulsified protein. By contrast, the control protein denatures quite rapidly. This suggests that once emulsified, the protein that is in contact with the silicone/water interface is relatively stabilized against denaturation. That is, the amount of damage by mechanical energy is a constant, and otherwise the proteins in the emulsion degrade at a very slow rate. Proteins can be stabilized by hydrophilic species. Thus, van Alstine and co-workers have shown that grafting PEO onto proteins retards denaturation, a process that has been commercialized.[16] The hydrophilic compounds act as osmolytes, controlling structure by controlling swelling by water. It is likely that such stabilization is also occurring at the PEO-protein/water interface. That is, the surface-active components, both protein and DC3225C, migrate to the interface. There, the protein sits against the PEO, rather than the silicone, and benefits from the stabilizing environment.

The situation is less clear for TES-PDMS. Such compounds should hydrolyze to initially give hydrophilic Si-OH (or Si-O⁻) groups that can behave

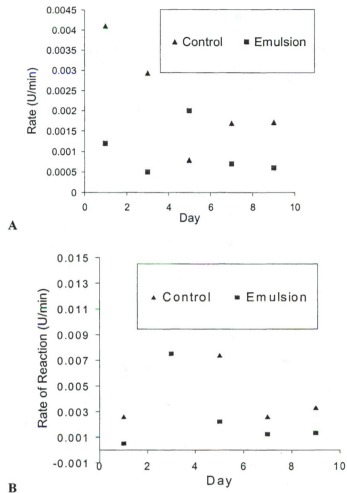

Figure 2: **A:** *Reaction of alkaline phosphatase with p-nitrophenyl phosphate;* **B:** *reaction of α-chymotrypsin with N-glutaryl-L-phenylalanine p-nitroanilide.*

in two important ways. First, they can anchor the TES-PDMS at the water/silicone interface (i.e., amplify the surface activity) and also stabilize the proteins through hydrophilic and, possibly, hydrophobic interactions (with adjacent Si-alkyl groups). However, it might be expected that $Si(OH)_3$ groups would undergo room temperature vulcanization (RTV) cure to give elastomers. We have seen no evidence for this. Future work will focus on preparing well-defined silicone/protein copolymers to see if such linked proteins are similarly stabilized against denaturation and to further probe the specific nature of the protein silicone interactions in TES-PDMS.

Conclusions

Enzymes and proteins denature over time. Denaturation is even more prevalent when the enzyme comes into contact with non-functional hydrophobic silicones and/or when subjected to shear stress during mixing. However, proteins residing at silicone oil interfaces bearing PEO- or triethoxysilyl-functional silicone groups are stabilized against denaturation. It appears that entrapping enzymes in water-in-silicone oil emulsions is an effective means of protecting these biomolecules from denaturation by the external environment.

Acknowledgements

Financial support of this work from the Natural Sciences and Engineering Research Council of Canada (NSERC) and the Canadian Institute for Health Research is gratefully acknowledged. We would like to thank Professor Paul Berti, McMaster University, for helpful discussions.

References

[1] Telefax: +1(905) 522 2509; email: mabrook@mcmaster.ca

[2] (a) Owen, M. J. *Siloxane Surface Activity*, In *Silicon-based Polymer Science: A Comprehensive Resource,* Zeigler, J. M.; Fearon, F. W. G., Eds., American Chemical Society (ACS Adv. Chem. Ser. 224): Washington, DC: 1990, Chap. 40, p. 705. (b) Owen, M. J. *Surface Chemistry and Applications*, In *Siloxane Polymers*, Clarson, S. J.; Semlyen, J. A.; Eds., Prentice Hall: Englewood Cliffs, NJ, 1993, Chap. 7, p. 309.

[3] Sun, L.; Alexander, H.; Lattarulo, N.; Blumenthal, N. C.; Ricci, J. L.; Chen, G. *Biomaterials* **1997**, *18*, 1593.

[4] Brook, M. A.; Loombs, L.; Heritage, P.; Jiang, J.; McDermott, M.; Underdown, B. *Microparticle Delivery System*, US Patent 5571531 (to McMaster University), Nov. 5, 1996.

[5] Heritage, P. L.; Loomes, L.; Jiang, J.; Brook, M. A.; Underdown, B. J.; McDermott, M. R. *Immunology* **1996**, *88*, 162.

[6] Bartzoka, V.; McDermott, M. R.; Brook, M. A. *Protein-Silicone Interactions at Liquid/Liquid Interfaces*, In *Emulsions, Foams and Thin Films*, Mittal, K. L.; Kumar, P., Eds., Dekker: New York, 2000, Chap. 21, pp. 371-380.

7 Bartzoka, V.; Chan, G.; Brook, M. A. *Langmuir* **2000**, *16*, 4589.

8 Brook, M. A.; Zelisko, P. M. *Polym. Prep. (Am. Chem. Soc., Div. Polym. Chem.)* **2001**, *42(1)*, 97.

9 Anderson, R; Vallee, B. *Proc. Nat. Acad. Sci. USA* **1975**, *72*, 394.

10 (a) Du, Y. J.; Cornelius, R. M.; Brash, J. L. *Colloids Surf. B*, **2000**, *17*, 59. (b) Sheardown, H.; Cornelius, R. M.; Brash, J. L. *Colloids Surf. B*, **1997**, *10*, 29.

11 Chapman, R. G.; Ostuni, E.; Takayama, S.; Holmlin, R. E.; Yan, L.; Whitesides, G. M. *J. Am. Chem. Soc.* **2000**, *122*, 8303.

12 Norde, W.; Favier, J. P. *Colloid Surf.* **1992**, *64*, 87.

13 With HSA, at least, confocal microscope studies of labelled protein indicated that the majority of protein resides at the silicone interface.

14 Hummel, B. *Can. J. Biochem. Physiol.* **1959**, *37*, 1393.

15 (a) Bloch, W; Schlesinger, M. *J. Biol. Chem.* **1974**, *249*, 1760. (b) Chlebowski, J; Coleman, J. *J. Biol. Chem.* **1974**, *249*, 7192.

16 (a) Burns, N. L.; van Alstine, J. M.; Harris, J. M. *Langmuir* **1995**, *11*, 2768. (b) www.shearwatercorp.com.

Chapter 20

Association of Siloxane Polymeric Surfactants in Aqueous Solution

Yining Lin and Paschalis Alexandridis*

Department of Chemical Engineering, University of Buffalo, The State
University of New York, Buffalo, NY 14260–4200

The micelle formation and structure of a siloxane polymeric
surfactant in water and in water mixed with ethanol has been
investigated. The formation of micelles (CMC) was
established by fluorescent probe molecules that detected
hydrophobic domains. The structure of the micelles was
determined by small angle neutron scattering (SANS). The
siloxane polymeric surfactant, consisting of a hydrophobic
polydimethylsiloxane (PDMS) backbone and hydrophilic
polyether grafts, forms micelles in water at a concentration of
0.05%. The micelles are spherical at room temperature and
consist of a compact siloxane core with a radius of 60 Å and a
relatively thin (20 Å) polyether corona. Addition of ethanol
suppresses the micelle formation and renders the micelles
smaller. The findings presented here are relevant to
waterborne coating and ink formulations, where siloxane
surfactants are often used in conjunction with cosolvents.

Introduction

The interest in the self-assembly of amphiphiles in water or in mixtures of water with less polar solvents is driven by both fundamental and practical considerations. In waterborne coating and ink formulations, polymeric surfactants are added in order to improve stabilization, solubilization, or provide surface modification. Water is the primary (and desirable) solvent, but cosolvents are needed in order to modulate the formulation performance. In order to improve existing formulations and to design new ones, we need fundamental information on association of polymeric surfactants in solution and on surfaces. In a recent study of Pluronic poly(ethylene oxide)-poly(propylene oxide)-poly(ethylene oxide) PEO-PPO-PEO block copolymers where water is typically used as a solvent (being selective for PEO) [1-4], we have shown that the addition to water of cosolvents, such as glycerol or ethanol, provides extra degrees of freedom in tailoring the solution properties [4-6]. Cosolvents also cause pronounced effects on the concentration range of stability of the different lyotropic liquid crystals formed by PEO-PPO-PEO block copolymers in water and on the characteristic length scales of the nanostructures [6-8].

Siloxane surfactants have unusually flexible polydimethylsiloxane (PDMS) backbones which may coil in the aggregates in aqueous solution [9]. According to some reports, close to the CMC, siloxane surfactants in aqueous solutions behave much like typical hydrocarbon surfactants [10,11]. However, published information about the association in solution of block and graft siloxane polymeric surfactants, such as the one we examine here, is very limited. Such lack of fundamental knowledge is in contrast with the current trend of increased use of functional polymers and surfactants in aqueous media and adsorbed on particles. Our desire to be able to tune association and adsorption in waterborne coating and ink formulations motivate us in studying interactions between their basic ingredients: polymeric surfactants, cosolvents and particles.

In this study, we highlight some of our recent finding on the topic of siloxane surfactant association in water and in water-cosolvent mixtures [12-16]. We first show how to determine the CMC using probe molecules such as DPH (1,6-diphenyl-1,3,5-hexatriene) and methyl orange, which undergo changes in their spectral properties as the environment changes from polar to hydrophobic. We then describe how the structure of micelles can be probed by a combination of SANS (that focuses primarily on the core of the micelle) and DLS (dynamic light scattering, that is sensitive to the solvated micelle). Finally, we discuss how the micelle formation and structure are affected in the addition of ethanol, a solvent fully miscible with, but less polar than water.

Materials and Methods

The siloxane graft copolymer examined here was provided by Goldschmidt AG, Essen, Germany and has the chemical formula $MD_{70}D'_5M$ (M: $Me_3SiO_{1/2}$-, D: -Me_2SiO-, D': $Me(R)SiO$-, R: PEO-PPO polyether copolymer) with MW=11500. The molecular weight per polyether group in the siloxane copolymer is 1200 and the composition of the polyether is 75% PEO and 25% PPO. 1,6-diphenyl-1,3,5-hexatriene (DPH) was obtained from Molecular Probes Inc., Eugene, OR. Methyl Orange 99.5% was purchased from Aldrich Co. Ethanol was purchased from Acros Co. All solutions were prepared with Milli-Q filtered water (18 $M\Omega.cm$) mixed with a cosolvent where appropriate. Deuterated water (D_2O) and deuterated ethanol (CD_3CD_2OD), used in the neutron scattering experiments, were purchased from Cambridge Isotope Laboratories.

Spectrophotometric measurements

DPH is a well-known probe of lipid membrane interiors and can also be used to detect the formation of micelles [17,18]. The fluorescence of DPH is minimal in water but is substantially enhanced by association with surfactants. Sample preparation was as follows: a stock solution of 0.4 mM DPH in methanol was prepared; 25μl of the DPH/methanol solution were added to 2.5 ml surfactant solution, so that the final surfactant solution contained 0.004 mM DPH and 250 mM methanol. UV-vis absorption spectra of the surfactant/DPH/water samples were recorded in the 320-400 nm range using a Beckman DU-70 UV-vis spectrophotometer. The main absorption intensity peak, characteristic of DPH, was at 354 nm.

Methyl orange has been used to investigate the formation of micelles as a solvatochromic dye molecule [12,13]. The non-covalent binding of methyl orange to the micelles is reflected by a hypsochromic shift of the long-wavelength absorption band of methyl orange in the presence of different concentrations of surfactants. The concentration of the dye (2.5×10^{-2} mM) was kept constant for all measurements.

Small angle neutron scattering

SANS measurements were performed at the National Institute of Standards and Technology (NIST) Center for Neutron Research, beam guide NG3. 5% siloxane surfactant solutions were prepared in D_2O or D_2O mixed with Dethanol, which provides good contrast between the micelle and the solvent. The experimental details are given in [6].

The absolute SANS intensity can be expressed as a product of $P(q)$ which is related to the form factor and the structure factor $S(q)$ [5,6,16]:

$$I(q)=NP(q)S(q)$$

where N is the number density of the scattered particles, in our case micelles, which depends on the copolymer concentration and the association number of micelles. The form factor $P(q)$, which takes into account the intramicelle structure, depends on the shape of the colloidal particle. A core-corona form factor has been proposed to describe the scattering generated from the contrast between the micelle core and corona [5,6,16], which have different solvent contents (the core is usually "dry" or has small amounts of solvent, whereas the corona is highly solvated), and the scattering due to the contrast between the micelle corona and the solvent phase:

$$P(q)=\{(4\pi R_{core}^{3}/3)(\rho_{core}-\rho_{corona})[3J_1(qR_{core})/(qR_{core})]+$$
$$(4\pi R_{micelle}^{3}/3)(\rho_{corona}-\rho_{solvent})[3J_1(qR_{micelle})/(qR_{micelle})]\}$$

where R_{core} and $R_{micelle}$ are the radii of the micelle core and whole micelle, respectively; ρ_{core}, ρ_{corona}, and $\rho_{solvent}$ are the scattering length densities (SLD) of the core, corona, and solvent (assuming a homogeneous solvent distribution in each of the domains). $J_1(y)$ is the first-order spherical Bessel function:

$$J_1(y)=[sin(y)-ycos(y)]/y^2$$

In fitting the core-corona model into the scattering data, we view the micelles as consisting of hydrophobic core composed of siloxane plus PPO segments (with little or no solvent present) and a relatively hydrated corona consisting of solvated PEO chains [5,6,16]. The SLD of the core, ρ_{core}, and the corona, ρ_{corona}, are a function of the average (over the core radius) volume fraction of siloxane in the core (α_{core}) and of the average volume fraction of PEO in the corona, respectively:

$$\rho_{core}=\alpha_{core}\rho_{hydrophobic}+(1-\alpha_{core})\rho_{solvent}$$
$$\rho_{corona}=\alpha_{corona}\rho_{PEO}+(1-\alpha_{corona})\rho_{solvent}$$

where $\rho_{hydrophobic}$ (=0.1 x 10^{10} cm^{-2}) and ρ_{PEO} (=0.547 x 10^{10} cm^{-2}) are the SLD of siloxane plus PPO (where $\rho_{siloxane}$=0.0658 x 10^{10} cm^{-2} and ρ_{PEO} =0.325 x 10^{10} cm^{-2}) and of PEO, respectively, and $\rho_{solvent}$ is the SLD of the water-cosolvent mixture (ρ_{D2O} = 6.33 x 10^{10} cm^{-2} and $\rho_{Dethanol}$ = 5.95 x 10^{10} cm^{-2}). α_{core} and α_{corona}

can be expressed in terms of the core and micelle radii and the micelle association number, $N_{assoc.}$, i.e., the number of block copolymer molecules which (on the average) participate in one micelle:

$$\alpha_{core} = 3N_{assoc.}V_{hydrophobic}/(4\pi R_{core}^3)$$
$$\alpha_{corona} = 3N_{assoc.}V_{PEO}/[4\pi (R_{micelle}^3 - R_{core}^3)]$$

where $V_{hydrophobic}$ is the volume of the siloxane plus PPO block (= 11904 Å^3) and V_{PEO} is the volume of the PEO blocks (= 7500 Å^3) of one siloxane molecule.

In summary, there are three fitting parameters in the core-corona form factor: R_{core}, $R_{micelle}$, and N_{assoc}. The volume fraction of polymer in the core and corona can be calculated on the basis of these three fitting parameters, according to the above equation. A core-corona form factor which accounts explicitly for the solvent content in the micelle core and corona has been found very useful in the case of PEO-PPO-PEO block copolymer micelles which undergo a progressive solvent loss in both the corona and the core when the temperature increases [5,6,16].

Dynamic light scattering

Dynamic light scattering measures the time-dependent scattering intensity emanating from the sample which leads to the correlation function $G^{(2)}(\Gamma)$ obtained by means of a multichannel digital correlator.

$$G^{(2)}(\Gamma) = A\,(1 + b\,|\,g^{(1)}(\tau)\,|^2)$$

where A, b, τ, Γ and $|g^{(1)}(\tau)|$ are the baseline measured by the counter, coherence factor, delay time, decay rate and normalized electric field correlation function, respectively. In our study $|g^{(1)}(\tau)|$ was analyzed by the EXPSAM (exponential sampling) method, yielding information on the distribution function of Γ from

$$|g^{(1)}(\tau)| = \int G(\Gamma)\exp(-\Gamma\tau)d\Gamma$$

EXPSAM works without any constraints and is based on the eigenvalue decomposition of the Laplace transformation of $G(\Gamma)$.

$G(\Gamma)$ can be used to determine an average apparent (or translational) diffusion coefficient, $D_{app} = \Gamma/q^2$, where $q = (4\pi n/\lambda)\sin(\theta/2)$ is the magnitude of the scattering wave vector. The apparent hydrodynamic radius R_h is related to

D_{app} via the Stokes-Einstein equation, $D_{app}=kT/6\pi\eta R_h$, where k is the Boltzmann constant, T is the absolute temperature, and η is the viscosity of the solvent.

We used a Lexel model 95 argon ion laser with Brookhaven BI-200SM goniometer to obtain the micelle size distribution from a plot of $G(\Gamma)$ versus R_h by the correlation data with the fitting EXPSAM routine, with $G(\Gamma_i)$ being proportional to the scattering intensity of particle i having an apparent hydrodynamic radius $R_{h,i}$. DLS provides information about the solvation of micelles, the possible presence of larger objects, and the size distribution of the micelles, whereas SANS gives better information about the micelle structure. In this sense, DLS and SANS are complementary. The surfactant concentrations studied at 24 °C were 1% and 5%, which are higher than the CMC (see discussion below).

Results and Discussion

Micelle formation

UV-vis spectra of aqueous siloxane solutions containing DPH with siloxane concentrations in the range 0.0001 to 10% w/v were recorded at 24 °C. At low concentrations the siloxane did not associate in aqueous solution and DPH was not solubilized in a hydrophobic environment, therefore, the UV-vis intensities of DPH were very low. At higher concentrations, the siloxanes formed micelles and DPH was solubilized in the hydrophobic micelle interior, giving a characteristic spectrum [17]. The CMC value for siloxane surfactant in aqueous solution was obtained from the first inflection of the absorption intensity at 354 nm vs. siloxane concentration plot for DPH probe (Figure 1). The arrow on the plot indicates the evaluated CMC (0.05%).

The spectra of methyl orange at a concentration 2.5×10^{-5} M for the siloxane surfactant over the 0.0001-10% (w/v) were used to construct λ_{max} versus siloxane concentration plots (Figure 1). The spectra of methyl orange remain unaltered in the presence of siloxane up to a certain siloxane concentration, but a progressive blue shift in λ_{max} is observed at high concentration also up to a certain value. In Figure 1, the absence of any change in λ_{max} in the dilute region indicates that the copolymer remains fully dissolved in water and does not affect the spectrum of the dye. At higher concentrations, the siloxane forms hydrophobic domains and the dye can be considered to partition from water to these domans resulting in decrease in λ_{max}. The CMC value for siloxane surfactant of various concentrations was obtained from the first inflection of the plot (Figure 1). The evaluated CMC value (0.05%), agrees very well with the one we obtained by DPH.

Micelle structure

Figure 2 shows the SANS data and model fits. We fit the data from 0.2%, 1% and 5% siloxane solutions using the core-corona form factor. As shown in Figure 2, this model fits well the scattering data in the q range 0.01 Å^{-1} to 0.1 Å^{-1}. The micelle core radius obtained from the fit is 60 Å and the micelle radius is 80 Å. The association number is 76. The micelle core radius, micelle radius and association number do not change as the siloxane concentration increase. The increase in the scattering intensity with increasing concentration is due to the increase in the number density of micelles. For comparison, Pluronic P105 PEO-PPO-PEO block copolymer (with molecular weight 6500 and 50% PEO) has core radius 46 Å, micelle radius 80 Å, and association number 78 [6].

Figure 3 shows the size distribution of 1% and 5% siloxane solutions in water obtained by DLS. The size distribution data show two well-distinguished peaks (fitted by Gaussians). The lower size distribution, centered around 130 Å, corresponds to the hydrodynamic radius of the siloxane surfactant micelles. The larger micelle radius obtained from DLS, compared to that obtained by SANS, results from the solvation of the siloxane micelles. The surfactant concentration does not show strong effect on the micelle radius, suggesting a closed-association process [17]. We speculate that the higher size distribution, centered around 500 Å (radius) with a standard deviation of about 80 Å, results from hydrolyzed siloxane impurities. The large aggregates contribute about 25 to 30% of the total scattered intensity but their concentration is very low: the number of particles corresponding to the higher peak is <0.01% of the total particles [12].

Cosolvent effects on micelle formation and structure

The CMC and structure of block or graft copolymer micelles will be affected by the addition of ethanol owing to its partition into the micelle core [6]. The CMC increases from 0.05% to ~ 0.3% [13] as determined by the dye-solubilized methods discussed above.

Figure 4 shows the effects of ethanol concentration on the micelle core and micelle radii, association number, polymer fraction in the core and corona, as detected by SANS. The micelle core and micelle radii decrease with increasing ethanol content. The micelle association numbers become smaller in the presence of ethanol. In the 0-40 vol % range, ethanol causes 50% reduction in $N_{assoc.}$ compared with the case of pure water, and 20% decrease in core and micelle radii.

The polymer volume fraction values in the core and corona can shed some light on the location of ethanol in the micelle. As seen in Figure 4, addition of 20 vol% ethanol decreases α_{core} from 1 to 0.96 and α_{corona} from 0.46 to 0.42.

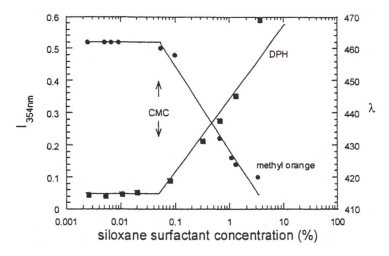

Figure 1. UV-vis spectra of DPH and methyl orange in aqueous siloxane solutions with concentrations in the range 0.001 to 10%. Above CMC, the siloxanes formed micelles and DPH was solubilized in the hydrophobic micelle interior, giving a characteristic spectrum with increasing intensity at 354 nm, while methyl orange showed a progressive blue shift in λ_{max}.

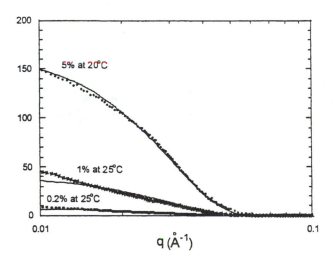

Figure 2. SANS pattern of siloxane surfactant in aqueous solution at different concentrations. As the siloxane concentration increases, the scattering intensity increases. The fitted curves were obtained by the core-corona form factor.

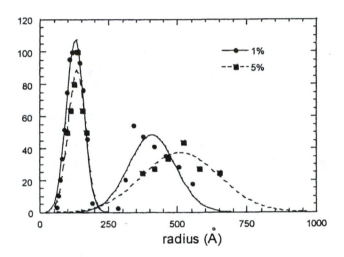

Figure 3. Dynamic light scattering data from 1% and 5% siloxane surfactants in water. The intensity-fraction distribution of micelle hydrodynamic radius was obtained by EXPSAM analysis. Two distributions can be seen. The micelle radius is 125 Å and the large aggregates have a hydrodynamic radius of 400–500 Å.

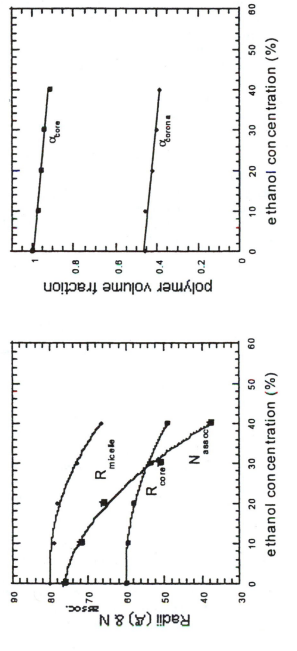

Figure 4. Structural information obtained from SANS in 5% siloxane solution at 20 °C, plotted as a function of the ethanol content. The micelle core radii, micelle radii, micelle association number and polymer volume fraction in the micelle core and corona decrease with added ethanol.

When the ethanol content in the mixed solvent increases to 40 vol %, α_{core} decreases to 0.92 (leaving a relatively high 8 vol % solvent content in the core), while α_{corona} becomes 0.39. The decreasing polymer volume fractions in the micelle core and corona with increasing ethanol content indicate that both siloxane and polyether are solvated to a higher degree in water-ethanol mixture. The addition of ethanol reduces the solvophobicity and increases the solubility of the siloxane parts, causing the lowering of the interfacial tension between the siloxane core and the solvent. In order to achieve thermodynamic equilibrium, the micelle size should be smaller than those in water. Ethanol also shows pronounced impacts on the micelle structure of Pluronic P105 PEO-PPO-PEO block copolymer [6].

We also measured by DLS the size distribution of siloxane micelles in water mixed with ethanol. The lower size distribution from micelle hydrodynamic radius decreases from 130 Å to 105 Å as the ethanol concentration increases from 0 to 20% in agreement with the trend observed in SANS. At the same time the impurity hydrodynamic radius decrease from 500 Å to 315 Å [14]. The fact that the "impurity" peak becomes smaller/weaker with added ethanol supports the idea that it originates from hydrophobic "impurities".

Conclusions

In this study, we focus on the formation of micelles in water by a rake-type siloxane polymeric surfactant, the structure of micelles (obtained by combination of SANS and DLS) and on the modulation of micelle formation and structure by the addition of ethanol. Such fundamental information is beneficial for applications of functional polymers in a variety of products, e.g., pharmaceutics, personal care products, detergents, coatings, and inks.

The CMC of this siloxane surfactant in aqueous solution has been determined by two methods: DPH UV-vis spectra and hypsochromic shift of methyl orange. The CMC value thus obtained is in good agreement in both experiments and equals 0.05%. The addition of 20% ethanol results in an increase of the CMC to 0.3%.

A core-corona form factor was used to extract information about the siloxane micelle size (R_{core}= 60 Å and $R_{micelle}$= 80 Å), and micelle association number ($N_{assoc.}$= 76). The polymer volume fractions in the core and the corona (α_{core} and α_{corona}) were also estimated on the basis of the above fitting parameters. The micelle core and micelle radii and the micelle association number decrease progressively upon the addition of ethanol, accompanying a decrease of the polymer volume fraction in the micelle core and corona. Ethanol offers better solvent conditions for the siloxane surfactant and leads to higher solvation of siloxane and polyether segments than water.

Acknowledgements: We thank Xerox Foundation for partial funding of this work and Dr. Thomas W. Smith (Xerox Corp., Ink-Jet Business Unit) for helpful discussions. We acknowledge the National Science Foundation (CTS-9875848 and CTS-0124848) for partial support of this research. We thank the support of the National Institute of Standards and Technology (NIST), U.S. Department of Commerce, in providing the neutron research facilities used in this work.

References

1. Alexandridis, P.; Hatton, T. A. "Poly(ethylene oxide)-block-Poly(propylene oxide)-block-Poly(ethylene oxide) Copolymer Surfactants in Aqueous Solutions and at Interfaces: Thermodynamics, Structure, Dynamics, and Modelling" *Colloids Surf. A* **1995**, *96*(1/2), 1-46.
2. Alexandridis, P. "Amphiphilic Copolymers and their Applications" *Curr. Opin. Colloid Interface Sci.***1996**, *1*(4), 490-501.
3. Alexandridis, P. "Poly(ethylene oxide)-Poly(propylene oxide) Block Copolymer Surfactants" *Curr. Opin. Colloid Interface Sci.* **1997**, *2*(5), 478-489.
4. Alexandridis, P.; Lindman, B. *Amphiphilic Block Copolymers: Self-Assembly and Applications*, Elsevier, Amsterdam, 2000.
5. Yang, L.; Alexandridis, P. "Polyoxyalkylene Block Copolymers in Formamide-Water Mixed Solvents: Micelle Formation and Structure Studied by Small-Angle Neutron Scattering" *Langmuir* **2000**, *16*(11), 4819-4829.
6. Alexandridis, P.; Yang, L. "SANS Investigation of Polyether Block Copolymer Micelle Structure in Mixed Solvents of Water and Formamide, Ethanol, or Glycerol" *Macromolecules* **2000**, *33*(15), 5574-5587.
7. Ivanova, R.; Lindman, B.; Alexandridis, P. "Evolution of the Structural Polymorphism of Pluronic F127 Poly(Ethylene Oxide)-Poly(Propylene Oxide) Block Copolymer in Ternary Systems with Water and Pharmaceutically Acceptable Organic Solvents: From 'Glycols' to 'Oils'" *Langmuir* **2000**, *16*(23), 9058-9069.
8. Alexandridis, P.; Ivanova, R.; Lindman, B. "Effect of Glycols on the Self-Assembly of Amphiphilic Block Copolymers in Water. II. Glycol Location in the Microstructure" *Langmuir* **2000**, *16*(8), 3676-3689.
9. Hill, R. M.; He, M.; Lin, Z.; Davis, H. T.; Scriven, L. E. "Lyotropic Liquid Crystal Phase Behavior of Polymeric Siloxane Surfactants" *Langmuir* **1993**, *9*, 2789-2798.
10. Hill, R. M. *Silicone Surfactants*. Marcel Dekker, Inc.: NY, 1999.

11. Gradzielski, M.; Hoffmann, H.; Robisch, P.; Ulbricht, W.; Gruening, B. "Aggregation Behavior of Silicone Surfactants in Aqueous Solutions" *Tenside, Surfactants, Detergents* **1990**, *27*(6), 366-379.

12. Lin, Y.; Alexandridis, P. "Association of Siloxane Polymeric Surfactants in Aqueous Solution" manuscript.

13. Lin, Y.; Alexandridis, P. "Microenvironment and Structure of Micelles Formed by a Polymeric Siloxane Surfactant in Aqueous Solutions" *Polymer* Preprints, **2001**, *42*(1), 229-230.

14. Lin, Y.; Alexandridis, P. "Cosolvent Effects on the Micellization of an Amphiphilic Siloxane Graft Copolymer in Aqueous Solutions" manuscript.

15. Lin, Y.; Smith, T. W.; Alexandridis, P. "Adsorption of an Amphiphilic Siloxane Graft Copolymer on Hydrophobic Particles" *Polymer Preprints*, **2001**, *42*(1), 245-246.

16. Lin, Y.; Alexandridis, P. "SANS Characterization of Micelles Formed by Polydimethylsiloxane-graft-Polyether Copolymers in Mixed Polar Solvents" manuscript.

17. Alexandridis, P.; Nivaggioli, T.; Hatton, T. A. "Temperature Effects on Structural Properties of Pluronic P104 and F108 PEO-PPO-PEO Block Copolymer Solutions" *Langmuir* **1995**, *11*(5), 1468-1476.

18. Alexandridis, P.; Holzwarth, J. F.; Hatton, T. A. "Micellization of Poly(ethylene oxide)-Poly(propylene oxide)-Poly(ethylene oxide) Triblock Copolymers in Aqueous Solutions: Thermodynamics of Copolymer Association" *Macromolecules* **1994**, *27*(9), 2414-2425.

Copolymers

Chapter 21

Poly(amidoamine organosilicon) Dendrimers and Their Derivatives of Higher Degree of Structural Complexity

Petar R. Dvornic and Michael J. Owen

Michigan Molecular Institute, 1910 West St. Andrews Road, Midland, MI 48640

Utilization of the copolymer concept in dendrimer chemistry has provided a variety of unique radially layered copolymeric dendrimers with hydrophilic polyamidoamine (PAMAM) interiors and hydrophobic (i.e., oleophilic) organosilicon (OS) exteriors. These globular, nano-scaled inverted unimolecular micelles with different functional end-groups, generically referred to as poly(amidoamine-organosilicon) (PAMAMOS) dendrimers, have been found exceptionally versatile as precise nanoscopic building blocks for the construction of more complex forms of the structural organization of matter. Here, we describe the hierarchy of some of these forms of structural complexity and propose several possible ways by which they may contribute to materials nanotechnology.

Dendrimers are spheroidal nano-scaled macromolecules with an exceptionally high density of functionality that may reach into hundreds and thousands of functional groups per molecule (see X in Figure 1) (*1*). By a variety of well controlled synthetic strategies they may be prepared with a very high degree of structural regularity that is unprecedented for conventional forms of macromolecular architecture, such as linear, randomly branched or cross-linked polymers. As a consequence, they represent unique nanoscopic (i.e., 1–15 nm in diameter) reactive building blocks that open up new avenues for synthetic chemistry of more complex forms of structural organization of matter, and for creative tailor-making of previously unattainable nano-structured materials.

Recently we showed that dendrimers may be prepared as compositionally copolymeric entities with significantly different repeat units (i.e., branch cells) (see Figure 1) (2). It was particularly pointed out that such a "copolymer approach" (2,3) would enable preparation of a practically unlimited variety of different dendrimers, and that by combining *both* architectural and compositional variations it should become possible to tailor-make a diversity of new, precisely nano-structured materials with "dialed-in" properties for specifically desired applications. In this report, we illustrate some concrete possibilities of this approach using the example of functionalized, radially layered copolymeric poly(amidoamine-organosilicon) (PAMAMOS) dendrimers as reactive building blocks for preparation of a variety of more complex forms of structural organization of matter (4).

Poly(amidoamine-organosilicon) (PAMAMOS) dendrimers

Poly(amidoamine-organosilicon) (PAMAMOS) dendrimers (2,5-7) are unique radially layered inverted unimolecular micelles that consist of a hydrophilic polyamidoamine (PAMAM), -[(CH$_2$)$_2$-C(O)-N(H)-(CH$_2$)$_2$-N]<, interior and a hydrophobic/oleophilic organosilicon (OS), -R-Si(R^1)(R^2)(R^3), exterior, as schematically illustrated in Figure 1. They may be prepared by various synthetic strategies from commercially available amine-terminated PAMAM dendrimers, including: (a) Michael addition of silicon-containing acrylates (5); (b) alkylation with haloalkylsilanes (5); (c) addition of epoxyalkylsilanes (2); or (c) addition of silylisocyanates. The first two of these strategies may be represented as shown in Reaction scheme 1.

The course of these reactions can be very accurately monitored by various spectroscopic and/or chromatographic techniques, providing for a high degree of synthetic control and enabling preparation of a variety of different PAMAMOS with precise molecular structure and predetermined degree of OS substitution (5-7). Selected examples of these dendrimers are represented in Figure 2.

Starting from the top left of this figure to its bottom right, these dendrimers can be broadly classified into three different groups: (*i*) chemically inert (i.e., non-reactive) PAMAMOS, such as those containing trimethylsilyl (TMS), triethylsilyl (TES), or trimethylsiloxy (TMSO) end-groups, and chemically reactive PAMAMOS such as: (*ii*) those containing unsaturations (i.e., vinylsilyl (VS), vinylsiloxy (VSO) or allylsilyl (AS) end-groups), and (*iii*) those containing alkoxysilyl end-groups, such as dimethoxymethylsilyl (DMOMS) or trimethoxysilyl (TMOS) units. All of these dendrimers can be prepared from

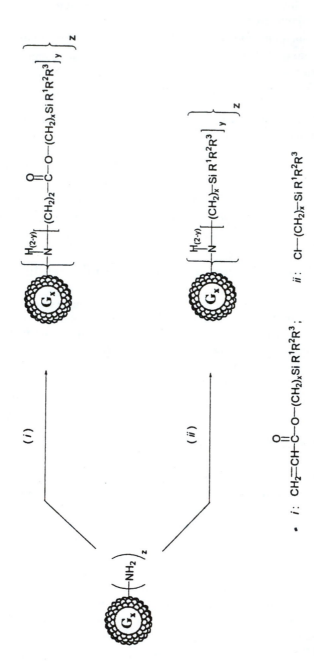

Reaction scheme 1

• i: $CH_2=CH-\overset{\overset{\displaystyle O}{\|}}{C}-O-(CH_2)_x Si\,R^1R^2R^3$;   ii: $Cl-(CH_2)_x-Si\,R^1R^2R^3$

as low as –60°C to above ambient temperature, the solubility dramatically changes within the same homologous series with the degree of OS substitution. For example, within the PAMAMOS-TMS series (top left in Figure 2), this change is from pronounced solubility of non-substituted amine-terminated PAMAMs in water and methanol (and insolubility in most organic solvents) (*5,8*), to solubility of the 57% TMS-substituted derivative in water, methanol and polar organics (such as chloroform and THF), to solubility of completely TMS-substituted PAMAMOS in methanol, polar organics and toluene, but insolubility in water (*5*). Computer modeling studies have revealed that this behavior is directly related to the degree of coverage of the PAMAM interior by OS exterior branch cells, and that only when this coverage becomes substantial enough to seriously restrict communication between the hydrophilic PAMAM and the surrounding environment, the oleophilic nature of OS composition starts dominating the resulting PAMAMOS dendrimer solubility. Hence, regardless of the fact that methylsilicones are generally insoluble in methanol (*9*), all of the PAMAMOS-TMS dendrimers with one layer of OS branch cells in their exterior are soluble in this solvent, indicating that their solubility must result from significant attractive interactions between the insufficiently "covered" PAMAM interior (see the top left structure in Figure 3) and surrounding solvent. If so, it would become necessary to further increase the degree of PAMAM coverage by OS branch cells in order to achieve methanol insolubility of the resulting PAMAMOS. For this, however, more complex dendrimers with more than one layer of exterior OS branch cells are needed.

Poly(amidoamine-organosilicon) dendrimers with more than one layer of exterior organosilicon branch cells

In principle, there are two different approaches that can be taken in order to prepare PAMAMOS dendrimers with more than one layer of exterior OS branch cells. One of these approaches would involve a step-wise addition of layer after layer of OS branch cells by some divergent growth strategy starting from an appropriately functionalized PAMAMOS of Figure 2, while the other utilizes convergent addition of preformed OS dendrons of desired generation directly to the PAMAM precursor. The dendrons must contain appropriate reactive functional groups at their focal point, and this approach may be illustrated as shown in Reaction schemes 2 and 3 (*4,10*).

It has been clearly shown (*5*) that, as expected, compositional characteristics indeed significantly influence properties of these PAMAMOS dendrimers, such as their glass transition temperature and solubility. While the former depends on the type of OS branch cells enclosing the PAMAM interior and may range from various generations of commercially available PAMAMs (*8*) and with different relative degree of their NH end-group conversion (i.e., OS substitution) (*5*).

240

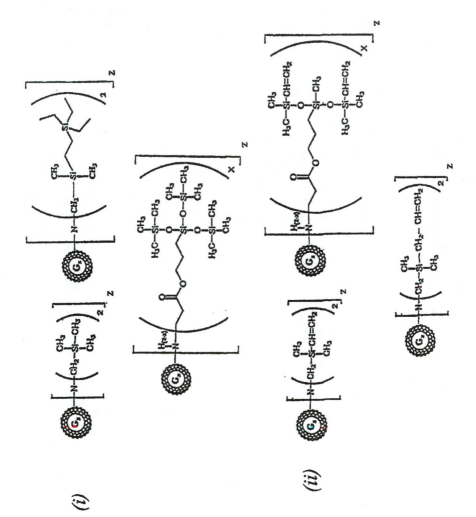

(i)

(ii)

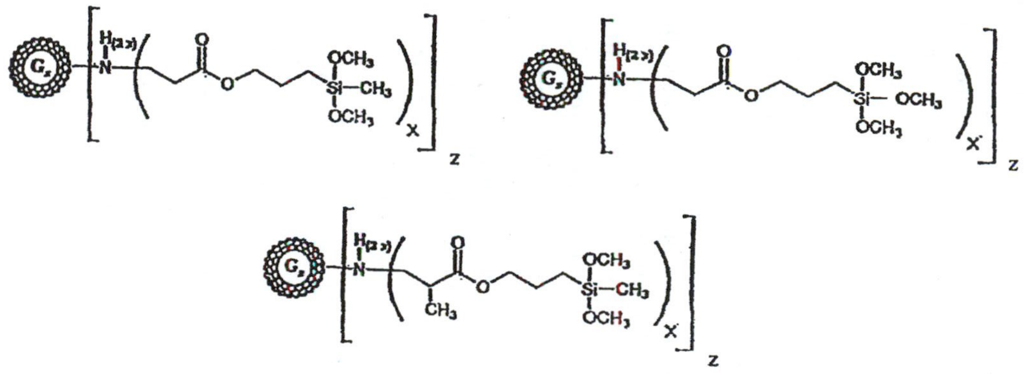

Figure 2: Selected examples of non-reactive and reactive PAMAMOS dendrimers. From the top-left down: (i) PAMAMOS-TMS; PAMAMOS-TMSO; PAMAMOS-TES; PAMAMOS-TMSO; (ii) PAMAMOS-DMVS; PAMAMOS-DMVSO; PAMAMOS-DMAS; (iii) PAMAMOS-DMOMS; PAMAMOS-TMOS; PAMAMOS-DMOMS.

In the first step of this example, polycarbosilane dendrons with haloalkyl focal points are prepared by a reiterative series of consecutive hydrosilylation and Grignard addition reactions, as shown for the polyethylsilane derivatives with chloromethyl focal points in Reaction scheme 2. The effectiveness of this route has been documented by many previous investigators and it routinely yields high quality products either with Karsted's or with Speier's hydrosilylation catalyst (11). The resulting dendrons may be isolated either with nonreactive ethylsilyl or with reactive vinylsilyl end-groups. The latter could be replaced by allyl units if an appropriate Grignard reagent is used in the synthesis.

The number of branch cells (i.e., triethylsilyl, TES, building blocks) in these OS dendrons will predetermine the number of OS layers surrounding the PAMAM interior in the PAMAMOS dendrimers resulting from Reaction scheme 3 (4,10). For clarity, these dendrimers are denoted PAMAMOS-[x,y], where x and y are integers that define the number of layers of PAMAM branch cells in their interiors and OS branch cells in their exteriors, respectively (see Figure 1). For example, PAMAMOS-[3,2] refers to a dendrimer having 3 layers of -[(CH_2)_2-C(O)-N(H)-(CH_2)_2-N]< branch cells (corresponding to a product of generation 2 amine-terminated PAMAM precursor) and 2 layers of −{Si[(CH_2)_2]_3}- branch cells (corresponding to a product prepared from G_2Et OS dendrons of Reaction scheme 2).

As expected, the properties of PAMAMOS-[x,y] dendrimers are dependent on the number of their OS layers (y) around the PAMAM interiors. For example, their solubility in methanol decreases with y, until eventually the PAMAMOS-[4,4] homologue, having 4 layers of OS branch cells built around a generation 3 PAMAM precursor (see structure at the bottom left of Figure 3), becomes insoluble just like a typical methylsilicone of traditional macromolecular architecture. As illustrated in Figure 3, computer modeling studies corroborate these experimental findings by showing that it is indeed at this generational stage (i.e., at 4 OS branch cell layers) that the degree of interior PAMAM coverage in PAMAMOS-[x,y] becomes practically complete and capable of preventing attractive interactions between the interior and methanol environment.

Poly(amidoamine-organosilicon) multi-arm star polymers

PAMAMOS dendrimers with reactive end-groups, such as those of groups *ii* and/or *iii* of Figure 2, enable various approaches to the preparation of multi-arm star polymers with hydrophilic/nucleophilic PAMAM interior and silicon-containing hydrophobic/oleophilic side arms. Two types of such polymers which represent the next step in the structural complexity of PAMAMOS organization have been recently described in more detail (4.12,13).

Reaction scheme 2

Reaction scheme 3

In these polymers the number of arms per star molecule is predetermined by the functionality of dendrimer precursor used, and by the degree of end-group conversion achieved in the synthesis. Thus, PAMAMOS star polymers containing up to 128 arms per molecule have been prepared either from generation 4 PAMAM dendrimers containing 64 NH$_2$ end-groups, or from a corresponding dimethylvinylsilyl (DMVS) PAMAMOS containing 128 Si(CH$_3$)$_2$CH=CH$_2$ functionalities (5,7). In the former case, the arms were polymethylene (PM) chains, $-$(CH$_2$)$_x$-CH$_3$, which were connected directly to the PAMAM cores via silicon-containing $-$[CH$_2$-Si(R)$_2$-(CH$_2$)$_2$-S]- bridges. In the latter case, the arms were polydimethylsiloxane (PDMS) chains, -[Si(CH$_3$)$_2$O]$_x$-, connected to the same PAMAM cores via $-$[CH$_2$-CH(OH)-CH$_2$-O-(CH$_2$)$_3$]- groups (4). In the former case, the length of the arms extended from 7 to 23 atoms from the outermost tertiary nitrogen dendrimer branch juncture, while in the latter, the PDMS chains had average degrees of polymerization of about 13 and 68, respectively.

The PM-armed stars were prepared via the azo*bis*isobutyronitrile (AIBN)-catalyzed thiol addition reaction on PAMAMOS-DMVS dendrimers as shown in Reaction scheme 4.

The syntheses were typically performed in THF or a THF/MeOH mixture at 45-65°C. The obtained products were insoluble in acetone and methanol, but soluble in toluene, *n*-hexane and chloroform, and their expected structures were confirmed by ^{1}H, ^{13}C and ^{29}Si NMR. The PM-stars having 12 or less carbon atoms per arm, showed T$_g$s ranging from $-$5°C to $-$30°C by DSC, while their derivatives having 18 carbon atoms per arm showed no T$_g$s but two distinct melting endotherms between 35°C and 50°C. These were identified by X-ray crystallography as originating from the side arms, since their diffraction patterns were identical to that of 1-octadecathiol which melts at 31°C.

The PDMS-armed star polymers were prepared by addition of monofunctional epoxypropoxypropyl-terminated PDMS to amine-terminated PAMAM dendrimers, as shown in Reaction scheme 5 (12).

The syntheses were performed in a THF/MeOH mixture at room temperature, and the composition of the obtained products was verified by ^{1}H, ^{13}C, ^{29}Si NMR and GPC. The products were soluble in THF and *n*-hexane, but insoluble in water and lower alcohols. In the case of epoxypropoxypropyl-PDMS of declared M$_w$ of 1,000, practically quantitative PAMAM substitution was

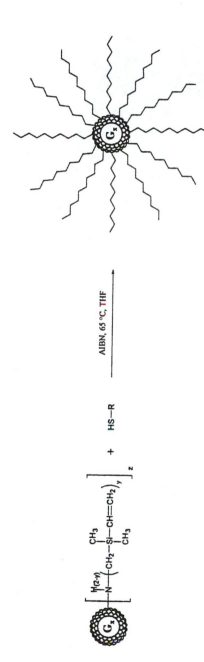

Reaction scheme 4

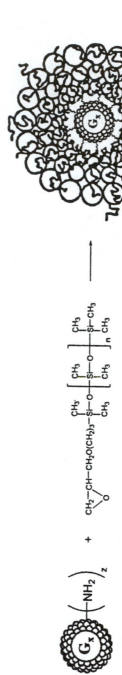

Reaction scheme 5

routinely achieved (*12*), while in the case of its M_w=5,000 derivative, serious steric hindrance was encountered at about 50 % NH substitution.

Poly(amidoamine-organosilicon) dendrimer-based networks

Functionalized PAMAMOS dendrimers are also extremely useful precursors for preparation of unique networks that represent the next level in the structural complexity of the PAMAMOS composition of matter. These networks are not only three-dimensionally organized structures, but they also exhibit a very high degree of nano-scale preciseness by which their building blocks, the well-defined hydrophilic PAMAM and hydrophobic OS domains are put together (*14-16*). While the preciseness of these domains (i.e., their sizes and size distribution) stems from the high degree of synthetic control embodied in the fundamental features of dendrimer synthetic chemistry (*1,5,8*), their topological disposition is predetermined by the selection of type of OS branch cells, the number of their layers in the PAMAMOS dendrimer precursor and by the selection of the cross-linking chemistry (*14,15*). As a consequence, all of these factors may be accurately engineered to enable formation of unique nano-templates with regularly distributed hydrophilic and hydrophobic nanoscopic domains.

From methoxysilyl-functionalized PAMAMOS-DMOMS or PAMAMOS-TMOS of Figure 2, such networks may be obtained as shown for the former in Reaction scheme 6 (2,14,15).

The process consists of two steps: (*i*) a water hydrolysis of methoxysilyl, Si-OCH$_3$, dendrimer end-groups into the corresponding silanols, Si-OH, and (*ii*) subsequent condensation of these silanol intermediates into siloxane, Si-O-Si, *inter*dendrimer bridges. The second reaction is in fact self-catalyzed by the basic PAMAMs, and is easily accomplished either by direct exposure of methoxysilyl-functionalized PAMAMOS-DMOMS or PAMAMOS-TMOS dendrimer precursors to atmospheric moisture, or by controlled addition of water, either in the form of vapor (for example: in a humidity chamber), or as liquid into an appropriate dendrimer solution (*14,15*). Since the process is a chain reaction in which a part of the water used in the methoxysilyl hydrolysis is regenerated in the silanol condensation step, a less than stoichiometric amount of water is needed. On the other hand, as long as methoxysilyl PAMAMOS dendrimer precursors are kept in dilute solution (preferably in methanol), their shelf life can be extended to over six months at room temperature.

For example, if methanol solution of a methoxysilyl, Si-OCH$_3$, functionalized dendrimer (either PAMAMOS-DMOMS or PAMAMOS-TMOS) is simply poured into an aluminum mold, covered with aluminum foil in order to prevent dust contamination, and left exposed to laboratory air for a period of time, slow curing will soon become evident as gradual densification of the initial liquid. The sample will also show a time-dependent loss of weight resulting from evaporation of methanol that is both present as a solvent and formed as a by-product of the cross-linking reaction 6. However, this change of weight becomes undetectable after a few days although under these conditions the curing reaction continues for several months, as can be easily seen by following the increase in glass temperature of the resulting product.

Reaction scheme 6

This long reaction time probably indicates that reactivity of the condensing methoxysilyl dendrimer groups is dependent on steric hindrance and that not all of the end-groups are always available for the cross-linking. If so, this would represent another architecturally-driven dendrimer property, which would create a favorable situation for promoting *intra*dendrimer condensation reactions that would in turn lead to the formation of closed siloxane loops between the cross-linking points at the outer surface of individual dendrimer domains.

Some Properties of PAMAMOS Dendrimer-Based Networks

As a consequence of their precise three-dimensional organization at the nanoscopic size level these PAMAMOS dendrimer-based networks are optically clear, perfectly transparent and colorless materials. In addition, if cast into films, sheets or coatings they generally have very smooth surfaces and are insoluble (but swell) in solvents, such as methanol, water, methylene chloride, THF, or acetone. Depending on the particular dendrimer precursor used, its density of functionality and composition of the curing reaction mixture, the mechanical properties of these networks may be varied at will to range from elastomers to plastomers, having T_gs from well below $-50°C$ to above room temperature. Generally, networks prepared from PAMAMOS-[x,1] dendrimers having only one layer of OS branch cells around the PAMAM interiors have glass temperatures that are very close to those of the PAMAM precursors from which they were derived while increase in the number of OS branch cell layers in the precursor dendrimer leads to the lowering of the resulting network's T_g.

The flexibility of the networks can also be affected by adding co-reagents into the cross-linking reaction mixture. For example, if curing is performed in the presence of a low molecular weight multifunctional cross-linker, such as $Si(OEt)_4$, it results in an increase in the obtained network's T_g due to an increase in its overall cross-link density. In contrast to this, introduction of α,ω-telechelic difunctional PDMS creates flexible linear spacers between the highly branched dendrimer domains and reduces the network's T_g by the extent dependent on the amount of PDMS added and the length of its chains (i.e., molecular weight).

Scanning electron microscopy (SEM) studies have shown featureless crosscuts of these networks, consistent with uniform spatial distribution of their nanoscopic PAMAM and OS domains; while atomic force microscopy (AFM) images showed a very high degree of smoothness where maximum roughness did not exceed a 2 nm range. Quasi-equilibrium advancing contact angle studies which are sensitive to the lowest surface energy component revealed methyl-like surfaces having solid surface energy of about 23 mN/m, similar to PDMS or paraffin waxes (*17*). This was reconfirmed by XPS which indicated only silicon, carbon and oxygen at the PAMAMOS network film surfaces.

Poly(amidoamine-organosilicon) dendrimer-based interpenetrating networks

PAMAMOS dendrimer-derived materials with the highest level of structural complexity achieved to date are interpenetrating networks (IPNs) with traditional linear polymers such as poly(methyl methacrylate) (PMMA) and cellulose acetobutyrate (CAB) (*18*).

250

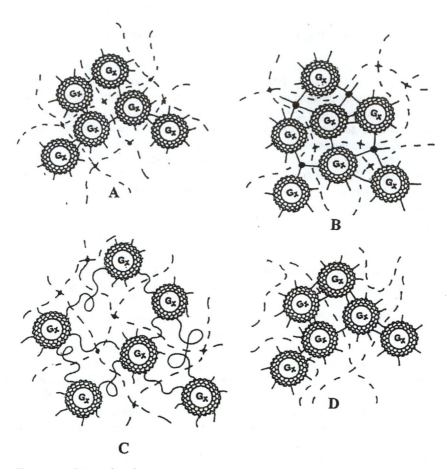

Figure 4: Generalized structures of three types of interpenetrating networks containing a dendrimer-based and a linear polymer-based partner component (A-C) and a semi-interpenetrating network containing a crosslinked dendrimer network enmeshed in a non-crosslinked linear polymer (D). Dendrimer partner networks are crosslinked without any additional co-reagent (A); with small molecular weight multi-functional crosslinker: large dark circles (B); and with high molecular weight linear polymer crosslinker (C). Small dark circles in A-C represent crosslinking points in the linear polymer networks.

An IPN is an intimate supramolecular combination of at least two polymers both of which are in network form, and wherein at least one of these polymer networks is synthesized in the immediate presence of the other. Generally, there are no induced covalent bonds between the two polymer networks, but due to the resulting interlocked configuration, the components of a true IPN cannot be separated without breaking their covalent bonds. In addition to this, the size scale of the potential phase separation in the case of mutually incompatible partner networks, is usually very small and frozen, so that a true IPN generally exhibits good dimensional stability, as well as a synergy of properties of the partner networks.

It should be noted that these PAMAMOS compositions represent the first truly interpenetrating networks involving a dendrimer-based partner component. Some previous reports that can be found in the literature mistakenly refer to linear polymer networks that are "filled" with isolated (i.e., non-crosslinked) dendrimer molecules as IPNs (*19*). In addition, it should also be noted that dendrimer-based IPNs represent a completely new class of macromolecular architecture in which dendritic and non-dendritic components are intimately and permanently combined at the supramolecular level of materials organization. Based on the dendrimer network alone, several sub-classes of this new class of macromolecular architecture may be envisioned as represented by the generalized structures shown in Figure 4. They may be formed by: (a) crosslinking dendrimers alone (i.e., without any additional co-reagent, as illustrated, for example, in Reaction scheme 6 and structure A in Figure 4); (b) crosslinking dendrimers together with small molecular weight di- or multi-functional crosslinker(s) (structure B in Figure 4); and (c) crosslinking dendrimers together with high molecular weight linear polymer crosslinker(s) (being either a di-functional α,ω-telechelic or multifunctional polymer with reactive side groups; the former is represented by structure C in Figure 4). Clearly, the selection of the crosslinking strategy will directly influence the segmental mobility, the available free volume and hence the macroscopic properties (such as mechanical properties, glass temperature, dimensional stability, ageing, etc.) of the resulting IPNs, even when the compositionally same (or very similar) partner networks are used.

PAMAMOS-PMMA IPNs may be prepared by a one-pot process in which all components (i.e. PAMAMOS dendrimer; dibutyltin dilaureate, a catalyst for the condensation reaction through alkoxysilyl groups; methyl methacrylate monomer (MMA); a free radical initiator, such as AIBN; and a free radical crosslinking agent) are mixed together and then the two networks are formed independently of each other at different crosslinking temperatures. Conveniently, the dendrimer network is formed first at room temperature according to the chemistry of Reaction scheme 6. As indicated above, in order to achieve desirable properties, this reaction may also include a co-reagent, such as a short telechelic PDMS disilanol (for example: $M_w = 400\text{-}700$) or a multi-functional small molecular crosslinker, such as tetraethoxysilane (TEOS). At the completion of the first stage, the process yields a dendrimer network swollen by the mixture of components that are subsequently used to form the second (i.e., linear polymer) network. The formation of the latter is readily achieved by heating the system under nitrogen to about 60°C to commence an AIBN-initiated MMA polymerization in the presence of ethylene glycol dimethacrylate as the crosslinking agent.

In contrast to this, PAMAMOS-CAB IPNs are prepared from preformed crosslinkable CAB polymer instead of a polymerizable monomer. The procedure is also a one-pot process wherein all components (including: lyophilized crosslinkable PAMAMOS dendrimer; CAB; tetramethoxysilane and dibutyltin dilaureate) are mixed under nitrogen in a common solvent, such as chloroform. The homogenized mixture is then cast onto a convenient substrate, such as a silylated glass plate, which is then heated at 50-60°C for about 25 h in nitrogen.

Specific properties of these IPNs depend on the relative amounts of component networks, and are described in more detail elsewhere (18). They are generally optically clear and perfectly transparent, easy to release from glass substrates and insoluble in organic solvents and water.

PAMAMOS dendrimer-based nanocomplexes and nanocomposites

Both PAMAMOS dendrimers and multi-arm star polymers behave as precise nano-scaled unimolecular inverted micelles that consist of hydrophilic PAMAM interiors and oleophilic OS exteriors. In their PAMAM interiors they posses a relatively high local concentration of nucleophilic tertiary amine branch junctures and neighboring amide groups that provide for pronounced ability to complex electrophiles (2,20-23). These may be either organic molecules (such as methylene blue, methyl red, etc.), or inorganic cations (such as Ag^+, Cu^+, Cu^{2+}, Ni^{2+}, Cd^{2+}, Fe^{2+}, Fe^{3+}, Au^{3+}, Co^{2+}, Pd^{2+}, Rh^{3+}, Pt^{2+}, Pt^{4+}, or lanthanides, such as Eu^{3+}, Tb^{3+}, etc.), or electrophilic organometallic compounds. Hence, these PAMAMOS species represent excellent substrates for preparation of a variety of different host-guest nanocomplexes which, because of the oleophilic character of their OS exteriors, are generally soluble in a variety of organic solvents in which their inorganic cationic guests would otherwise not be (12).

In addition to this, the complexed electrophiles remain chemically active for further transformation while retaining their established location within the dendritic interior (2,20-23). For example, complexed $CuSO_4$, $Cu(Ac)_2$ or $Cd(Ac)_2$ can be easily converted into CuS, CuSe, CdS or CdSe by exposing a salt-"impregnated" PAMAMOS-TMS to H_2S or H_2Se gas, or into dendrimer-encapsulated metal by reduction with an appropriate reducing agent, such as hydrazine, sodium borohydride, ascorbic acid, etc. The resulting zerovalent metal particles remain entrapped (i.e., encapsulated) inside the highly branched dendritic molecular interior and hence "solubilized" in otherwise hostile organic media. Furthermore, because of the truly nanoscopic size of these particles that are smaller than the wavelength of light, they color these organic "solutions" but do not disturb their optical clarity and transparency. This behavior may be illustrated by an example of copper transformation in a PAMAMOS multi-arm star polymer prepared from generation 4 PAMAM and linear mono-functionalized PDMS having M_w of about 1000 (12), as shown in Figure 5.

The same ability to complex electrophiles and encapsulate nanoparticles is also exhibited by PAMAMOS dendrimer-based networks, regardless of their physical form (i.e., films, sheets or coatings) (2). All of these forms can easily absorb (i.e., by simple immersion into a water or methanol solution of the desired electrophile), complex and retain (i.e., when dried) guest species, acting in a manner that is reminiscent of "molecular sponges". In addition, the guest species can also be subject to chemical transformation (as in confined "nano reactor") to yield (from inorganic cations) a variety of elastomeric or plastomeric nanocomposites containing nano-sized metallic particles regularly distributed throughout the three-dimensionally crosslinked PAMAMOS template/matrix. This ability to template such nano-particles by a variety of different physical forms of optically clear and transparent networks opens up unprecedented possibilities for applications in a variety of fields in electronics, opto-electronics, sensors, catalysis, biomedical materials, etc. It should be particularly noted that with respect to this PAMAMOS and their nanocomplexes and nanocomposites are unique among dendrimers and they offer great promise for nanomaterials engineering of tractable elsatomeric or plastomeric films, sheets, membranes, coatings or otherwise shaped objects (14,15). Schematically, this assortment of possibilities may be illustrated as shown in Figure 6.

PAMAMOS dendrimer-based coatings

PAMAMOS dendrimer-based coatings are of particular interest because not only may they be cast on a variety of different surfaces, but if the substrate contains reactive surface groups (such as Si-OH groups in various glasses) they can also establish covalent bonds during the cross-linking reaction, resulting in a "permanent" coat that can not be "peeled off" the substrate surface by simple mechanical action (see Figure 7). In other cases, such as substrates having negatively charged surfaces/groups that are not reactive with alkoxysilyl units, secondary bonds can be established between the substrate and protonated tertiary amines of the PAMAM domains. Either way, the coated substrates can be effectively surface-modified to behave as the self-sustaining networks described above, including the ability to complex electrophiles and encapsulate nano-sized (organic or inorganic) guest species.

Conclusion

In summary, PAMAMOS dendrimers not only represent a striking example of exceptional possibilities that are offered to dendrimer chemistry by applying the compositional copolymer approach, but they are also extraordinarily versatile building blocks for preparation of a variety of more complex types of the

254

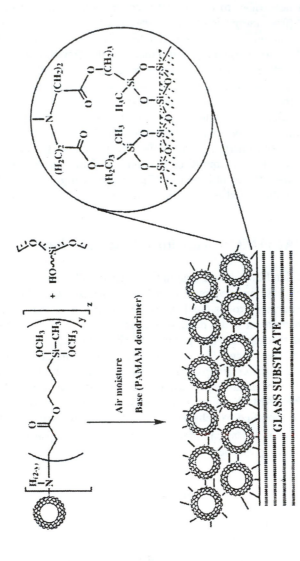

Figure 7: Formation of covalently bonded PAMAMOS dendrimer network coat on a glass surface having silanol surface groups.

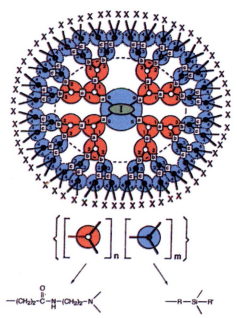

Figure 1: Schematic representation of a radially layered poly(amidoamine-organosilicon) (PAMAMOS) dendrimer having two layers of red (i.e., PAMAM) branch cells and two layers of blue (i.e., OS) branch cells.

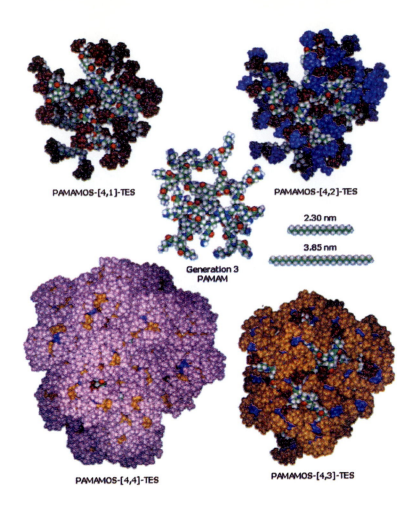

PAMAMOS-[4,1]-TES

PAMAMOS-[4,2]-TES

2.30 nm

3.85 nm

Generation 3
PAMAM

PAMAMOS-[4,4]-TES

PAMAMOS-[4,3]-TES

Figure 3: Computer models of PAMAMOS-[4,y]-TES dendrimers. From top-left clockwise: y=1, 2, 3, 4. Generation 3 PAMAM from which these PAMAMOS are generated is shown in the middle together with two extended polyethylene (PE) chain segments that serve as "molecular size rulers". Atom color code for PAMAM and PE: carbon (green), hydrogen (white), oxygen (red) and nitrogen (blue). OS branch-cell layers color code: first innermost TES layer (burgundy), second layer (navy), third layer (gold), forth outermost layer (pink).

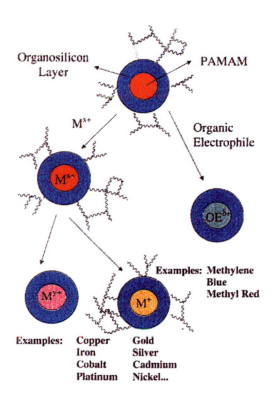

Organosilicon
Layer

PAMAM

M^{x+}

Organic
Electrophile

M^{x+}

$OE^{\delta+}$

M^{y+}

M^+

**Examples: Methylene
Blue
Methyl Red**

Examples: **Copper   Gold
Iron   Silver
Cobalt   Cadmium
Platinum   Nickel...**

*Figure 6: Templating ability of PAMAMOS dendrimer-based networks to act as
nanoscopic "molecular sponges" and "chemical reactors".*

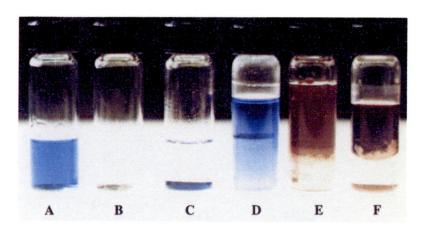

Figure 5. Different phases in the preparation of a hexane-soluble copper nanocomplex and nanocomposite with PAMAMOS-PDMS 128-arm star polymer. A: water solution of $Cu(Ac)_2$; B: hexanes solution of PAMAMOS-PDMS star polymer; C: $Cu(Ac)_2$ is insoluble in water; D: two-phase system obtained after mixing A and B, thorough shaking of the resulting mixture and subsequent separation of immiscible layers (upper layer: organic; lower layer: aqueous); E: D after copper reduction with hydrazine to form zerovalent metal; F: E after decantation.

structural organization of matter. Some of these types, including (in the order of increasing degree of structural complexity) the higher generation multi-layer copolymeric dendrimers, multi-arm star polymers, three-dimensionally cross-linked regularly nano-domained networks, and unprecedented interpenetrating networks, are described in this contribution. Also, some possible applications of these systems, such as formation of nanocomplexes and nanocomposites, as well as shaping of different forms, including films, sheets, membranes and/or coatings are also described. Schematically, the versatile scope of this unique dendrimer family and the flexibility of synthetic strategies based on them as nanoscopic building blocks for nanomaterials engineering may be illustrated as shown in Figure 8.

Acknowledgement

Many colleagues have significantly contributed to this work. We particularly wish to acknowledge Drs. Agnes de Leuze-Jallouli, Scott D. Reeves, Jieming Li, Jin Hu, Robert A. Bubeck and Steven E. Keinath of Michigan Molecular Institute, Frederic Vidal, Isabelle Hemonic and Dominique Teyssie of Unverste de Cergy-Pontoise, Cergy-Pontoise cedex, France, Sylvie Boileau of CNRS, Thiais, France, as well as Ms. Susan V. Perz and Mr. Lee W. Hoffman of Dow Corning Corporation, Midland, Michigan and Mr. Paul L. Parham of Michigan Molecular Institute. Financial support for parts of this program was provided by Dow Corning Corporation and Dendritech Inc., both of Midland, Michigan, and their contribution is gratefully acknowledged.

References

1. See for example: (a) Vögtle, F.; Gestermann, S.; Hesse, R.; Schwierz, H,; Windisch, B., *Prog. Polym. Sci.*, **2000**, *25*, 987. (b) Bosman, A.; Janssen, H.M.; Meijer, E.W., *Chem. Rev.*, **1999**, *99*, 1665.(c) Fischer, M.; Vögtle, F., *Angew. Chem., Int. Ed. Engl.* **1999**, *38*, 884. (d) Majoral, J.-P.; Caminade, A.M., *Chem. Rev.* **1999**, *99*, 884. (e) Archut, A.; Issberner, J.; Vögtle, F., in *Organic Synthesis Highlights III;* Mulzer, J.; Waldman, H., Eds., Wiley-VCH: Weinheim, **1998**; pp 391-405. (f) Frey, H.; Lach, C.; Lorenz, K., *Adv. Mater.* **1998**, *10*, 279. (g) Zeng, F.; Zimmerman, S.C., *Chem. Rev.* **1997**, *97*, 1681. (h) Newkome, G.R.; Moorefield, C.N.; Vögtle, F., *Dendritic Molecules: Concepts, Synthesis, Perspectives;* VCH Verlagsgesellschaft: Weinheim, **1998**. (i) Dvornic, P.R.; Tomalia, D.A., *Curr. Opin. Colloid Interface Sci.* **1996**, *1*, 221. (j) Voit, B.I., *Acta. Polym.* **1995**, *46*, 87. (k) Ardoin, N.; Astruc, D., *Bull. Soc. Chim Fr.* **1995**, *132*, 875. (l) Hawker, C.J.; Frechet, J.M.-

256

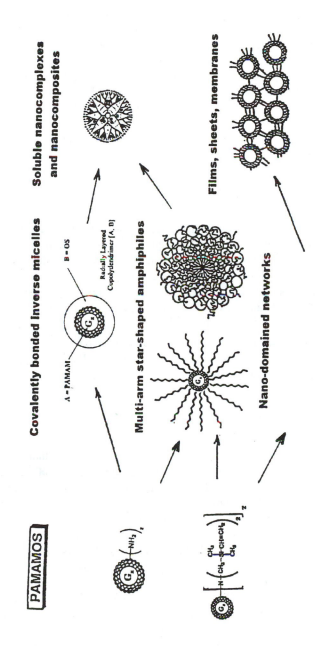

257

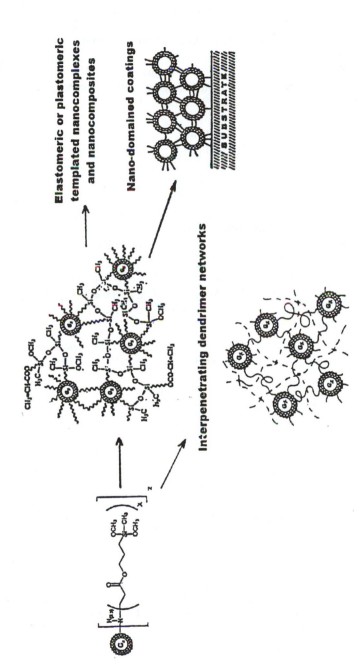

Figure 8: Schematic representation of the versatility of PAMAMOS dendrimers as nanoscopic synthetic building blocks for preparation of various structures of higher degree of structural complexity. The two functional PAMAMOS dendrimers shown are selected as representatives of their respective families (ii and iii of Figure 2) and other memvers of these families may be used instead.

258

J., *Three-dimensional dendritic macromolecules: design, synthesis and properties* in *New Methods of Polymer Synthesis*, Ehdon, J.R.; Eastmond, G.C., Eds., Blackie Academic and Professional: Glasgow, **1995**, Vol. 2, pp. 290-330.

2. Dvornic, P.R; de Leuze-Jallouli, A.M; Owen, M.J; Perz, S.V, Chapter 16 in *"Silicones and Silicone-Modified Materials"*, Clarson, S.J; Fitzgerald, J.J; Owen, M.J, Smith, S.D, Eds., ACS Symp. Ser. 729, ACS, **2000**, pp.241-269.

3. Hawker, C.J.; Frechet, J.M.J., *J. Am. Chem. Soc.*, **1992**, *114*, 8405.

4. Dvornic, P.R.; Owen, M.J.; Keinath, S.E.; Hu, J.; Hoffman, L.W.; Parham, P.L., *Polym. Preprints.*, **2001**, *42(1),* 126.

5. Dvornic, P.R; de Leuze-Jallouli, A.M; Owen, M.J; Perz, S.V, *Macromolecules*, **2000**, *33*, 5366.

6. de Leuze-Jallouli, A.M; Swanson, D.R; Perz, S.V; Owen, M.J; Dvornic, P.R., *Polym. Mater. Sci. Eng.*, **1997**, *77*, 67.

7. Dvornic, P.R; de Leuze-Jallouli, A.M; Swanson,D.R; Owen, M.J; Perz,S.V, *US Patent*, 5,739,218, **1998**.

8. Dvornic, P.R.; Tomalia, D.A., *Dendritic Polymers, Divergent Synthesis (Starburst® Polyamidoamine Dendrimers)* in *The Polymeric Materials Encyclopedia*, Salamone, J.C., Ed., CRC Press, Boca Raton, Vol. 3, pp. 1814-1830, **1996**.

9. Kuo, A.C.M., *Poly(dimethylsiloxane)* in *Polymer Data Handbook*, Mark, j.E., Ed., Oxford University Press, New York-Oxford, **1999**, pp. 411-435.

10. Dvornic, P.R; de Leuze-Jallouli, A.M; Owen, M.J; Perz, S.V, *US Patent*, 6,077,500, **2000**.

11. (a) van der Made, A.W; van Leeuwen, P.W.N.M, *J. Chem. Soc, Chem. Commun.*, **1992**, 1400. (b) van der Made, A.W; van Leeuwen, P.W.N.M; de Wilde, J.C; Brandes, R.A.C, *Adv. Mater*, **1993**, *5*, 466. (c) Zhou, L.-L; Roovers, J, *Macromolecules*, **1993**, *26*, 963. (d) Seyferth, D; Son, D.Y; Rheingold, A.L; Ostrander, R.L, *Organometal.*, **1994**, *13*, 2682.

12. Dvornic, P.R.; Hu, J.; Reeves, S.D.; Owen, M.J., *Silicon Chem.*, **2002**, *1*, 0000.

13. Dvornic, P.R.; de Leuze-Jallouli, A.M.; Perz, S.V.; Owen, M.J., *Mol. Cryst. Liq. Cryst.,* **2000**, *353*, 223.

14. (a) Dvornic, P.R; de Leuze-Jallouli, A.M; Owen, M.J; Perz, S.V, *Polym. Preprints*, **1998**, *39(1),* 473. (b) Dvornic, P.R; de Leuze-Jallouli, A.M; Owen, M.J; Perz, S.V, *Polym. Preprints*, **1999**, *40(1),* 408. (c) Dvornic, P.R; de Leuze-Jallouli, A.M; Owen, M.J; Dalman, D.A.; Parham, P.; Pickelman, D.; Perz, S.V, *Polym. Mat. Eng. Sci.,* **1999**, *81,* 187.

15. Dvornic, P.R; de Leuze-Jallouli, A.M; Owen, M.J; Perz, S.V, *US Patent*, 5,902,863, **1999**.
16. Ruckenstein, E; Yin, W., *J. Polym. Sci., Polym. Chem.*, **2000**, *38*, 1443.
17. de Leuze-Jallouli, A.M; Dvornic, P.R; Perz, S.V.; Owen, M.J., *Polym. Preprints*, **1998**, *39(1), 475.
18. Vidal, F; Hémonic, I; Teyssie, D; Boileau, S; Reeves, S.D; Dvornic, P.R; Owen, M.J, *Polym. Preprints*, **2001**, *42(1),* 128.
19. Ottaviani, M.F.; Montali, F.; Turro, N.J.; Tomalia, D.A., *J. Phys. Chem. Chem. B.,* **1997**, *101*, 158.
20. Balogh, L.; Tomalia, D.A., *J. Am. Chem. Soc.*, **1998**, *120*, 7355-7356.
21. Zhou, L. Sun; Crooks, R.M., *J. Am. Chem. Soc.*, **1998**, *120*, 4877-4878.
22. Balogh, L.; de Leuze-Jallouli, A.M.; Dvornic, P.R.; Owen, M.J.; Perz, S.V.; Spindler, R., *U.S. Patent* 5,938,934, **1999**.
23. Bubeck, R.A.; Bauer, B.J.; Dvornic, P.R.; Owen, M.J.; Reeves, S.D.; Parham, P.L.; Hoffman, L.W., *Polym. Mater. Sci. Eng.,* **2001**, *84*, 866.

Chapter 22

Surfactant Properties of Poly(dimethylsiloxane)-Containing Block Copolymers from Living Radical Polymerization

Laurence Bes[1], Kim Huan[1], David M. Haddleton[1,*], and Ezat Khoshdel[2]

[1]Department of Chemistry, University of Warwick, Coventry CV4 7AL, United Kingdom
[2]Unilever Research, Port Sunlight, Quarry Road East, Bebington, Wirral CH63 3JW, United Kingdom

Transition metal mediated living radical polymerization has been utilized to prepare amphiphilic PDMS containing block copolymers. The PDMS based macroinitiator has been synthesized for allow the polymerization of 2-dimethylaminoethyl methacrylate (DMAEMA) monomers with Cu(I)Br/npropyl-2-pyridinalmethanimine complex. Well-defined block copolymers have been prepared with different molecular weights of DMAEMA block. Dynamic surface tension of aqueous solutions of poly(DMAEMA-DMS-DMAEMA) were investigated at different pH. These block copolymers were found to be highly surface active in aqueous media, particularly in basic pH where the charge density (degree of protonation) of the DMAEMA is low. An improvement in aggregation formation is established by the addition of salt (NaCl) in the aqueous solution. Aggregates formed with these polymers have a large hydrodynamic diameter.

Surface active polymers, or polymeric surfactants, have gained in interest over the last two decades. They are now used commercially in many different applications, among which stabilization is the most widespread. An excellent way to design surface-active polymers is to alternate hydrophilic and hydrophobic segments in block copolymers. Poly(dimethylsiloxane) (PDMS) has been chosen as hydrophobic segment as it is extremely hydrophobic being completely insoluble in water. Poly(dimethylsiloxane) also lowers surface tension (down to around 20 mNm^{-1}), is an excellent wetting agent for low energy surfaces and is a constituent of powerful antifoamers [1].

Most siloxane surfactants are copolymers of PDMS and poly(alkylene) oxides of intermediate molecular weight. They are, generally, prepared by condensation polymerization, i.e., by coupling alkoxymethylsiloxane polymers and hydroxy-terminated poly(oxyalkylene)s using a transetherification reaction [2], or by the hydrosilylation of methyl siloxanes containing Si-H groups with vinyl functional poly(oxyalkylenes) [3].

PDMS-based block copolymers have also been prepared using living polymerization techniques. PS-PDMS block copolymers were first synthesized by anionic polymerization by Saam *et al* in 1970 [4]. Amphiphilic PDMS-based block copolymers were then prepared by living anionic polymerization such as poly(4-vinylpyridine)-poly(dimethylsiloxane) block copolymers [5], poly(dimethylsiloxane)-poly[alkyl(meth)acrylic acid] block copolymers [6] and poly(dimethylsiloxane)-poly[2-(dimethylamino)ethyl methacrylate] block copolymer [7]. Recent advances in polymerization chemistry has seen the emergence of transition-metal mediated living radical polymerization (TMMLRP)[8,9], which has been used to achieve poly(dimethylsiloxane) containing block copolymers with precise control of architectural design [10,11].

In earlier work we have demonstrated that poly(dimethylsiloxane) block copolymers have been synthesized by using PDMS macro-initiators with copper mediated living radical polymerization. The use of mono-and di-functional carbinol hydroxyl functional initiators led to AB and ABA block copolymers with narrow PDI and controlled M_n. Polymerization with MMA and DMAEMA is reported with a range of molecular weights produced. Polymerizations proceed with excellent first order kinetics indicative of living polymerizations. Preliminary works on amphiphilic block copolymers showed that they formed aggregates in aqueous solution [11].

In this present work we report the surface activity and the aggregation behavior of the resulting amphiphilic PDMS-based triblock copolymers prepared by TMMLRP. These studies were performed using dynamic surface tensiometry, ^{1}H NMR and dynamic light scattering at different pH and on different molecular weight copolymers.

Experimental

Materials

Carbinol (hydroxyethoxypropyl) terminated poly(dimethylsiloxane) was purchased from ABCR Gelest Chemical with a molecular weight of approximately 5000 g mol^{-1} (PDI = 1.70 as determined by ourselves using SEC calibrated with PMMA standards). 2. N-(n-Propyl)-2-pyridylmethanimine ligand was synthesized as previously described, [12] toluene (BDH, 99%) was degassed and stored under nitrogen. Copper (I) bromide (Aldrich, 98%) was purified as described previously [12]. 2-(Dimethylamino)ethyl methacrylate (DMAEMA) and methyl methacrylate (MMA) (Aldrich, 99%) was filtered through a column of basic alumina to remove inhibitors and stored under nitrogen. Immediately prior to polymerization, all solvents, monomers and other reagents were degassed via a minimum of three freeze-pump-thaw cycles. All manipulations were carried out under nitrogen atmosphere using standard Schlenk or syringe techniques.

Measurements

^{1}H-NMR spectra were recorded on a Bruker-DPX 300 MHz spectrometer. Molecular weight and molecular weight distribution were measured by SEC on a system equipped with a guard column, 2 mixed D columns (Polymer Laboratories) using PMMA standards, with DRI detectors and eluted with tetrahydrofuran at 1 mL min^{-1}.

Preparation of difunctional PDMS macroinitiators

The synthesis of difunctional PDMS macroinitiators has been described previously [11] following an esterification reaction between 2-bromo-2-

methylpropionyl bromide and carbinol (hydroxyethoxypropyl) terminated PDMS.

^{1}H NMR (δ): 0.00 (m, 6H), 0.55 (m, 4H), 1.56 (m, 4H), 1.87 (s, 6H), 3.35 (t, 4H), 3,64 (t, 4H), 4.24 (t, 4H).

Polymerization of 2-dimethyl(amino)ethyl methacrylate with PDMS macroinitiator.

The molar amount of each reagent used were [DMAEMA]/[I]/[Cu]/[Ligand] = 64/1/1/2 for copolymer **A**, [DMAEMA]/[I]/[Cu]/[Ligand] = 287/1/1/2 for **B** and [DMAEMA]/[MMA]/[I]/[Cu]/[Ligand] = 120/24/1/1/2 for **C**.

A general polymerization procedure was carried out as follows. Cu^{I}Br, along with a magnetic stirrer bar, was placed in a dry Schlenk flask, which was evacuated and flushed with nitrogen three times. 2-Dimethyl(amino)ethyl methacrylate, toluene 50% v/v and PDMS initiator were added to the Schlenk using degassed syringes. The solution was subsequently de-oxygenated by three freeze-pump-thaw cycles. Finally, once the flask had reached 90°C the ^{n}propyl-2-pyridinalmethanimine ligand was added with stirring. The reaction mixture immediately turned dark brown in color on addition of the ligand. The reaction was left for a period of three hours to reach near complete conversion, after which the product was purified by passing the solution over basic alumina.

Quaternization

Quaternization was carried on copolymer **A**. A 3-fold excess of methyl iodide was added in a solution of the copolymer in THF. The solution was left stirring 24 hours at room temperature. The precipitate was filtered and purified by Soxhlet extraction with THF and dried under vacuum for 24 hours.

Solution preparation and characterization

All samples were prepared by the dilution of dry copolymers with doubly distilled water. Some samples were prepared with 0.1 mol L^{-1} of NaCl where stated. The pH was adjusted by addition of 1 M solution of HCl or NaOH. Surface tension measurements were carried out on a dynamic surface tensiometer DST9005 (Nima Technology) using du Nouy ring flamed after each experiment, at 20°C. The surface tension of the doubly distilled water was checked regularly. Dynamic light scattering measurements were carried out using the Malvern Zetasizer 3000 spectrometer equipped with a 5mW-helium neon laser operating at 633 nm and a 7132 correlator operating in 8 x 8 groups. Determination of the hydrodynamic diameter was done using the CONTIN algorithm. All measurements were carried out at a scattering angle 90° and a temperature of 25°C.

Results and Discussion

Polymer Synthesis

Four copolymers were prepared using PDMS macroinitiators with copper mediated living radical polymerization. The PDMS based macroinitiator was synthesised via the esterification of 2-bromoisobutyryl bromide with a hydroxyethoxypropyl (carbinol)difunctional PDMS prepolymer. It has previously been demonstrated that this α-bromoester PDMS macroinitiator is suitable to initiate a polymerization with MMA and DMAEMA monomers with excellent first order kinetics indicative of living polymerizations [11].

Polymerization of 2-dimethyl(amino)ethyl methacrylate with the PDMS macroinitiator

Triblock copolymers **A** and **B** were synthesized with different molecular weights of DMAEMA according to the ratio [Monomer]/[initiator] : **A** containing 49 %-molar of DMAEMA and **B** containing 81 %-molar of DMAEMA. ^{1}H NMR (figure 2) confirms the incorporation of the DMAEMA block within the polymer by appearance of the broad signal typical of DMAEMA (CH$_2$ group , d-e, at 4.05 and at 2.54 ppm and N(CH$_3$)$_2$, f, at 2.18 ppm). The polymerization of DMAEMA with PDMS macroinitiators has already been described [11]. In all cases first order kinetic plots gave excellent linear plots, indicative of living polymerization. And M_n obtained by ^{1}H NMR was very close to that predicted, table 1.

Copolymer **A** was quaternized with methyl iodide to obtain polycationic triblock copolymers. The degree of quaternization was determined by ^{1}H NMR with the disappearance of the $N(CH_3)_2$ signal at 2.18 ppm and the appearance of $N^+(CH_3)_3$ signal at approximately 3.3 ppm. The quaternization is quantitative. Incorporation of methyl methacrylate into the poly 2-(dimethylamino)ethyl methacrylate segment to form a statistical segment copolymer was also studied in order to varying degrees of hydrophilicity of the hydrophilic block. The incorporation of MMA in the hydrophilic segment is showed in ^{1}H NMR by the appearance of two signals at 3.3 and 3.5 ppm corresponding from the OCH_3 of the MMA group. The first signal corresponding to a MMA close to a MMA unit and the second one to MMA close to a DMAEMA unit, the presence on two signals shows statistical incorporation.

Table 1. Molecular characterization of PDMS-based copolymers.

	%-molar fraction [a]			M_n of copolymer (g/mol) [a]	PDI [b]
	DMS	DMA	MMA		
A	51	49	0	15 000	1.30
B	19	81	0	47 800	1.30
C	32	56	12	24 700	1.30
AQ [c]	51	49	0	15 000	1.30

a) determined by ^{1}H NMR, b) determined by SEC, c) A quaternized with MeI

Surface tension properties

An important feature of a surfactant from a physico-chemical point of view is that the molecule is able to orient itself so as to expose hydrophilic regions into a polar environment and hydrophobic segment into a lipophilic phase. This process results in a reduction in the interfacial tension. The surface-active character of siloxane surfactants is due to the methyl groups with the -(Si-O)$_n$- backbone serving as a flexible framework on which to attach these groups [9-12]. The surface energy of a methyl-saturated surface is approximately 20 dyne cm^{-1} and this is also the lowest surface tension achievable using siloxane surfactant.

Poly(dimethylsiloxane) has a low T_g and thus the self association of the molecules are expected to have a high mobility at 25°C, resulting in a dynamic equilibrium between aggregations and unimers. This should improve the surface activity and aggregation formation.

Surface tension measurements were carried out on aqueous solutions of **A** at different concentrations and at different pH. The solution's pH was adjusted by addition of HCl or NaOH. Figure 1 shows that the surface tension of an aqueous solution of **A** is pH dependent. At pH = 3 the surface tension of the aqueous solution is reduced by only 10 mN m^{-1} whereas at basic pH the surface tension of the aqueous solution is lowered by over 30 mN m^{-1} to 40 mN m^{-1}. This observation suggests that the chain conformation at the air/water interface differs considerably as a function of pH as result of protonation of the PDMAEMA blocks. In order to confirm this result, surface tension measurements were carried out on **A** quaternized with methyl iodide. No change in surface tension was observed. Milling *et al* [13] have reported that for a aqueous solution of poly(DMAEMA-*b*-MMA) at pH ≥ 6, the absorption at the surface appears to be micellar. They observed a slow disaggregation of the micelles at pH = 6 and 6.5. The surface tension eventually rises to a high value at pH = 5 solution when the surface is populated by unimers. The same phenomenon has also been observed with poly(DMAEMA-*b*-MMA) diblock copolymers quaternized by either HCl or ethyl bromide [14, 15]. This was explained in terms of an electrostatic repulsion between the charged PDMAEMA blocks leading to a low packing density at the air/water interface and to a small decrease in the water surface tension.

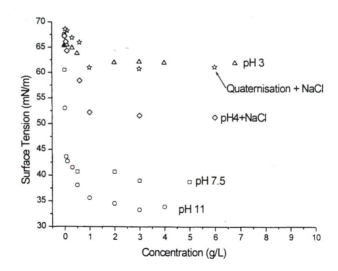

Figure 1. Concentration dependence of the surface tension of aqueous solution of copolymers AQ and A at different pH

In order to investigate what happens in solution, ¹H NMR analysis of **A** at basic pH and acid pH and on quaternized **A** was used, figure 2. At basic pH the signal corresponding at the PDMS block in **A** is insignificant. This suggests the formation of aggregates with a siloxane core and the PDMAEMA segment acting as a solvated corona. At acidic pH, and when quaternized, the signal of the PDMS (approximately 0 ppm) in **A** becomes significant corresponding to solvation of the PDMS block. This result lead us to conclude that when the PDMAEMA is protonated there is less tendency for self-association of the amphiphilic copolymer.

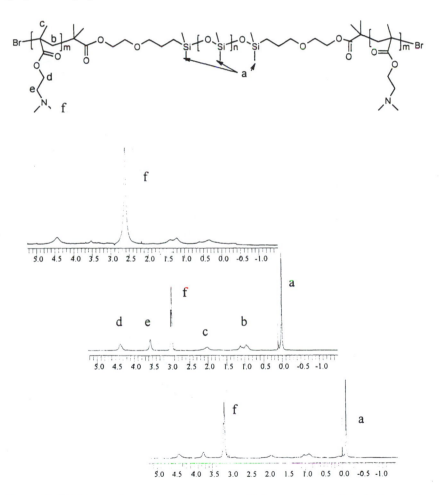

Figure 2. *¹H NMR of copolymer A at basic pH (top), acidic pH (middle) and AQ (quaternized, bottom) in D₂O.*

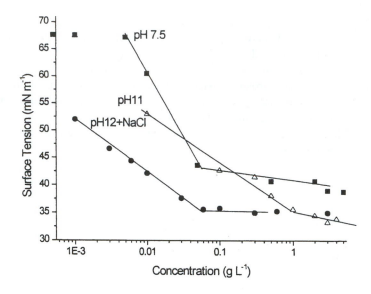

Figure 3. Concentration dependence of the surface tension of aqueous solution of copolymer A at different basic pH.

Thus a reduction in the pH increases the extent of ionization of the DMAEMA block, and the solubility of this block increases. At the maximum ionization of the DMAEMA block the predominant species in solution appears to be unimers and the surface activity is commensurately low.

At basic pH **A** behaves as a surfactant, figure 3. There is a break in the surface tension versus concentration curve reflecting the onset of self-association (such as micelle or aggregation formation). The surface tension falls linearly until the break identified as critical aggregation concentration (cac). At pH = 7.5 the cac is identified at 0.06 g L^{-1}. The surface tension above this concentration remains constant at about 43 mN m^{-1}. When the solution is more basic, pH = 11 the surface tension after the cac is lower, 33 mN m^{-1}, the charge density (degree of protonation) of the DMAEMA block is low which increases the hydrophobicity of this block. However, the cac of 1 g L^{-1} at this pH is higher than at pH = 7.5. In order to improve the self-association formation, salt (NaCl 0.1 mole/L) was added to the solution which resulted in a decrease of the cac to 0.6 g L^{-1} at pH = 12.

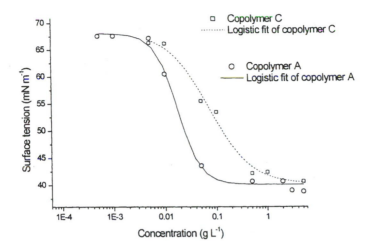

Figure 4. Concentration dependence of the surface tension of aqueous solution of A and C at the same pH.

Copolymer **C** with 12 %-molar of methyl methacrylate statistically incorporated into the poly(2-(dimethylamino)ethyl methacrylate) segment was also investigated. The concentration dependence of the surface tension of aqueous solutions of **A** and **C** at the same pH are shown in figure 4. The incorporation of MMA into the hydrophilic segment seems not to influence the surface activity of the copolymers in aqueous solution. The lowering of surface tension, as expected, by varying the hydrophilicity of the PDMAEMA block did not take place.

Figure 5 shows a comparison of the concentration dependence of the surface tension of aqueous solution between two copolymers having different %-molar fraction of DMAEMA. Copolymer **B**, with the higher %-molar fraction of DMAEMA, has a higher CAC. Critical aggregation concentrations vary with molecular structure, within a homologous series, proportionately larger hydrophobic groups lead to smaller values for CAC.

Dynamic light scattering characterization.

Dynamic light scattering experiments were carried out for **A** at different basic pH, **B** and **C**, in order to determine the size of the aggregates formed in aqueous solution. In all cases the hydrodynamic diameter is greater than that calculated for a fully extended chain. Implying a more complex aggregate structure. This same phenomenon is found in the case of the salt solution of **A**. This ensure an extended double electric layer does not artificially increase the

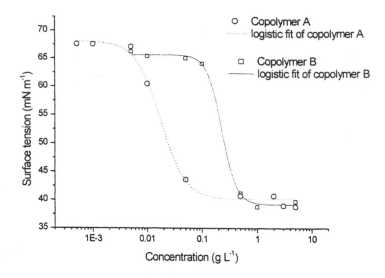

Figure 5. Concentration dependence of the surface tension of aqueous solutions of A at B at the same pH .

size by a few nanometers. The fact that the PDI is also quite high lead us to conclude that aggregation of the micelles occurs. However, when we attempted to remove this aggregation with ultrasound we observed the same result. One explanation is that P(DMAEMA-*b*-DMS) triblocks copolymers tend to self-assemble in aqueous solution to give aggregates with a hollow center, a type of vesicle.

Table 2. Hydrodynamic diameters of the aggregates for the copolymers A, B, C.

		CAC (g L-1) a)	*Hydrodynamic diameter (nm)*	*PDI*	*D theory* b)
	pH 7.5	0.06	85	0.3	50
A	pH 11	1	44	0.5	50
	pH 12 + NaCl	0.06	90	0.4	50
B	pH 7.5	1	224	0.4	115
C	pH 7.5	1	149	0.4	71

a) from tensiometry b) corresponding at the end to end distance calculated for fully stretched chain.

Conclusions

Transition metal mediated living radical polymerization has been used to prepare some DMAEMA-DMS-DMAEMA triblock copolymers. Poly(dimethylsiloxane) based macroinitiators have been used to initiate the polymerization of DMAEMA with a Cu(I)Br/^{n}propyl-2-pyridinalmethanimine complex. Well-defined block copolymers have been prepared with different molecular weights of the DMAEMA block. These DMAEMA-DMS triblock copolymers have been found to be highly surface active in aqueous media, particularly in basic pH where the charge density (degree of protonation) of the DMA is low. The aggregation formation for **A** is by the addition of salt (NaCl) to the aqueous solution. Aggregates are formed with a high hydrodynamic diameter, which seem to be hollow vesicles.

We thank Unilever (KH) and the EC Marie Curie Fellowship Scheme (LB, IHP-MCFI-99-1) for funding this work.

References

(1)   J. Lindman, H. Kronberg, *Surfactant and Polymers in Aqueous Solution*, Wiley, **1998**.

(2)   Snow, S. A.; Fenton, W. N.; Owen, M. J., Langmuir **1990**, *6*, 385.

(3)   Clarson, S. J.; Semlyen, J. A., eds., Siloxane Polymers, PTR Prentice Hall, New York, 1993.

(4)   Saam, J. C.; Gordon, D. J.; Lindsey, S., Macromolecules **1970**, *3*, 1.

(5)   Nugay, N.; Kücükyavuz, Z.; Kücükyavuz, S., Polym. Int. **1993**, *32*, 93.

(6)   Lim, K. T.; Webber, S. E.; Johnston K. P., Macromolecules **1999**, *32*, 2811.

(7)   De Paz Bañez, Robinson, K. L., Armes, S. P., Macromolecules **2000**, *33*, 451.

(8)   Haddleton, D. M.; Jasieczek, C. B.; Hannon, M. J.; Shooter, A. J. *Macromolecules* **1997**, *30*, 2190.

(9)   Wang, J. S.; Matyjaszewski, K. *J. Amer. Chem. Soc.* **1995**, *117*, 5614.

(10)   Miller, P. J., Matyjaszewski K., Macromolecules **1999**, *32*, 8760.

(11)   Huan, K.; Bes, L.; Haddleton, D. M.; Khoshdel, E.; J. Pol. Sci., Polym. Chem., 2001, 39, 1853.

(12) Haddleton, D. M.; Crossman, M. C.; Dana, B. H.; Duncalf, D. J.; Heming, A. M.; Kukulj, D.; A. J. Shooter, Macromolecules **1999**, *32*, 2110.

(13) A. J. Milling, R. W. Richards, F. L. Baines, S. P. Armes, N. C. Billingham, Macromolecules, **2001**, 12, 4173.

(14) F. L. Baines, N. C. Billingham, S. P. Armes, Macromolecules **1996**, *29*, 3416.

(15) S. Antoun, J-F Gohy, R. Jerome, Polymer **2001**, *42*, 3641.

Chapter 23

Organic–Inorganic Hybrid Materials from Polysiloxanes and Polysilsesquioxanes Using Controlled/Living Radical Polymerization

Jeffrey Pyun, Jianhui Xia, and Krzysztof Matyjaszewski*

Center for Macromolecular Engineering, Department of Chemistry, 4400 Fifth Avenue, Pittsburgh, PA 15213

The synthesis of polysiloxane materials containing well-defined organic polymers is discussed using controlled/living radical polymerization techniques. In particular, the use of atom transfer radical polymerization (ATRP) enables the synthesis of a wide range of organic/inorganic hybrid materials utilizing polysiloxane macroinitiators, or macromonomers in the polymerization of organic vinyl monomers. Hybrid block and graft copolymers have been prepared containing poly(dimethylsiloxane) (pDMS) segments, or polyhedral oligomeric silsesquioxane (POSS) groups. Hybrid nanoparticles composed of an inorganic colloidal core and an outer shell of tethered organic polymers have been synthesized via ATRP of vinyl monomers from polysilsesquioxane nanoparticle surfaces.

The synthesis of organic/inorganic hybrid materials from polysiloxanes and organic polymers has been achieved from various techniques. However, the preparation of hybrid copolymers with both well-defined inorganic and organic segments remains a challenge. Methodologies for the synthesis of linear, cyclic (oligomeric) and network polysiloxanes are well established. Specifically, living ionic polymerizations have been used to make linear polysiloxanes, while cyclic oligomers and networks have been made from the hydrolysis and condensation of multifunctional alkoxysilanes.[1,2] By combining these methods with controlled/living polymerization of organic monomers, well-defined hybrids have been prepared. In particular, the use of siloxane based (macro)initiators and (macro)monomers in atom transfer radical polymerization (ATRP)[3-5] has enabled the synthesis of hybrid block/graft copolymers and functional colloids (Figure 1) [6,7]

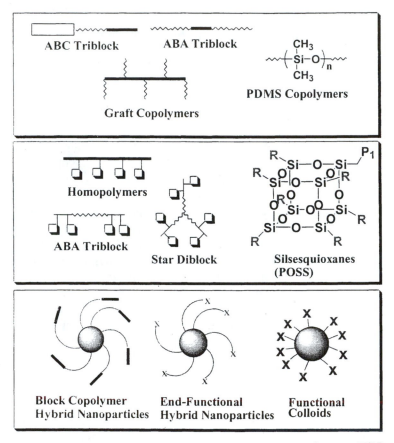

Figure 1: Hybrid materials from polysiloxanes prepared using ATRP

Results and Discussion

Poly(dimethylsiloxane) Block and Graft Copolymers from ATRP. The synthesis of block and graft copolymers from poly(dimethylsiloxane) (pDMS) was conducted using ATRP. Previously, we reported the synthesis of AB diblock, ABA triblock and ABC triblock copolymers.[8,9] In these materials, condensation or anionic polymerization was used to prepare the pDMS segments. ABA triblock copolymers of pSty-*b*-pDMS-*b*-pSty ($M_{n\ SEC}$ = 20;700 M_w/M_n = 1.6) were prepared from a difunctional pDMS ($M_{n\ SEC}$ = 9,800 M_w/M_n = 2.4) which was used to initiated the ATRP of styrene. For the ABC triblock copolymer, living anionic pSty chains ($M_{n\ SEC}$ = 4,600; M_w/M_n = 1.09) was chain extended with hexamethylcyclotrisiloxane and quenched with chlorodimethylsilane. The silane terminated pDMS-*b*-pSty ($M_{n\ SEC}$ = 7,760; M_w/M_n = 1.15) was then functionalized with alkyl halides (e.g., benzyl chloride, 2-bromoisobutyrate) by hydrosilation with a 2-bromoisobutyrate functional alkene. These copolymers were then used as macroinitiators in the ATRP of methyl methacrylate ($M_{n\ SEC}$ = 10,100 M_w/M_n = 1.21) and *n*-butyl acrylate ($M_{n\ SEC}$ = 10,200; M_w/M_n = 1.18). By the use of polysiloxane macroinitiators in the ATRP of vinyl monomers, well-defined block copolymers where prepared without the need for coupling techniques.

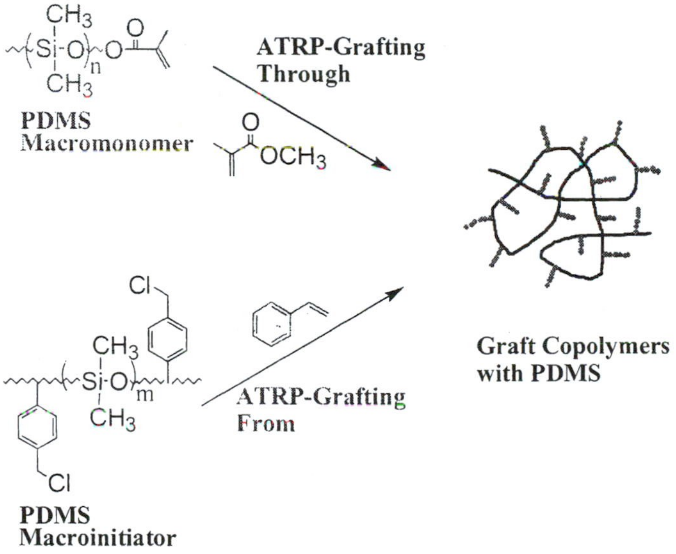

Scheme 1: Synthesis of poly(dimethylsiloxane) graft copolymers using ATRP

Graft copolymers were also prepared using ATRP using both "grafting from" and "grafting through" approaches (Scheme 1). In the former system, a copolymer of poly(dimethylsiloxane-*r*-methylvinylsiloxane) ($M_{n\ SEC}$ = 6,600; M_w/M_n = 1.76) was functionalized with benzyl chloride groups via hydrosilation. Subsequent ATRP of styrene yielded a graft copolymer with a soft pDMS backbone and hard pSty grafts ($M_{n\ SEC}$ = 14,800; M_w/M_n = 2.10).[8]

Alternatively, graft copolymers were prepared by the ATRP copolymerization of a methacrylate-functional pDMS macromonomer and methyl methacrylate (MMA).[10] Methacrylate pDMS macromonomers were prepared by the anionic ring-opening of hexamethylcyclotrisiloxane, followed by quenching of the lithium silanoate chaing end with 3-(methacryloxy)propylchlorodimethylsilane ($M_{n\ SEC}$ = 2,200; M_w/M_n = 1.18). ATRP of the pDMS macromonomer with MMA yielded a graft copolymer with a relatively uniform distribution of grafts throughout the copolymer. In this graft copolymer, the hard pMMA segments were in the backbone, while the soft pDMS segments are the pendant grafts. Figure 2 shows the Jaacks plot for the ATRP copolymerization of MMA and the pDMS macromonomer. As shown in the plot, the reactivity ratio of MMA using ATRP (r_{MMA} = 1.24) revealed that a steady consumption of the pDMS macromonomer occurred throughout the reaction, unlike in the conventional radical process where homopropagation of pMMA radicals to MMA dominated over the cross-propagation to the pDMS macromonomer (r_{MMA} = 3.07) .

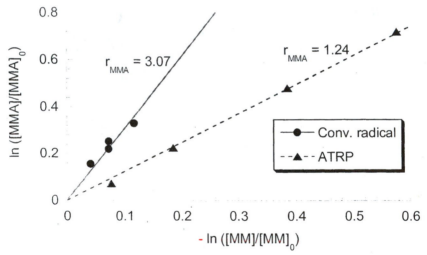

Figure 2. Jaacks plots for the copolymerizations of MMA with pDMS-macromonomer (M_n=2200, 5 mol % in the feed) in xylene solution (MMA/xylene=1/1 by weight) using the conventional radical polymerization (RP) and ATRP.[10] Conditions; RP: [MMA]₀/[macromonomer]₀/[AIBN]₀= 380/20/1, 75°C. ATRP: [MMA]₀/[macromonomer]₀/[initiator]₀/[CuCl]₀/ [dNbpy]₀= 285/15/1/1/2, 90°C.

(Co)polymers from Polyhedral Oligomeric Silsesquioxane (POSS) Monomers using ATRP. Another class of siloxane containing hybrid materials are copolymers prepared from silsesquioxane (POSS) monomers. POSS segments are cubic siloxane octamers prepared from the hydrolysis and condensation of functional trialkoxysilanes. The synthesis of linear homopolymers from methacrylate and styryl functional POSS monomers was conducted via conventional radical polymerization.[11] Despite the bulkiness of POSS monomers (corner to corner distance = 1.5 nm) radical homopolymerizations proceeded to high monomer conversions, yielding high molar mass polymers with broad molecular weight distributions. The presence of bulky POSS groups in polymeric systems has been found impart enhanced thermal and mechanical properties to these materials.[12,13]

Recently, a methacrylate functional POSS monomer (MA-POSS, cyclopentyl corner groups) was polymerized by ATRP to prepare homopolymers and ABA triblock, AB star diblock copolymers.[14] In these block copolymers, low T_g macroinitiators were used to prepare materials with both soft polyacrylate segments and hard p(MA-POSS) segments.

Homopolymers of p(MA-POSS) prepared by ATRP did not reach high degrees of polymerization (DP$_n$), as for the conventional radical polymerization systems. Despite modification of reaction conditions, p(MA-POSS) with DP$_{n\ NMR}$ < 15 were obtained. SEC of the p(MA-POSS) revealed that monomodal, narrow molecular weight distributions were obtained (M$_{n\ SEC}$ = 9,590; M$_w$/M$_n$ = 1.14. Figure 3) despite the low DP$_n$ of the polymer.

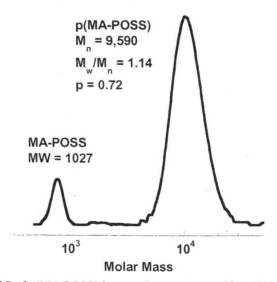

Figure 3. SEC of p(MA-POSS) homopolymer prepared by ATRP. Conditions for the ATRP were the following: [MA-POSS]:[EBiB]:[Cu(I)Cl]:[PMDETA] = 0.38 M: 0.019 M: 0.009 M: 0.009 M, 60 °C in o-xylene. Conversion of MA-POSS determined from ^1H NMR (p = 0.72).

The SEC and ^{1}H NMR analysis of the molar mass indicated that homopolymerization of MA-POSS proceeded until a limiting DP_n was reached. This observation can be explained by the steric bulk of the POSS groups which may bury alkyl halide end-groups from the ATRP catalyst preventing further polymerization after a certain molar mass of the polymer was achieved.

An additional advantage of controlled/living radical polymerization is the ability to prepare statistical copolymers while maintaining control of molar mass and macromolecular architecture. In the synthesis of POSS containing copolymers, this concept was demonstrated by the ATRP of methyl methacrylate (MMA), benzyl methacrylate (BzMA) and MA-POSS (cyclopentyl) using a tetrafunctional alkyl halide initiator (tetrakis(2-bromoisobutyryloxy)erythritol, TBiB). From this approach, a four armed star copolymer was prepared, possessing a random sequence distribution of MMA, BzMA and MA-POSS. In the polymerization a catalyst system of Cu(I)Cl and 2 equivalents of 4,4'-di-(5-nonyl)-2,2'-bipyridine was employed at 90°C, while monomer feed ratios were the following: f_{MMA} = 48-mol%, f_{BzMA} = 48%, $f_{MA-POSS}$ = 4%. The copolymerization by ATRP of these monomers proceeded to high monomer conversion as determined from ^{1}H NMR (p_{MMA} = 0.91; p_{BzMA} = 0.90; $p_{MA-POSS}$ = 0.70, Figure 4) and yielded well-defined star copolymers possessing low polydispersity ($M_{n\,SEC}$ = 26,970; M_w/M_n = 1.12).

The versatility of controlled/living radical polymerizations was also demonstrated by the synthesis of well-defined copolymers containing both POSS and carboxylic acid groups. Two approaches were taken in synthesis of these materials using either reversible addition chain transfer polymerization (RAFT),[15] or ATRP (Scheme 2). In the RAFT system, methacrylic acid (MAA, 55-mol%), *tert*-butyl methacrylate (BMA, 18-mol%), hexafluoroisopropyl acrylate (HFA, 18 mol%) and MA-POSS (cyclopentyl, 9-mol%) were directly polymerized in the presence of 2-phenylprop-2-yl dithiobenzoate (CDB) and AIBN to high conversion of each monomer (p_{MAA} = 0.70; p_{BMA} = 0.80; p_{HFA} = 0.95 ; $p_{MA-POSS}$ = 0.90; $M_{n\,SEC}$ = 15,000; M_w/M_n = 1.37).

In the preparation of POSS and carboxylic acid containing copolymers using ATRP, BzMA was copolymerized with MA-POSS and then deprotected via catalytic hydrogenation to impart acid functionality to the POSS copolymer. Unlike the RAFT approach, methacrylic acid could not be directly copolymerized by ATRP due to interactions of the monomer with the Cu(I)Cl/dNbpy catalyst system. Similar to the p(MMA-*r*-BzMA-*r*-(MA-POSS)) four-armed star copolymer (Figure 4), a tetrafunctional initiator (TBiB) was used in the copolymerization of BzMA (80-mol%) and MA-POSS(cyclopentyl, 20-mol%) by ATRP. Monomer conversions proceed to p_{BzMA} = 0.80 and $p_{MA-POSS}$ = 0.68 and a well-defined star copolymer was obtained ($M_{n\,SEC}$ = 23,840; M_w/M_n = 1.12). Benzyl groups were then removed from the star copolymer by

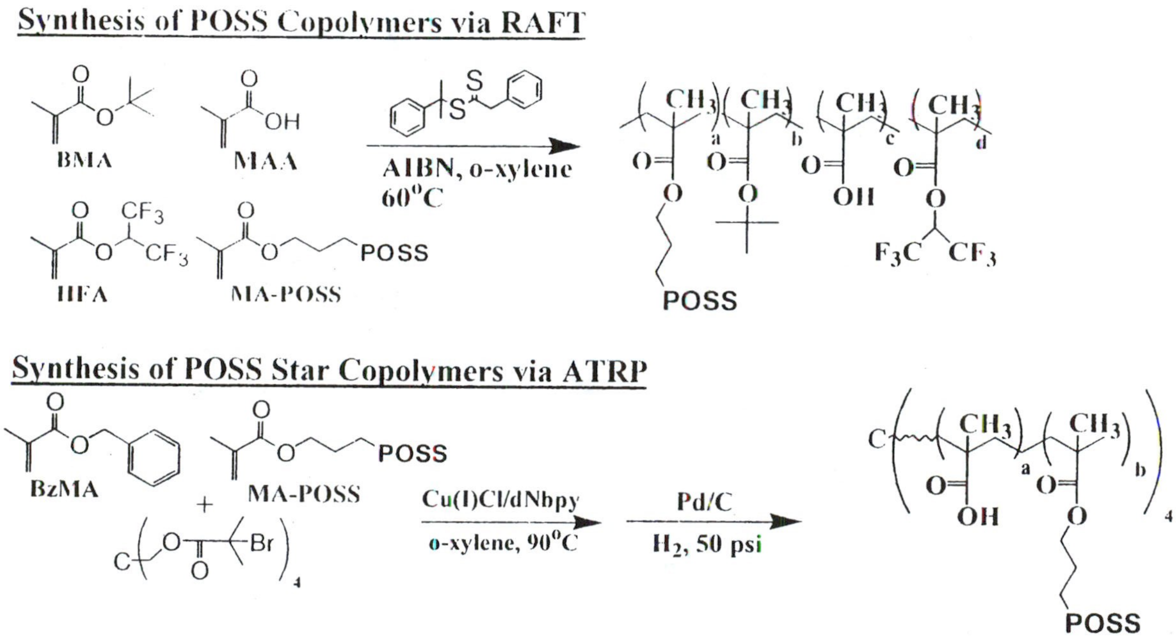

Synthesis of POSS Copolymers via RAFT

Synthesis of POSS Star Copolymers via ATRP

Scheme 2: Synthesis of POSS and carboxylic acid containing copolymers via RAFT and ATRP. (Top) RAFT copolymerization of MAA, BMA, HFA, MA-POSS. Conditions for RAFT were the following: [MAA]:[BMA]:[HFA]:[MA-POSS]:[CDB]:[AIBN] = 1.15 M: 0.37 M: 0.37 M: 0.19 M: 0.017 M: 0.008 M, 65°C in THF. (Bottom) ATRP copolymerization of BzMA and MA-POSS in the presence of tetrafunctional initiator.

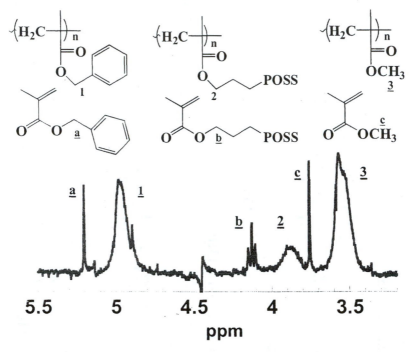

Figure 4. ^1H NMR of four-armed star of p(MMA-r-BzMA-r-(MA-POSS)) and residual monomers. Conditions for the ATRP terpolymerization were the following: [BzMA]:[MMA]:[MA-POSS]:[TBiB]:[Cu(I)Cl]:[dNbpy] = 0.72 M: 0.72 M: 0.036 M: 0.014 M: 0.0179 M: 0.035 M, 90 °C in o-xylene. Conversion were determined to be the following: $p_{MMA} = 0.91$; $p_{BzMA} = 0.90$; $p_{MA-POSS} = 0.70$. Assignments of resonances are the following: **1**, methylene protons from pBzMA, **a**, methylene protons from BzMA; **2**, methylene protons from pMA-POSS, **b**, methylene protons from MA-POSS; **3**, methyl protons from pMMA, **c**, methyl protons from MMA

catalytic hydrogenation in the presence of H_2 and palladium on carbon (10-wt% Pd). ^1H NMR analysis of the star copolymer after hydrogenation indicated that the deprotection reaction proceeded to high conversion (p = 0.99), as evidenced by the complete consumption of benzylic proton resonances from pBzMA (Figure 5) in the spectrum.

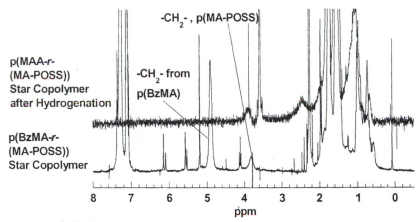

Figure 5: ^1H NMR of four-armed star copolymer of p(MAA-r-(MA-POSS)) synthesized using ATRP with the following conditions [BzMA]:[MA-POSS]:[TBiB]:[Cu(I)Cl]: [dNbpy] = 1.38 M: 0.34 M: 0.013 M: 0.017 M: 0.034 M, 90 °C in o-xylene. Conversion were determined to be the following: p_{BzMA} = 0.87; $p_{MA-POSS}$ = 0.68. Catalytic hydrogenation conducted in the presence of Pd/C (10-wt%) proceeded to high conversion as determined by consumption of pBzMA methylene protons at δ = 5.0

(Co)Polymers Tethered to Polysilsesquioxane Nanoparticles. The synthesis of hybrid materials was also done using colloidal initiators for ATRP reactions.[16,17] In these systems, spherical polysilsesquioxane networks were prepared via a microemulsion[17,18] process where 2-bromoisobutyrate groups were condensed onto the colloidal surfaces . Analysis of these functional colloids by light scattering (dynamic (DLS), static (SLS)) and atomic force microscopy (AFM) indicated that networks of high molar mass were formed (M_w = 4 x 10^6 g/mol) having nearly unifrom particle sizes ($D_{effective-DLS}$ = 27 nm; $D_{effective-AFM}$ = 19 nm, Figure 6).

These colloidal initiators where then used in the ATRP of styrene and benzyl acrylate to prepare hybrid nanoparticles with either homo-, or block

SiO$_{1.5}$ Colloidal Initiator: D$_{eff}$ = 19 nm, h$_{bearing}$ = 17   0.7 nm

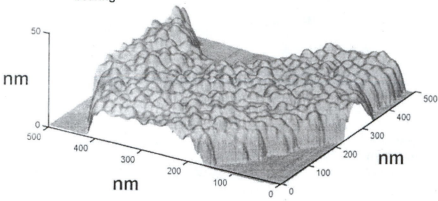

pSty hybrid nanoparticle: D$_{eff}$ = 27 nm, h$_{bearing}$ = 18.2   0.5 nm, pSty M$_n$ = 5,230; M$_w$/M$_n$ = 1.22

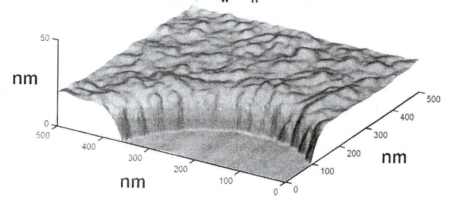

Figure 6: AFM height image of 2-bromoisobutyrate functional nanoparticles and polystyrene hybrid nanoparticles. Molar mass of pSty after HF core destruction, M_n = 5,230; M_w/M_n = 1.2.

copolymers tethered to the surface.   In the ATRP of styrene from the 2-bromoisobutyrate functional nanoparticles, DLS and AFM (Figure 6) measurements of the pSty hybrid particle effective diameter confirmed an increase in particle size ($D_{eff\ DLS}$ = 54 nm; $D_{eff\ AFM}$ = 27 nm) relative the colloidal initiator. The molar mass of pSty (M_n = 5,230; M_w/M_n = 1.22) tethered to the particle was determined after particle destruction by treatment with hydrofluoric acid (HF). Chain extension of the pSty hybrid particles was then conducted with benzyl acrylate to synthesize a material with both hard and soft segments tethered to a central particle. DLS and AFM again indicated that a further increase in particle size was observed ($D_{eff\ DLS}$ = 106 nm; $D_{eff\ AFM}$ = 56 nm). These block copolymer hybrid particles are anticipated to possess interesting viscoelastic properties and are currently the subject of future investigations.

Conclusions

The synthesis of organic/inorganic hybrid materials containing polysiloxanes has been successfully done using ATRP. By the functionalization of a polysiloxane with an activated alkyl halide or polymerizable group, these materials were able to be incorporated into the polymerization of organic vinyl monomers using ATRP.

Acknowledgements

The authors wish to thank Dr. J.J. Schwab (Hybrid Plastics) for supply of POSS monomers and Dr. S. Rubinsztajn (the General Electric Corporation) for supply of PDMS (co)polymers. The National Science Foundation (DMR-0131093) and the ATRP Industrial Consortium are also gratefully acknowledged for financial support.

References

(1) Novak, B. M. *Adv. Mater.* **1993**, *5*, 422-433.

(2) Wen, J.; Wilkes, G. L. *Chem. Mater.* **1996**, *8*, 1667-1681.

(3) Wang, J. S.; Matyjaszewski, K. *J. Am. Chem. Soc.* **1995**, *117*, 5614.

(4) Patten, T. E.; Xia, J.; Abernathy, T.; Matyjaszewski, K. *Science* **1996**, *272(5263)*, 866-868.

(5) Matyjaszewski, K.; Xia, J. *Chem. Rev.* **2001**, *010*, 2921-2990.

(6) MacLachlan, M. J.; Manners, I.; Ozin, G. A. *Adv. Mater.* **2000**, *12*, 675-681, and other reports in this issue.

(7) Pyun, J.; Matyjaszewski, K. *Chem. Mater.* **2001**, *13*, 3436-3448.

(8) Nakagawa, Y.; Miller, P. J.; Matyjaszewski, K. *Polymer* **1998**, *39*, 5163-5170.

(9) Miller, P. J.; Matyjaszewski, K. *Macromolecules* **1999**, *32*, 8760-8767.

(10) Shinoda, H.; Miller, P. J.; Matyjaszewski, K. *Macromolecules* **2001**, *34*, 3186-3194.

(11) Lichtenhan, J. D.; Yoshiko, A.; Otonari, M. J. C. *Macromolecules* **1995**, *28*, 8435.

(12) Romo-Uribe, A.; Mather, P. T.; Hadda, T. S.; Lichtenhan, J. D. *J. Polym. Sci., Part B: Polym. Phys.* **1998**, *36*, 1857.

(13) Mather, P. T.; Jeon, H. G.; Romo-Uribe, A.; Haddad, T. S.; Lichtenhan, J. D. *Macromolecules* **1999**, *32*, 1194.

(14) Pyun, J.; Matyjaszewski, K. *Macromolecules* **2000**, *33*, 217-220.

(15) Chiefari, J.; Chong. Y. K.; Ercole. F.; Krstina, J.; Jeffery, J.; Le, T. P. T.; Mayadunne. R. T. A.; Meijs, G. F.; Moad, C. L.; Moad, G.; Rizzardo, E.; Thang, S. H. *Macromolecules* **1998**, *31*, 5559-5562.

(16) von Werne, T.; Patten, T. E. *J. Am. Chem. Soc.* **1999**, *121*, 7409-7410.

(17) Pyun, J.; Matyjaszewski, K.; Kowalewski, T.; Savin, D.; Patterson, G.; Kickelbick, G.; Huesing, N. *J. Am. Chem. Soc.* **2001**, *123*, 9445-9446.

(18) Baumann, F.; Deubzer, B.; Geck, M.; Dauth, J.; Schmidt, M. *Adv. Mater* **1997**, *9*, 955.

Chapter 24

Preparation and Properties of Silicone–Urea Copolymers Doped with Cobalt(II) Chloride

E. Yilgör, M. Gördeslioğlu, B. Dizman, and I. Yilgör

Chemistry Department, Koç University, Sariyer 80910 Istanbul, Turkey

Incorporation of metal salts into polymeric materials leads to improvements in thermal, mechanical and electrical properties of the system. Depending on the structure and the oxidation state of the metal ion, these materials also display interesting optoelectronic properties. Silicone-urea copolymers containing various amounts of $CoCl_2.6H_2O$ were prepared by solution blending. The level of metal incorporation ranged from 10% to 100% on molar basis. Products obtained were characterized by infrared spectroscopy, thermal analyses and stress-strain tests. In order to understand the mechanism of complex formation, quantum mechanical calculations were carried out on model 1,3-dimethylurea/Co(II) systems. Blends of 1,3-dimethylurea and $CoCl_2.6H_2O$ were also prepared and analyzed by infrared spectroscopy.

Metal containing polymers, which have been studied since 1950s, received increasing attention during the last two decades, mainly due to need for high performance polymers for electrical, electronic, optoelectronic and space applications (1-5). This is due to unique set of properties that can be imparted to the polymeric systems through the incorporation of pure metal particles or metal salts. Depending on the type and nature of the polymer, structure and oxidation state of the metallic additive and its concentration, the polymer properties that can be modified and improved include higher surface or bulk conductivity, semiconductor properties, thermal and environmental stability, catalytic activity, electro optical and other physicochemical properties (1,2).

Metal containing polymers can be prepared by the direct polymerization of metal containing monomers or through the incorporation of metal particles or metal salts into preformed polymers via different techniques, such as, solution blending, melt compounding, plasma polymerization or mechanochemical synthesis (4). An interesting class of metal containing polymers is ionomers, which contain low levels of metal ions. These systems display highly improved mechanical properties, melt viscosities, overall durability and membrane performance (6). Two comprehensive reviews on the preparation, properties and applications of metal containing polymeric systems have recently been published (3,4). In addition, several new articles on metal salt containing block or segmented copolymers, including polyurethanes have also been reported (7-9). When polyether polyurethanes are doped with metal salts, there is a strong competition between hard (urethane) and soft (polyether) segments to form complex with the metal ion (10). Urethane groups may interact through (N-H), (C=O) or (C-O) groups, whereas ether interacts through the oxygen (-O-). This leads to the phase mixing and deterioration of the elastomeric properties of the system.

In this investigation polydimethylsiloxane-urea copolymers were used as the base resin and their interaction with cobalt(II) chloride was investigated. It is well known that polydimethylsiloxane (PDMS) has a very non-polar backbone and siloxane oxygen does not show strong interaction with other polar species (11). As a result, it is expected that the interaction between Co(II) ions and the polymer will be limited to urea hard segments only. In order to better understand the interaction between urea and Co(II) ions, 1,3-dimethylurea (DMU) was used as a model compound. In addition, quantum mechanical calculations were performed to determine the geometries of most stable complexes formed between DMU and Co(II) and their energy of formation. Theoretical vibrational spectrum of DMU and DMU-Co(II) complexes were also calculated and compared with experimental spectra.

EXPERIMENTAL

Materials

α,ω-Aminopropyl and N-methylaminopropyl terminated PDMS oligomers with number average molecular weights of 2500 g/mole were kindly provided by Th. Goldschmidt AG, Essen, Germany. Bis(4-isocyanatocyclohexyl)methane (HMDI) was a product of Bayer AG, Leverkusen, Germany. Reagent grade 1,3-dimethylurea (DMU) was obtained from Aldrich. $CoCl_2.6H_2O$ was purchased from Riedel de Haen and dehydrated under vacuum at 100°C for 24 hours before use. Complete removal of water was confirmed by FTIR spectroscopy. Reagent grade tetrahydrofuran (THF) was a product of Merck. All starting materials had purities better than 99.5% and were used as received.

Preparation of siloxane-urea (PSU) copolymer

Polymerization was carried out in a 500 mL, 3-neck round bottom reaction flask fitted with an overhead stirrer, nitrogen inlet and a dropping funnel. 16.90 g of HMDI (64.4 mmole) was weighed into the flask and was dissolved in 100 g of THF. 164.90 g of amine terminated PDMS (64.4 mmole) was dissolved in 100 g THF and introduced into the dropping funnel. PDMS solution was added onto HMDI solution dropwise, at room temperature. As the viscosity of the reaction mixture increased it was incrementally diluted with a total of 220 g more THF. Progress and the completion of the reaction were monitored by FTIR spectroscopy, following the disappearance of very strong isocyanate peak at 2270 cm^{-1}.

Preparation of metal salt complexes with PSU and DMU

Calculated amounts of silicone-urea copolymer, PSU, and $CoCl_2.6H_2O$ were weighed into separate flasks and were dissolved in THF to produce homogeneous solutions with concentrations around 5-10% by weight. Two solutions were then mixed together and a clear blue solution was obtained. A film was cast from this solution into a glass Petri dish. The solvent, THF was evaporated at room temperature overnight. The sample was then placed in a vacuum oven at 100°C and kept there until a constant weight is reached. Clear blue films thus obtained were kept in a desiccator until further characterization. A control PSU film was also prepared using the same procedure. DMU and $CoCl_2.6H_2O$ blends were also prepared following a similar procedure, except the drying temperature, which was kept at 50°C, to prevent the sublimation of DMU.

Characterization methods

FTIR spectra of the samples were obtained on a Nicolet Impact 400D spectrometer, with a resolution of 2 cm^{-1}, using thin films cast on KBr discs from THF solution. GPC curves were obtained on a Polymer Laboratories PL-110 GPC, equipped with PL-gel columns of 500, 1000 and 10000Å and a refractive index detector. Measurements were done at 23°C, with a flow rate of 1 mL/min. DSC thermograms were obtained on a Rheometric Scientific PL-DSC Plus, under dry nitrogen atmosphere, with a heating rate of 10°C/min. Thermogravimetric analysis (TGA) of the samples were obtained under nitrogen atmosphere, with a heating rate of 10°C/min, using a Shimadzu TGA-50H. Tensile behavior of the polymers were studied on an Instron Model 4411 tensile tester, at room temperature, with a crosshead speed of 2.0 cm/min. Dog-bone samples for tensile tests were cut from the films by using a standard ASTM die.

Quantum mechanical calculations

Theoretical calculations were performed within Density Functional Theory (DFT) using Gaussian98 (12). The geometry of monomers has been optimized within DFT approach utilizing the Becke three-parameter exchange-correlation-functional (B3LYP) (13) with the basis sets of 3-21g, 6-311g, 6-31g*, TZV and LanL2DZ.

RESULTS AND DISCUSSION

Silicone-urea copolymers display excellent combination of properties due to very good microphase separation between hard and soft segments and very strong hydrogen bonding between the urea hard segments (14). These include excellent low temperature flexibility, high tensile strength, good elastomeric properties, good oxidation, UV and heat resistance and very good processability by solution or melt techniques.

In this project we wanted to utilize several critical properties of siloxane-urea copolymers, which are; (i) complete phase separation between soft (PDMS) and hard (urea) phases (14,15), (ii) very strong hydrogen bonding capacities of urea groups, and (iii) acidic nature of the urea protons. As a result of these factors, it is expected that only urea groups will have interaction (or complex formation) with the transition metal ions incorporated into these copolymers, but not the PDMS backbone. In order to better understand the interaction mechanisms between urea groups and transition metal salts, we have also performed control experiments by using dimethylurea.

Transition metal salts may interact with urea groups in several different ways. One possibility is the formation of complexes through only electrostatic interactions between highly polar urea groups and transition metal salts. This may lead to the formation of clusters, size and geometry of which depend on the concentration of urea and metal salts. Another possibility is the formation of covalently bonded new compounds due to reaction between urea hydrogens and the transition metal chlorides, acetates, etc., with the release of respective acids.

Reactions between DMU and transition metal salts have been demonstrated (16). When aqueous solutions containing equimolar amounts of DMU and $CoCl_2.6H_2O$ or $Zn(CH_3COO)_2$ were heated to 80°C a sudden drop in the pH of the mixture was observed, clearly showing the release of HCl and CH_3COOH respectively (16). Completion of the reactions were indicated when the pH of the systems reached a constant value. At the end of each reaction formation of a precipitate was also observed from otherwise homogeneous solutions, clearly indicating the formation of new compounds.

Quantum mechanical calculations (QMC)

In order to understand the energetics of the model reactions between DMU and Co(II) chloride, chemical structures and vibrational spectra of the products formed, QMC were performed. These theoretical studies suggested the possibility of formation of two different products, which are shown in Figure 1. Energy of formation for product shown in Figure (1-a), the intermolecular product, was determined to be 351 kJ/mole, whereas for the intramolecular product (Figure 1-b), it was calculated to be 376 kJ/mole. N—Co bond distances for intermolecular (1-a) and intramolecular (1-b) compounds were calculated to be 1.91 and 1.86Å respectively, by QMC.

When the urea groups are a part of a polymer chain, as in the case of siloxane-urea copolymers, then intramolecular reactions are expected to lead to linear systems, whereas intermolecular reactions will result in branching and cross linking. As will be discussed later, all siloxane-urea copolymers containing Co(II) chloride were soluble in THF, indicating mainly intramolecular reactions.

FTIR studies

FTIR spectroscopy was also used to investigate the interaction between $CoCl_2$, DMU and PSU, where the changes in peak shapes and the peak positions for N–H and C=O stretchings were determined.

Pure DMU gives a fairly symmetrical and sharp N–H peak centered at 3341 cm^{-1}. As shown in Figure 2, N–H peaks become broader and shift to higher wavenumbers in DMU/$CoCl_2$ complexes with increasing Co(II) concentrations. DMU also has a very sharp, strongly hydrogen bonded C=O peak at 1624 cm^{-1} with a shoulder at 1588 cm^{-1}. Upon mixing with $CoCl_2$, the peak at 1624 cm^{-1} becomes weaker as the concentration of Co(II) in the complex increases. On the other hand, a new peak is formed at 1582 cm^{-1}. As shown in Figure 3, this new peak becomes stronger as the concentration of Co(II) in the complex increases.

Similar behavior was also observed for silicone-urea copolymers cured with varying amounts of $CoCl_2$. As shown in Figure 4, silicone-urea copolymer shows a sharp C=O peak at 1621 cm^{-1}, indicating very strong hydrogen bonding. As the system is doped with $CoCl_2$, C=O peak becomes weaker with increasing Co(II) concentration and for the system containing equimolar amounts of urea and Co(II) 1621 cm^{-1} peak disappears completely . Silicone-urea copolymer also shows a very sharp C–N peak at 1529 cm^{-1}. Upon doping with Co(II) this peak shifts slightly to higher wavenumbers, as clearly seen in Figure 4. In addition to these changes, in $CoCl_2$ doped systems a new peak at 1574 cm^{-1} forms and becomes stronger as the concentration of Co(II) in the system increases. These changes in FTIR spectra of the system clearly indicate strong interaction between urea groups and $CoCl_2$ as was also demonstrated by quantum calculations.

Stress-strain tests

Tensile properties of siloxanê-urea copolymers doped with Co(II) chloride were substantially higher than the base polymer. The improvement in both the tensile modulus and the ultimate tensile strength was directly proportional with the amount of metal salt incorporated as shown in Table 1. This was interesting because complex formation with Co(II) reduces the hydrogen bonding capacity of the urea groups, since urea protons are lost. However, as shown in Figure 1, if intermolecular complexes are formed an improvement in the tensile properties can be expected. Intramolecular complex may also improve the tensile properties due to the possibility of strong electrostatic interaction between Co(II) in the complex and other urea groups in the close vicinity, leading to clusters or physical crosslinking.

Thermal properties

DSC and thermogravimetric analysis were used to study the thermal properties of silicone-urea/$CoCl_2$ complexes. There was no noticeable

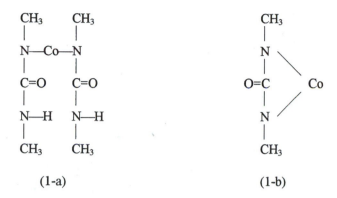

Figure 1. Possible structures of the compounds formed through reactions between DMU and Co(II) chloride, determined by QMC.

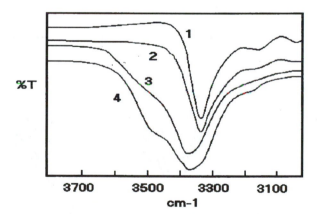

Figure 2. FTIR spectra of N–H region for DMU/CoCl$_2$ complexes with different molar compositions (1) DMU, (2) 10% CoCl$_2$, (3) 50% CoCl$_2$, (4) 100% CoCl$_2$

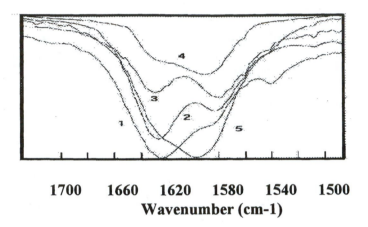

1700   1660   1620   1580   1540   1500
Wavenumber (cm-1)

Figure 3. FTIR spectra of C=O region for DMU/CoCl$_2$ complexes with different molar compositions (1) DMU, (2) 10%, (3) 25%, (4) 50%, (5) 100% CoCl$_2$

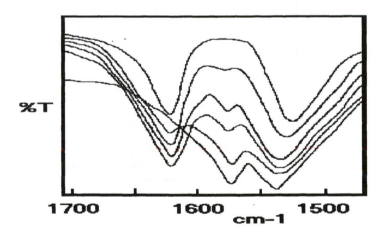

Figure 4. FTIR spectra of C=O region of PSU/CoCl$_2$ complexes with different compositions. From top to bottom: Pure DMU and Co(II)/DMU complexes with 10, 25, 50, 75 and 100% by mole of CoCl$_2$.

Table 1. Compositions and Tensile Properties of CoCl₂ Containing Silicone-urea Copolymers

Sample Description	CoCl₂ content (weight %)	Modulus (MPa)	Tensile Str. (MPa)	Elong. (%)
PSU-0	--	3.68	1.80	520
PSU-10 400	10.0	17.9		5.00
PSU-15 320	15.0	25.4		5.80

difference in the Tg of PDMS in the parent or the doped systems, which were determined to be –118°C. As expected, these results also prove that there is no interaction between Co(II) and PDMS. No high temperature transition was observed in the CoCl₂ doped silicone-urea systems. Difficulties in detecting a high temperature transition in silicone-urea copolymers, in DSC, have already been reported (15). This may be due to very small amount of urea groups present in these systems or due to very high melting points of urea groups, which is above the degradation temperature of the system. Incorporation of CoCl₂ into silicone-urea copolymers reduces the thermal stability of the system. TGA curves for silicone-urea copolymer and samples doped with 10 and 50 mole percent of CoCl₂ are given in Figure 5. While pure silicone-urea copolymer is thermally stable up to 275°C, Co(II) doped systems start degrading at around 225°C. This behavior is a direct result of Co(II) being a Lewis acid, which is an effective catalysts in cleaving the siloxane (Si-O) bonds in the PDMS backbone of these copolymers (17).

CONCLUSIONS

Interaction of Co(II) chloride with silicone-urea copolymers were investigated. Incorporation of transition metal salts into polymers may lead to improvements in mechanical, electrical and thermal properties of the system. Depending on the structure and the oxidation state of the metal ion, these materials may also display interesting optoelectronic properties (9). In this study, it has been demonstrated that protons on urea groups react with CoCl₂ through the release of HCl. Products obtained show improved mechanical properties. Our preliminary studies indicate that Co(II) doped silicone-urea

copolymers also show substantial improvements in the electrical conductivity of the system. Thermal stabilities of Co(II) containing silicone-urea copolymers display slightly reduced thermal stability when compared with the parent copolymer.

294

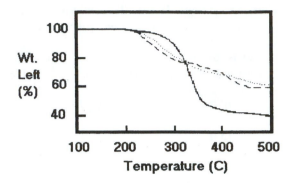

Figure 5. Thermogravimetric analysis of silicone-urea copolymer (——) and copolymer doped with 10% (········) and 50% (– – –) by mole of $CoCl_2$.

REFERENCES

1. Carraher, J. E. Jr.; Sheats, C. E.; Pittman, C. U. Jr. *Organometallic Polymers*, Academic Press, New York **1978**.
2. Carraher, J. E. Jr., Sheats, C. E.; Pittman, C. U. Jr. Eds., *Metal Containing Polymer Systems*, Plenum Press, New York **1985**.
3. Pomogailo, A. D.; Sarost'yanov, V. S. *Synthesis and Polymerization of Metal-Containing Monomers*, CRC Press, Boca Raton, Florida, **1994**.
4. Biswas M.; Mukherjee, A. *Adv. Polym. Sci*, **1994**, *115*, 89.
5. Symposium on *"Metal Containing Polymers"*, ACS Spring 2000 Meeting, San Francisco, Ca, *Polym. Prepr.*, **2000**, *41(1)*, 0000.
6. Eisenberg, A.; King, M. *Ion-containing Polymers*, Academic Press, New York, **1977**.
7. Wang, Y.; Goethals, E. J. *Macromolecules*, **2000**, *33*, 808.
8. Shen, Q.-D.; Chen, L.; Hu, T.-D.; Yang, C.-Z. *Macromolecules*, **1999**, *32*, 5878.
9. Bermudez, V. Z.; Ferreira, R. A.; Carlos, L. D.; Molina, C.; Dahmouche, K.; Ribeiro, S. J. L. *J. Phys. Chem.*, *B* **2001**, *105*, 3378.
10. Wang, H.-L.; Kao, H.-M.; Wen, T.-C. *Macromolecules*, **2000**, *33*, 6910.
11. Yilgor, E.; Burgaz, E.; Yurtsever, E.; Yilgor, I. *Polymer*, **2000**, *41(3)*, 849.

12. Frisch, M. J.; Trucks, G. W.; Schlegel, H. B.; Scuseria, G. E.; Robb, M. A.; Cheeseman, J. R.; Zakrzewski, V. G.; Montgomery, J. A. Jr.,; Stratmann, R. E.; Burant, J. C.; Dapprich, S.; Millam, J. M.; Daniels, A. D.; Kudin, K. N.; Strain, M. C.; Farkas, O.; Tomasi, J.; Barone, V.; Cossi, M.; Cammi, R.; Mennucci, B.; Pomelli, C.; Adamo, C.; Clifford, S.; Ochterski, J.; Petersson, G. A.; Ayala, P. Y.; Cui, Q.; Morokuma, K.; Malick, D. K.; Rabuck, A. D.; Raghavachari, K.; Foresman, J. B.; Cioslowski, J.; Ortiz, J. V.; Stefanov, B. B.; Liu, G.; Liashenko, A.; Piskorz, P.; Komaromi, I.; Gomperts, R.; Martin, R. L.; Fox, D. J.; Keith, T.; Al-Laham, M. A.; Peng, C. Y.; Nanayakkara, A.; Gonzalez, C.; Challacombe, M.; Gill, P. M. W.; Johnson, B.; Chen, W.; Wong, M. W.; Andres, J. L.; Gonzalez, C.; Head-Gordon, M.; Replogle, E. S.; Pople, J. A. *Gaussian 98*, Revision A.6, Gaussian, Inc., Pittsburgh PA, **1998**.
13. Becke, A. D. *J. Chem. Phys.*, **1993**, *98*, 564.
14. Yilgor, E.; Yilgor, I.; *Polymer*, **2001**, *42,* 7953.
15. Tyagi, D.; Yilgor, I.; McGrath, J. E.; Wilkes, G. L. *Polymer*, **1984**, *25*, 1807.
16. Yilgor, E.; Gordeslioglu, M.; Dizman, B.; Kaya, E.; Yilgor, I. *Polym. Prepr.,* **2001**, *42(1)*, 211.
17. Voronkov, M. G.; Mileshkevich, V. P.; Yuzhelevskii, Yu. A. *The Siloxane Bond*, Consultants Bureau, New York **1978**.

Chapter 25

Block and Graft Copolymers Containing Poly(dimethylsiloxane) and Poly(4-vinyl-pyridine) Segments by Free Radical Polymerization

Yongsin Kim[1], Daniel Graiver[2], Gary T. Decker[2], Fernando J. Hamilton[1], and H. James Harwood[1,*]

[1]Maurice Morton Institute of Polymer Science, The University of Akron, Akron, OH 44325-3909
[2]Dow Corning Corporation, 2200 West Salzburg Road, Midland, MI 48686–0994

Polysiloxanes with terminal or pendent aldehyde functionality can be used as components of copper salt-based redox initiation systems to prepare block, graft and crosslinked copolymers containing poly(4-vinylpyridine) and polysiloxane segments. The block and graft copolymers form very viscous solutions in dilute HCl and have useful anti-foam activity.

Introduction

Block and graft copolymers containing polysiloxane backbones and terminal or pendent hydrophilic polymer segments can be expected to have useful properties, particularly as antifoaming or defoaming agents, surfactants, wetting agents and hydrogels. This paper concerns polysiloxanes containing terminal or pendent poly(4-vinylpyridine) segments.

Previously, block copolymers containing polysiloxanes and poly(2-vinylpyridine) or poly(4-vinylpyridine) segments have been prepared using anionic polymerization techniques (*1-5*). Synthesis via controlled radical polymerizations of 2-vinylpyridine or 4-vinylpyridine initiated by poly(dimethylsiloxane) macroinitiators containing bis(silyl pinacolate) groups has also been reported (*6*).

We have previously reported that polysiloxanes with terminal or pendent aldehyde functionality can be prepared by ozonolysis of polysiloxanes containing terminal or pendent hexenyl groups and then used as macromolecular reducing agents for the synthesis of block and graft copolymers (*7, 8*). In this paper, we report the synthesis of both block and graft copolymers containing poly(dimethylsiloxane) and poly(4-vinylpyridine) segments by polymerizations in which polysiloxanes with terminal or pendent aldehyde groups serve as macroreductants for redox-initiated reactions.

Experimental

Polysiloxanes with Aldehyde Functionality

The preparation of aldehyde-functional polysiloxanes by ozonolysis of polysiloxanes containing hexenyl groups was described in previous patents, publications and abstracts (*7-10*). Table I provides information about the materials employed in this investigation. Polymers I-III contain terminal aldehyde groups and can be used to prepare block copolymers. Polymer IV contains two mole percent pendent aldehyde functionality and can be used to prepare graft or crosslinked copolymers.

4-Vinylpyridine Polymerization Procedure

All polymerizations were conducted in 1 ounce bottles under an argon atmosphere using benzene as a solvent to minimize chain transfer reactions. A typical reaction mixture contained 10 ml benzene that had been distilled from CaH_2, copper octanoate, triphenylphosphine, triethylamine, pyridine, 4-vinylpyridine and aldehyde-functional poly(dimethylsiloxane). The amounts of aldehyde-functional poly(dimethylsiloxane) and copper(II) salts were varied to increase the amount of polysiloxanes in the copolymers. Tables II-III provide recipes for the polymerization mixtures.

After being heated at 70°C for 21 hours, the reaction mixtures were cooled to room temperature and added to hexane or ether. The emulsions that formed were placed in a hood until the solvents had evaporated. The products were then dried for several days under vacuum at 55°C and ground to fine powders. They were then extracted with THF to remove catalyst components and residual polysiloxanes and were analyzed by ^{1}H-NMR. THF extractions were performed by mixing 0.5 g samples with 10 ml THF and by shaking constantly overnight. Then the supernatent THF layers were removed and the polymer precipitates were washed with additional THF three times. A control experiment in which aldehyde-functional polysiloxane was omitted from the formulation failed to yield polymer (#14-2). Homopolymers of 4-vinylpyridine were also prepared by initiation using AIBN (#14-1) or butyraldehyde (#14-4).

Table I. Aldehyde-Functional Poly(dimethylsiloxanes) Employed in This Study

Polysiloxanes Containing Aldehyde-Functional Groups at Chain Ends
(I, II and III)

$$\underset{\underset{\displaystyle CH_3\,CH_3\ CH_3}{|\quad|\quad|}}{\overset{\overset{\displaystyle O}{||}\qquad\quad CH_3\,CH_3\ CH_3\qquad\overset{\displaystyle O}{||}}{HC(CH_2)_4Si(OSi)_nOSi(CH_2)_4CH}}$$

Polysiloxanes Containing Aldehyde-Functional Groups at Branch Sites (IV)

$$\underset{\underset{\displaystyle O}{\overset{\displaystyle ||}{CH_3\ (CH_2)_4CH}}}{\overset{\overset{\displaystyle CH_3\quad CH_3}{|\qquad|}}{(CH_3)_3Si(OSi)_n(OSi)_mOSi(CH_3)_3}}$$

Polymer	n	Mn Calculated from n	Mn GPC	Mw/Mn
I	30	2,522	4,800	1.88
II	100	7,702	10,000	1.81
III	200	15,102	20,400	1.69
IV	50	3,306	3,800	2.12

Table II. Polymerization Recipes for Controls and Block Copolymers

Experiments	Controls					Polysiloxane-poly(4-vinylpyridine) block copolymers				
Sample #	14-1	14-2	14-4	14-3	17-3	83-3	19-1	19-2	19-3	19-4
Cu(II)Octanoate										
(g)	0	0.13	0.13	0.13	0.05	0.10	0.10	0.10	0.10	0.10
(mole,$\times 10^{-4}$)		3.58	3.58	3.58	1.43	2.86	2.86	2.86	2.86	2.86
Triphenylphosphine										
(g)	0	0.38	0.38	0.38	0.15	0.30	0.30	0.30	0.30	0.30
(mole,$\times 10^{-3}$)		1.43	1.43	1.43	0.57	1.14	1.14	1.14	1.14	1.14
Triethylamine										
(g)	0	0.13	0.13	0.13	0.05	0.10	0.10	0.10	0.10	0.10
(mole,$\times 10^{-3}$)		1.24	1.24	1.24	0.49	0.99	0.99	0.99	0.99	0.99
Pyridine										
(g)	0	0.63	0.63	0.63	0.25	0.50	0.50	0.50	0.50	0.50
(mole,$\times 10^{-3}$)		7.90	7.90	7.90	3.16	6.32	6.32	6.32	6.32	6.32
4-Vinylpyridine										
(g)	5.00	5.00	5.00	5.00	5.00	5.00	5.00	5.00	5.00	5.00
(mole,$\times 10^{-2}$)	4.76	4.76	4.76	4.76	4.76	4.76	4.76	4.76	4.76	4.76
Initiator	AIBN	none	butyral-dehyde	I	I	I	I	I	II	III
(g)	0.01		0.03	0.40	0.40	0.40	0.80	1.60	2.00	4.00
(mole,$\times 10^{-4}$)	0.48		3.61	1.59	1.59	1.59	3.17	6.34	2.60	2.65
Benzene (ml)	10	10	10	10	10	10	10	10	10	10

Table III. Polymerization Recipes for Graft Copolymers

Experiments	Polysiloxane-poly(4-vinylpyridine) graft copolymers						
Sample #	27-1	27-2	27-3	27-4	27-5	27-6	27-7
Cu(II) Octanoate							
(g)	0.10	0.05	0.10	0.15	0.20	0.40	0.80
(mole,x10^{-4})	2.86	1.43	2.86	4.29	5.71	11.4	22.9
Triphenylphosphine							
(g)	0.30	0.15	0.30	0.45	0.60	1.20	2.40
(mole,x10^{-3})	1.14	0.57	1.14	1.72	2.29	4.58	9.15
Triethylamine							
(g)	0.10	0.05	0.10	0.15	0.20	0.40	0.80
(mole,x10^{-3})	0.99	0.49	0.99	1.48	1.98	3.95	7.91
Pyridine							
(g)	0.50	0.25	0.50	0.75	1.00	2.00	3.00
(mole,x10^{-3})	6.32	3.16	6.32	9.49	12.7	25.3	38.0
4-Vinylpyridine							
(g)	5.00	1.25	1.25	1.25	1.25	1.25	1.25
(mole,x10^{-2})	4.76	1.19	1.19	1.19	1.19	1.19	1.19
Initiator	IV	IV	IV	IV	IV	IV	IV
(g)	0.66	0.32	0.64	0.97	1.28	2.56	5.09
(mole,x10^{-4})	1.74	0.84	1.68	2.55	3.37	6.74	1.34
Benzene (ml)	10	5	5	5	5	5	5

Solution Properties

To study solution properties of polysiloxane-poly(4-vinylpyridine) block or graft copolymers, the copolymers (0.1 g) were mixed in distilled water (5 ml) and 5% HCl was gradually added to the mixtures until gels formed. The addition of 5% HCl was continued until the polymer gels dissolved. The amounts of HCl added at the points of hydrogel formation and polymer dissolution were noted. Once polymer dissolution occurred, the reversion of their solutions to hydrogels was tested by adding NaOH solution.

Anti-foaming Experiments

To evaluate the potential use of polysiloxane-poly(4-vinylpyridine) block or graft copolymers as anti-foaming agents, the ability of the copolymers to break soap emulsions was tested. This was performed by mixing 0.5 ml – 1 ml of polymer solutions in 5% HCl, obtained as above, with 5 ml soap water (0.5 g Dial hand soap containing mostly sodium laurel sulfate dissolved in 500 ml distilled water) in test tubes and shaking them vigorously. The heights of foam formed were measured after five minutes and compared to those formed by soap water alone, by soap water containing poly(4-vinylpyridine) solution in 5% HCl and by soap water containing only 5% HCl.

Results and Discussion

Polysiloxane-Poly(4-vinylpyridine) Block Copolymers

Table IV provides information about 4-vinylpyridine polymerizations that were initiated by Polymers I, II and III in combination with copper octanoate, triphenylphosphine, triethylamine and pyridine. Included in the table are yield, % conversion, the molecular weights that can be calculated from the 4-vinylpyridine unit/siloxane unit ratios determined by NMR (Mn) along with wt. % of polysiloxane in the copolymers and the percent polysiloxane incorporated in the copolymer.

In efforts to isolate the copolymers by pouring the polymerization mixtures into hexane or ether, emulsions formed that had to be concentrated to dryness to obtain the polymers. These were then purified by extraction with THF. Figure 1 shows the ^{1}H-NMR spectrum of polymer 19-3 after purification. Typically, the polysiloxane-poly(4-vinylpyridine) block copolymers contained 2-10 wt. % polysiloxane segments and the peak molecular weights of the polymers obtained

Table IV. Block Copolymers Containing Polysiloxane and Poly(4-vinylpyridine) Segments

Sample #	14-3	17-3	83-3	19-1	19-2	19-3	19-4
Starting polysiloxane	I	I	I	I	I	II	III
Polysiloxane (g)	0.4	0.4	0.4	0.8	1.6	2.0	4.0
4-Vinylpyridine (g)	5.0	5.0	5.0	5.0	5.0	5.0	5.0
Copolymer yield (g)	4.2	4.3	4.4	4.4	5.0	4.3	5.4
VP conversion (%)	77.8	79.6	81.5	75.9	75.8	61.4	59.6
M_n copolymer (^{1}H-NMR)	98K	115K	115K	85K	70K	107K	148K
Wt. % polysiloxane in copolymer	2.6	2.2	2.2	3.0	3.6	7.2	10.2
Polysiloxane incorporation (%)	26.9	23.5	24.2	16.3	11.3	15.5	13.8
M_p copolymer (GPC)	---	383K	429K	526K	402K	---	---

by GPC, using DMF as the solvent, M_p, were approximately 4xMn. This suggests that the copolymers have multiblock structures similar to the polysiloxane-polystyrene block copolymers previously obtained with the same initiating system. Such structures are believed to result from termination of growing poly(4-vinylpyridine) segment radicals by combination.

Polysiloxane-Poly(4-vinylpyridine) Graft Copolymers

Table V lists the results of 4-vinylpyridine grafting onto Polymer IV. The products obtained were only partially soluble in chloroform and were mostly swollen gels. The gel contents of the copolymers increased as the amount of polysiloxane used in the polymerizations increased (27-1 < 27-5 << 27-6 << 27-7). The strong tendency of propagating poly(4-vinylpyridine) radicals to terminate by combination is believed to be responsible for the crosslinking reaction that occurs during these polymerizations. NMR analysis of swollen polymer gels allowed us to calculate the Mn's of the PVP segments from the 4-vinylpyridine unit/siloxane unit ratios. The polysiloxane contents of the copolymers were in the range of 2-12 wt. % and increased with the amount of Polymer IV employed.

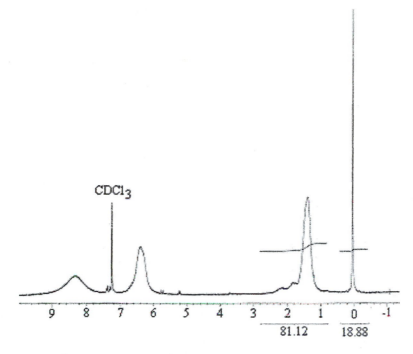

CDCl₃

81.12   18.88

Figure 1. ¹H-NMR Spectrum of Purified Polysiloxane-Poly(4-vinylpyridine) Block Copolymer (19-3) in CDCl₃.

Solution Properties

Polysiloxane-poly(4-vinylpyridine) block copolymers (#19-2, 19-3 and 19-4) were tested for solubility in dilute HCl solution. Solutions of copolymers (2 wt %) in 0.1% HCl were very viscous and gel-like but became milky solutions of very low viscosity in 5% HCl. When base was added to one polymer solution (#19-3), the polymer reprecipitated. The NMR spectrum of the recovered polymer was the same as that of the original polymer.

Anti-foaming Properties

Table VI compares the relative heights of the foams present five minutes after soap solutions containing small amounts of the polymer solutions in 5% HCl were vigorously shaken. The results indicate that the block and graft copolymers have useful anti- or de-foaming properties and that their effectiveness depends on both the sizes of the poly(4-vinylpyridine) and polysiloxane segments and on whether they have block or graft architectures.

Table V. Graft Copolymers Containing Polysiloxane and Poly(4-vinylpyridine) Segments

Sample #	27-1	27-2	27-3	27-4	27-5	27-6	27-7
Starting polysiloxane	IV	IV	IV	IV	IV	IV	IV
Polysiloxane (g)	0.66	0.33	0.64	0.97	1.28	2.56	5.09
4-Vinylpyridine (g)	5.0	1.25	1.25	1.25	1.25	1.25	1.25
Copolymer yield (g)	5.3	1.3	1.4	1.3	1.8	1.6	2.0
VP conversion (%)	93.6	80.6	72.5	56.8	72.3	42.5	31.5
Mn of PVP segment (^{1}H-NMR)	187K	99K	51.7K	36.6K	45.2K	33.2K	28.4K
Wt. % polysiloxane in copolymer	2.0	3.7	6.9	9.4	7.8	10.3	11.8
Polysiloxane incorporation (%)	16.0	14.6	15.0	12.6	10.9	6.4	4.6

Conclusions

Polysiloxanes with pendent aldehyde functional groups can be used to synthesize block and graft copolymers bearing poly(4-vinylpyridine) segments. The copolymers appear to have useful anti-foam activity.

Table VI. Relative Foam Heights After Vigorous Shaking

Copolymer	Polysiloxane Segment Type	Relative Foam Height (%)
None	-----	50*
14-1	None	12
19-2	I	6
19-3	II	12
19-4	III	0
27-4	IV	0

*100x Foam Height/Combined Height of Foam and Solution

References

1. Dean, J.W. U.S. Patent 3,673,272, 1972.
2. Dean, J.W. U.S. Patent 3,875,254, 1975.
3. Lee, J.A.; Hogen-Esch, T.E. *Polymer Preprints* **1993**, *34(1)*, 556.
4. Lee, J.A.; Hogen-Esch, T.E. *Polymer Preprints* **1996**, *37(1)*, 591.

5. Nugay, N.; Kucukyavuz, Z.; Kucukyavuz, S. *Polym. Internat.* **1993,** *32,* 93.
6. Crivello, J.V.; Lee, J.L.; Conlon, D.A. *J. Polym. Sci. Part A Polym. Chem.* **1986,** *24,* 1251.
7. Graiver, D.; Nguyen, B.; Hamilton, F.J.; Kim, Y.; Harwood, H.J. In *Silicones and Silicon-Modified Materials;* Clarson, S.J.; Fitzgerald, J.J.; Owen, M.J.; Smith, S.D., Eds.; ACS Symp. Series 2000; American Chemical Society, 2000; Vol. 729, Chapter 30.
8. Graiver, D.; Decker, G.T.; Tselepsis, A.J.; Hamilton, F.J.; Harwood, H.J. *Polymer Preprints* **1999,** *40(2),* 146.
9. Graiver, D.; Khieu, A.Q.; Nguyen, B.T. U.S. Patent 5,739,246, 1998.
10. Graiver, D.; Khieu, A.Q.; Nguyen, B.T. U.S. Patent 5,789,516, 1998.

Chapter 26

Poly(dimethylsiloxane)-*graft*-oligo(hexafluoropropeneoxide) Via Ring-Opening Polymerization

Mark A. Buese, Jose F. Gonzalez, Curtis G. Harbaugh,
Garrett M. Stearman, and Michael S. Williams

Clariant LSM (Florida) Inc., P.O. Box 1466, Gainesville, FL 32602

Cyclotetrasiloxanes with an oligo(hexafluoropropeneoxide)
substituent were prepared and ring-opening polymerized. The
synthesis of the cyclosiloxanes with three different linkages
between the siloxane and fluoroether is described. The
cationic and anionic polymerizations were examined. The
equilibrium cyclosiloxanes were extracted from the polymer
and characterized. The copolymerization with a polycyclo-
siloxane resulted in the formation of cross-linked networks.

Various fluorosilicone homopolymers and copolymers are quite useful as
lubricating oils for severe environmental conditions, crude-oil antifoams,
magnetic media lubricants, pressure-sensitive adhesive release-liners, selective
oxygen permeable membranes, very low temperature elastomers, low refractive
index elastomers, and oil and soil resistant rubbers, sealants, and coatings.
Polytrifluoropropylmethylsiloxane, PTFPMS, is the most widely available
fluorosilicone and is used in applications where resistance to hydrocarbons and
other organic compounds is required. Although it has a lower affinity for

hydrocarbons, PTFPMS displays a higher liquid surface tension than PDMS (*1*). Surface properties are determined primarily by the composition and orientation of the atoms at the surface layer. Although the CF_3 group intrinsically donates the lowest surface energy of any fragment, it cannot dominate the surface of PTFPMS and cannot overwhelm the contribution from the ethylene link between the CF_3 and siloxane backbone (*2*). To achieve a fluorosilicone with lower surface tensions than PDMS, the polymer must have a high density of aliphatic fluorine-containing groups. The group must not affect the backbone flexibility of the polysiloxane in a detrimental manner and the linking group between the fluorocarbon and silicone must be sufficiently removed from the surface such that its contribution is negligible (*3*). This requires a larger fluoroorgano substituent on the siloxane unit.

The polymerization of cyclic oligomers is the primary route to the majority of silicone polymers and copolymers. The polymerization of octamethylcyclotetrasiloxane, D_4, to polydimethylsiloxane, PMDS, and other unstrained cyclosiloxanes, those larger than the trimer, with relatively small substituents at silicone has been extensively documented (*4*). The process results in a mixture of linear and cyclic structures. The proportions of these structures are relatively easily predicted from statistical considerations where the linear homologues displaying a Flory-Schultz distribution (*5*) and the cyclic homologues display a modified Jacobson-Stockmayer distribution (*6*). The experimentally observed equilibrium cyclic oligomer concentrations for the cyclic tetramer through hexamer are higher than those predicted by Jacobson-Stockmeyer theory, and the cyclic octamer through approximately the 21-mer are lower than predicted (*7,8,9*). The larger the substituents on the silicon, the more pronounced is the deviation from the theoretical equilibrium cyclization constants for any given siloxane homopolymer (*10*).

The equilibrium constants for the cyclic tetramer, pentamer, and hexamer of PTFPMS are high, with the tetramer and pentamer nearly an order of magnitude greater than that of their counterpart in equilibrium with PDMS. The summation of the product of the equilibrium constants for the tetramer through pentadecamer and their molecular weight indicate that linear polymer cannot exceed 26% by weight in an attempt to prepare high polymer (*10*). The critical volume concentration for PTFPSO is approximately 10%, which virtually excludes typical condensation polymerizations such as the hydrolysis of dichlorosilanes, as the leaving group is a sufficient diluent to preclude linear polymer formation (*4*). For these reasons PTFPS is prepared by kinetically controlled ring-opening polymerization of the strained cyclic trimer. Even under these conditions, if the catalyst is not destroyed immediately upon the attainment of high polymer, depropogation to a mixture dominated by cyclic oligomers results (*11*). The ring-opening polymerization of unstrained cycloperfluoroorganosiloxane with large perfluoroalkyl substituents is even less successful than the polymerization of the PTFPS cyclic tetramer, as implied during the condensation of 1H,1H,2H,2H-perfluorooctylmethyldichlorosilane with water (*12*).

The use of cyclic trimers has been extended to the preparation of dimethylsiloxane copolymers where the ring contains a single silicon with one or two large perfluoroorgano substituents (*13,14*). In both cases high polymers were prepared by cationic and anionic polymerization. The patent literature describes a trimer containing one to three fluoroalkoxyalkyl groups (*15*) and a single hexafluoropropylene oxide oligomer attached to the ring (*16,17*). They were claimed to be polymerizable. The preparations of the cyclic trimers were not trivial, requiring the use of solvents and tetramethyldisiloxane-1,3-diol.

Polytrifluoropropylmethyl-siloxane-co-dimethylsiloxane has been prepared from cyclic tetramers where the weight of linear polymer prepared in solution increased from 0 to 31 to 45% as the proportion of trifluoropropylmethylsiloxy units decreased from 100 to 50 to 25% by mole (*18*). The preparation of copolymers of larger perfluoroalkylsiloxane and dimethylsiloxane via copolymerization of octamethylcyclotetrasiloxane and unstrained perfluoro-alkylmethylsiloxane cyclic oligomers has resulted in commercialized fluids (*12*). The weight fraction of perfluoroalkyl groups can be increased without significantly increasing the magnitude of the equilibrium constants for the cyclosiloxanes from that of those from dimethylsiloxane by copolymerization.

This report describes the preparation and ring-opening polymerization of an oligo(oxyperfluoropropylene)heptamethylcyclotetrasiloxane, using acidic and basic initiators, to poly(dimethylsiloxane)-*graft*-oligo(oxyperfluoropropylene) and to networks. The properties of polyfluoroalkylpolyethers are complimentary to properties of silicones desirable for many applications such as high shear and thermal stability, nonflammability, and high compressability. They have the potential to significantly enhance some properties of silicones, such as low surface energy and low refractive index. Perhaps most importantly, they have the potential to significantly improve some of the shortcomings of silicones, such as their solubility in common organic solvents. The cyclotetrasiloxanes had a single large oligomer of hexafluoropropeneoxide attached via a propylamide or a propylester linkage. This route was chosen as all the starting materials for the cyclic tetramer are commercially available and the intermediates could be prepared in high yield without the use of solvents. The polymerization of the tetramer resulted primarily in a linear polymer and cyclosiloxanes. The cyclosiloxanes could be extracted with hydrocarbons. The extracted cyclosiloxanes could be polymerized in like manner to the starting tetramer, resulting in essentially the same polymers. In this manner the effective yield of these copolymer can be quite high.

Experimental

All chemicals were purchased from Lancaster Synthesis with the exception of methylallylamine, which was purchased from Aldrich. Heptamethylcyclo-tetrasiloxane, **1**, was prepared as previously described (*19*). 1,3,5,7-Tetra-(2-heptamethylsiloxane-ylethyl)-1,3,5,7-tetramethylcyclotetrasiloxane, **9,** was prepared and isolated as previously described (*20*).

Gas Chromatography was carried out using a Hewlett-Packard 5890 Series II Chromatograph with a thermal conductivity detector and a SPB-1 30 m x0.53 mm capillary column with a 0.5 mm thick film. IR spectra were recorded using a Nicolet 5PC FTIR Spectrometer. NMR spectra were recorded using a JEOL JNM GX400 FT NMR Spectrometer.

Synthesis of [3-(perfluoro-2,5,8-trimethyl-3,6,9-trioxadodecan-oyl)-oxy-propyl]heptamethylcyclotetrasiloxane

As illustrated in Figure 1, a 1L round bottom flask was charged with 500 g of **1** and was heated to 80°C and 200 µL of Pt 1,3-di-vinyltetramethyldisiloxane complex in xylene (3% Pt) was added. The mixture was stirred and 125 g of allyl alcohol was added dropwise. The reaction was exothermic. Addition was carried out at 90-120°C over a period of 2 hours. A GC trace indicated the presence of (3-hydroxy)heptamethylcyclotetrasiloxane, **2**, at 84%. Distillation at 70°C and 0.2 mmHg gave 450g (75% yield) with 98% purity by GC analysis.

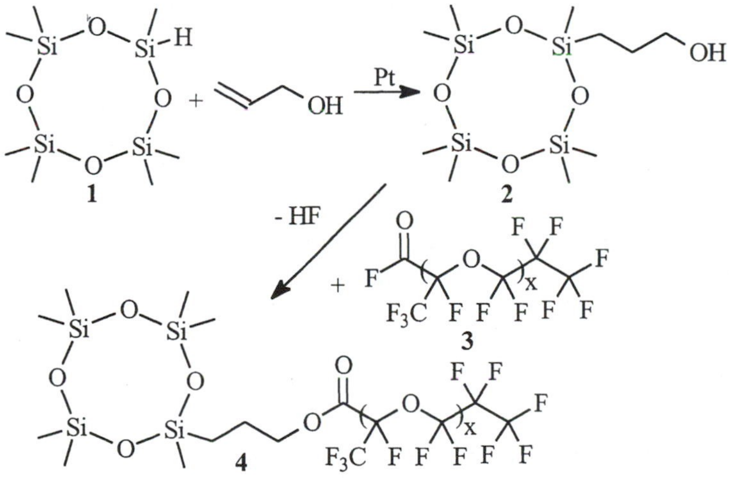

Figure 1. Preparation of [3-(perfluoro-2,5,8-trimethyl-3,6,9-trioxadodecan-oyl)-oxy-propyl]heptamethylcyclotetrasiloxane where x=3

As illustrated in Figure 1, a 50 mL round bottom flask was charged with 7.7 g of **2** and 2.3 g of triethylamine. The mixture was stirred and 15.0 g of 97% perfluoro-2,5,8-trimethyl-3,6,9-trioxadodecanoyl fluoride, **3** where x = 3, was added dropwise. A GC trace indicated that all of **3** and most of **2** were consumed with predominately the formation of [3-(perfluoro-2,5,8-trimethyl-3,6,9-trioxadodecan-oyl)-oxypropyl]heptamethylcyclotetrasiloxane, **4** where x = 3. The product was washed once with dilute HCl, and twice with water. The crude product was distilled at 110°C and 0.3 mmHg resulting in 14.7 g of **4** where x = 3 (66%) which was 97% pure by GC analysis. 1 H NMR 400 MHz, CDCl3 δ 0.1(s 21 H), 0.6 (t 2 H), 1.8 (p 2 H), 4.4 (m 2 H); 19 F NMR 376 MHz, CDCl3: δ -81 (CF3). -82 (CF3), -83 (CF3), -85 (CF3), -130 (CF2), -132 (CF2), -144 (CF); IR (neat liquid on NaCl): cm^{-1}: 2970 (m), 1780 (s), 1245 (vs), 1200 (s), 1145 (s), 1070 (vs), 995 (s), 970 (s), 810 (vs) 750 (m). In like manner **4** where x = 4, 5, 6, 7, and x averages 8 were synthesized.

Synthesis of *N*-(3-heptamethylcyclotetrasiloxan-yl)-propylperfluoro-2,5,8,-trimethyl-3,6,9,-trioxadodecanamide

As illustrated in Figure 2, a 100 mL round bottom flask equipped with a magnetic stirring bar and an addition funnel was charged with 9.0 g of allyl amine. The liquid was stirred and 40 g of 97% **3** where R = H, x = 3 was added dropwise. The addition of dilute HCl resulted in two phases. The amide layer was washed with dilute HCl and twice with water. Distillation at 101°C and 3 mmHg yielded 21.6 g (52%) of >99% *N*-allyl-perfluoro-2,5,8,-trimethyl-3,6,9,-trioxadodecanamide, **5** where R = H, x = 3.

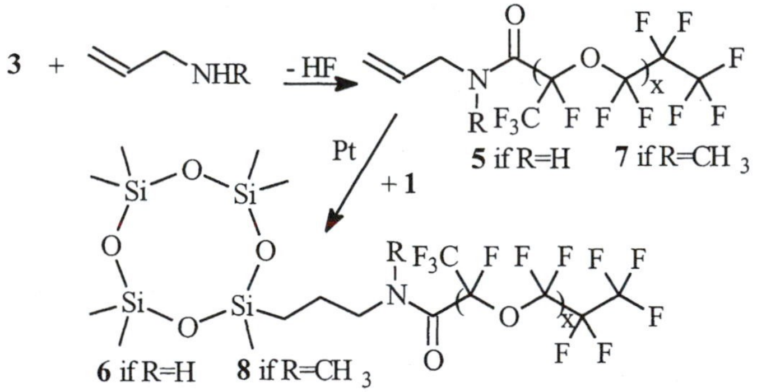

Figure 2 Preparation of N-(3-heptamethylcyclotetrasiloxan-yl)-propylperfluoro-2,5,8,-trimethyl-3,6,9,-trioxadodecanamide where R=H and x=3 or N-methyl-N-(3-heptamethylcyclotetrasiloxan-yl)-propylperfluoro-2,5,8,-trimethyl-3,6,9,-trioxadodecanamide where R=Me and x=3

As illustrated in Figure 2, a 3-necked 25 mL round bottom flask was equipped with a magnetic stirring bar, a condenser, and a temperature probe. The flask was charged with 10.0 g of >99% **5** and 4.0 g of **1**. The mixture was heated to 100°C and 10 μL of a Pt 1,3-di-vinyltetramethyldisiloxane complex in xylene (3% Pt) was added. A very exothermic reaction occurred. A gas chromatographic analysis indicated a high conversion to a single product. Distillation at 116-20°C and 0.04 mmHg resulted in 5.7 g (43% yield) of >99% *N*-(3-heptamethylcyclotetrasiloxan-yl)-propyl-perfluoro-2,5,8,-trimethyl-3,6,9,-trioxadodecanamide **6** where R = H, x = 3. ^{1}H NMR 400 MHz, CDCl$_3$: δ 0.1 (m 21 H), 0.6 (t 2 H), 1.7 (p 2 H), 3.4 (m 2 H), 6.7 (s 1 H); IR (neat liquid on NaCl): cm^{-1}: 3460 (m), 2970 (m), 1705 (s), 1550 (m), 1310 (m), 1245 (vs), 1200 (s), 1150 (s), 1070 (vs), 995 (s), 970 (s), 810 (vs), 750 (m). In like manner runs using **6** where x = 4, 5, and 6 were synthesized.

Synthesis of *N*-methyl-*N*-(3-heptamethylcyclotetrasiloxan-yl)-propylper-fluoro-2,5,8,-trimethyl-3,6,9,-trioxadodecanamide

As illustrated in Figure 3, in an equivalent manner to the preparation of **5**, 5.0 g of *N*-methylallylamine and 17.5 g of 97% **3** reacted and were distilled at 98°C and 2.4 mmHg to give 14.3 g (77% yield) of >99% *N*-methyl-*N*-allyl-perfluoro-2,5,8,-trimethyl-3,6,9,-trioxa-dodecanamide, **7** where R = Me, x = 3.

As illustrated in Figure 3, in an equivalent manner to the preparation of **6**, 7.0 g of of >99% **7** and 2.8 g of **1** reacted in the presence Pt 1,3-di-vinyltetra-methyldisiloxane complex and were distilled at 100-5°C and 0.04 mmHg to yield a 5.2 g (53% yield) of >99% *N*-methyl-*N*-(3-heptamethylcyclotetrasiloxan-yl)-propyl-perfluoro-2,5,8,-trimethyl-3,6,9,-trioxa-dodecanamide, **8** where R = Me, x = 3. The product was analyzed by spectroscopy with the following results: ^{1}H NMR 400 MHz, CDCl$_3$: δ 0.1 (m 21 H), 0.5 (m 2 H), 1.7 (m 2 H), 3.1 (m 3 H), 3.5 (m 2 H)); IR (neat liquid on NaCl): cm^{-1}: 2970 (m), 1685 (s), 1410 (w), 1300 (m), 1245 (vs), 1200 (s), 1140 (m), 1080 (vs), 995 (s), 970 (s), 810 (vs), 750 (m). In like manner syntheses using **8** where x = 4, 5, 6, 7, and with mixtures where x averages 7.3 were carried out.

Cationic Polymerization

As illustrated in Figure 4, a 1.5 dram vial containing 1.0 g of **4** where x = 3, was injected 2.0 μL of trifluoromethanesulfonic acid and the mixture was shaken to form a polymer. The vial was warmed for 10 minutes. A viscous oil resulted upon cooling to room temperature. The oil was warmed and allowed to stand for 4 hours with little or no apparent change. To the vial was added 0.01 g of MgO and the oil was warmed to facilitate dispersion of the salt. Upon cooling, 1 mL of pentane was added to the vial and the mixture shaken. Two liquid phases and a solid phase were apparent. The liquid layers were filtered using a 3 mL syringe equipped with a 0.45 micron filter. A series of peaks was observed by GC analysis of the freshly formed suspension. After separation, the upper layer was removed and placed in a vial. The polymer was extracted with additional pentane. Evaporation of the pentane from the extract resulted in a liquid residue of 0.2 g. The GC trace of this residue displayed the same signals for siloxanes observed for the polymer suspension. The copolymer layer was heated with a stream of N_2 to remove the pentane. The resulting 0.6 g of a gum displayed no flow. A small portion was shaken with pentane and the suspension immediately analyzed by GC. Only a peak for pentane was observed. Similar results were observed for all of the polymers.

In like manner, polymers were prepared from 1.0 g of **6** where x = 3 and 1.0 g of **8** where x = 3 by the injection of 2.0 µL of trifluoromethanesulfonic acid. A viscous linear-cyclic mixture, similar to that of **4**, resulted in the case of **8** and the mixture from **6** was much more viscous than the fluid from **4**.   The polymerizations of **8** where x = 4 and **8** where x = 7.3 were examined.

Anionic Polymerization

As illustrated in Figure 3, a 1.5 dram vial containing 2.0 g of **6** where x = 3.4 was injected with 19.0 µL of 1M tetrabutylammonium fluoride in tetrahydrofuran and the mixture shaken to form a polymer. The vial was warmed gently over 10 minutes and upon cooling increased in viscosity. The heating was repeated until, upon cooling, it resulted in a heavy oil that displayed almost no flow when the vial was inverted at room temperature. No apparent increase in viscosity was observed upon subsequent heating. After 24 hours the vial was heated strongly with the formation of bubbles, presumably from the decomposition of the tetrabutylammonium salt. A linear-cyclic mixture similar to that observed in the cationic polymerization was observed by GC analysis.

In like manner, polymers were prepared from **8** where x = 4 and **8** where x = 7.3 by the addition of tetrabutylammonium fluoride solution. Viscous linear-cyclic mixtures resulted but the were of notably lower viscosity than that from **6** and lower in viscosity than those from cationic polymerization. The miscibility of these polymers in acetone, cyclohexane, and toluene is shown in Table I.

Table I Miscibilities of solvents with polymers from **8** where x = 4 and x = 7.3

x =	% Increase in Volume of Polymer after 24 hr		
	Cyclohexane	Toluene	Acetone
4	6.1	9.4	12.7
7.3	5.4	5.3	8.5

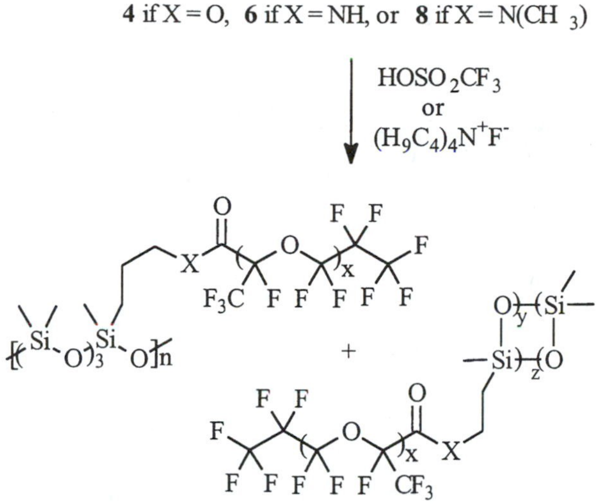

4 if X = O, **6** if X = NH, or **8** if X = N(CH$_3$)

$$HOSO_2CF_3$$
or
$$(H_9C_4)_4N^+F^-$$

Figure 3 Polymerization of oligo(oxyperfluoropropylene)heptamethylcyclotetrasiloxanes

Polymerization to Networks

As illustrated in Figure 4, cationic copolymerizations of **9** and **8** where x=4 and x = 7.3 were carried out in proportions calculated to give similar low cycle ranks networks as given in Table 2 (*21*). Placing the rubbers in MgO terminated the siloxane redistribution. For comparison, rubbers were prepared by the copolymerization of D$_4$ and **9**, D$_4$, **8** where x = 7.3 and **9**, and **8** where x = 7.3, **8** where x = 4 and **9**. In all cases, soft rubbers were formed. Swelling of the rubbers was examined in cyclohexane and toluene as given in Table 2.

Table II Preparation and swelling studies of networks from 8

	Mass in g		MW	Cycle	MW	% Mass Increase[b]	
x =	8	9	RU	Rank[a]	J to J[a]	C$_6$H$_{12}$	C$_7$H$_8$
0[c]	10.0007	0.1515	74.15	30.9	11,000	310	710
4	10.0026	0.1548	290.47	28.8	9,500	8.5	8.7
7.3	10.0062	0.1977	429.66	28.8	7,100	6.0	7.6
6.6[d]	10.0239	0.1803	346.43	30.7	10,000	6.1	8.1
4.0[e]	1.1490	0.0156	290.91	26.2	16,000	8.9	

[a] Calculated, [b] 100(Final Wt – Initial Wt)/Initial Wt, [c] D$_4$, [d] Mix of x = 4 and x = 7.3, [e] Mix of x = 7.3 and D$_4$

Figure 4 Copolymerization with 1,3,5,7-Tetra-(2-heptamethylsiloxane-ylethyl)-1,3,5,7-tetramethylcyclotetrasiloxane to Networks

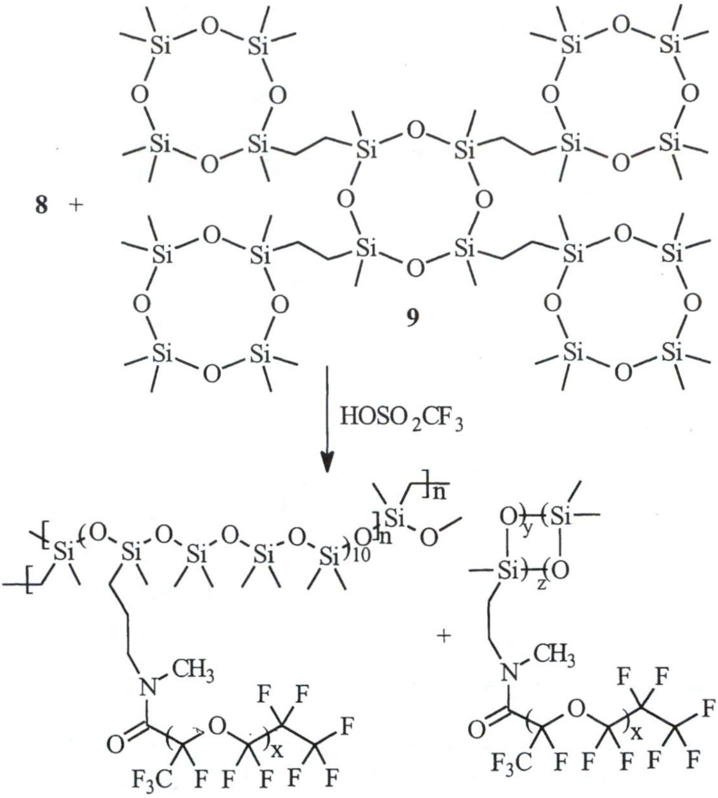

Results and Discussion

The synthesis of **4** occurred in high conversion with nearly complete consumption of **3**. Likewise, the preparations of **6** and **8** were carried out in high conversion. Although **6** could be prepared by either of the synthetic routes (Figures 1 and 2), **4** was prepared in only modest yield via the route given in Figure 2 and attempts to produce **8** via the route given in Figure 1 were unsuccessful. In all cases, the Si was linked to the carbonyl through a three-carbon bridge with no evidence of a two-carbon bridge. All the cyclosiloxanes were dense (d >1.4), low viscosity liquids with low refractive indices (RI <1.35 at 25°C). The viscosity increased with x, and copolymers from **6** were notably more viscous than copolymers from **4** and **8**. The weight percent of oligo(oxy-perfluoropropene) in the cyclosiloxane was high in all cases. The lowest weight percent was 65% for **8** where x = 3, and the highest weight percent was 80% for **8** where x = 7.3.

Polymerization of **4**, **6**, and **8** resulted in polymers of apparent high molecular weight. The gum had been extracted of compounds volatile enough to be analyzed by GC. These compounds were observed as four series of four peaks of descending intensity. As in the case of the polymerization of **4** where x = 3, the first group of compounds displayed the retention times for standards of the dimethylcyclosiloxane tetramer (D_4) through the heptamer (D_7). The second sequence of GC peaks was more intense than the other series and started with a retention time identical to that of the starting **4** where the oligo-(oxyperfluoropropylene) was a single homolog. This series was consistent with a series of cyclosiloxanes with only one perfluoroether substituted siloxy unit and three (D_3D^F) through six (D_6D^F) dimethylsiloxy units. The third series was less intense and each peak was relatively broad. This is consistant with a cyclic tetramer ($D_2D^F_2$) through heptamer ($D_5D^F_2$) with two perfluoroether substituted siloxy units in the ring. The last series was of very low intensity, broad, and poorly resolved. It is reasonable to assume that this is a series of cyclo tetramers (DD^F_3) through heptamer ($D_4D^F_3$) where there are three perfluoroether substituted siloxy units in the ring.

For the polymerization of **4** where x = 3, the molar concentrations for D_4 through D_6 should equal the cyclic equilibrium constant for dimethylsiloxanes modified by the mole fraction of dimethylsiloxy units of all siloxy units in the mixture (*19*). Therefore the theoretical molar ratio of D_4 to D_5 to D_6 should equal K_4/K_4 to $0.75K_5/K_4$ to $0.75^2K_6/K_4$ if polymerization occurred with the complete and random redistribution of all siloxy units in the mixture. The observed molar ratio of D_4 to D_5 to D_6 calculated from the area counts divided by the molecular weight of the cyclic was 1.0 to 0.4 to 0.1, which agreed with the theoretical ratio of 1.0 to 0.4 to 0.1. The theoretical molar ratio of D_3D^F to D_3D^F to D_3D^F is given by $4K_4/4K_4$ to $5(0.75)K_5/4K_4$ to $6(0.75)^2K_6/4K_4$. The observed ratio by GC was 1.0 to 0.5 to 0.2 agreed with the theoretical ratio of 1.0 to 0.5 to 0.2. The agreement of these two sets of ratios not only indicated formation of a random redistribution, but also that a high molecular weight copolymer was formed. If a low molecular weight polymer resulted, the ratios of larger to smaller cyclics would be smaller, reflecting the additional factor of the molecular weight of the linear polymer at equilibrium with the cyclic oligomers.

The polymers after extraction and removal of the extracting solvent, usually pentane, were isolated as 60 to 70% of the mass of the starting 3-(Hepta-methylcyclotetrasiloxane-yl) terminated oligo(oxyperfluoropropene) that was polymerized. This far exceeds the 10 to 26% polymer that results in the equilibrium polymerization of cyclotrifluoropropylmethylsiloxanes. This occurs in spite of the fact that less than 45% by weight of poly(trifluoropropyl-methylsiloxane) is fluorocarbon. The polymers in this study all exceeded 60% fluoroether and resulted in more than 60% linear polymer at equilibrium. Furthermore, the cyclic oligomers were easily extracted from the linear polymer with a volatile hydrocarbon, and the extracted cyclic oligomers were recyclable in a subsequent polymerization after removing the hydrocarbon.

The solubility of these polymers in solvents which will readily dissolve the oligo(oxyperfluoropro-pene) substituted cyclotetrasiloxane is illustrated for **8** where x = 3 and **8** where x = 7.3 in Table 1. The pentane extracted and vacuum dried polymers were allowed to stand in the presence of the solvents for more than 24 hours, and the increases in volume of the polymers were recorded. The polymers chosen for the study were the lowest viscosity polymers produced. In all cases, a dramatic difference in solubility from the monomer was observed, and smaller amounts of solvent were absorbed by the polymer made from **8** with the larger oligo(oxyperfluoropropene) substituent.

The ability to make rubbers of low cycle rank suggested that it was reasonable to assume that the magnitude of the cyclic equilibrium constants was similar to those for the cyclodimethylsiloxane. The dimethylsiloxane unit constituted 75 mole percent of the repeating units for these mixture. Although all formed soft rubbers and had similar cycle ranks, they behaved differently depending upon the average molecular weight of the repeating unit. The higher the molecular weight, the more pronounced the flow at elevated temperatures. This is reasonable due to the greater propensity for a repeating unit to be part of a cyclic structure, either free or incorporated into the network, as the average repeating unit size increases. The rate of apparent network formation was slower for the networks formed from **8** versus those prepared from D_4. The cure was homogeneous in nature for networks from **8** while networks from D_4 cured most rapidly at the air interface. Furthermore, the termination of the catalyst, via migration from the network into the MgO, differed for the copolymer networks from the termination of the network from D_4. The surface would quickly form a skin with networks from D_4, which would resist flow upon heating, but the inside would liquefy. The termination of the networks from **8** cured much more slowly and the formation of a skin was less pronounced. These results are consistent with the surface being dominated by the perfluoroether moieties. The swelling of the networks, Table 2, gave results consistent with the solubility studies with the linear polymers in Table 1.

Conclusions

A relatively efficient route to the preparation of fluorosilicone with a high fluorocarbon content was demonstrated via the ring-opening polymerization of 3-(Heptamethylcyclotetrasiloxane-yl) terminated oligo-

(oxyperfluoropropene)s. Analysis of the equilibrium cyclosiloxanes and the nature of the formation of networks indicate that the process is random and can be treated as any copolymerization of cyclosiloxanes where the equilibrium cyclic concentrations can be reasonably approximated by those for cyclodimethylsiloxanes. The use of a cyclic tetramer facilitates not only the synthesis of the monomer, but also the control of the polymer's size and structure within the limitations of a random system.

References

1. Owen, M.J. *Appl. Polm. Sci.*, **1988**, *35*, 895.
2. Zisman, W.A., in *Advances in Chemistry 43*, **1964**, A.C.S., Washington, DC, 1.
3. Owen, M.J., in *Siloxane Polymers*, Clarson, S.J. and Semlyen, J.A. ed., **1993**, Prentice Hall, Engelwood Cliffs, NJ, 309.
4. Beervers, M.S.; Semlyen, J.A. *Polymer*, **1971**, *12*, 373.
5. Flory, P.J. *J. Am. Chem. Soc.*, **1946**, *68*, 2294.
6. Jacobson, H.; Stockmayer, W.H. *J. Chem. Phys.*, **1950**, *18*, 1600.
7. Brown, J.F.; Slusarczuk, C.M. *J. Am. Chem. Soc.*, **1965**, *87*, 931.
8. Carmichael, J.B.; Winger, R. *J. Polym. Sci., Part A*, **1965**, *3*, 971.
9. Semlyen, J.A.; Wright, P.V. *Polymer*, **1969**, *10*, 543.
10. Wright, P.V.; Semlyen, J.A., *Polymer*, **1970**, *11*, 462.
11. Yu, A.; Yuzhelevskii, E.B.; Kagan, E.G.; Fedoseeva, N.P., *Vysokomol. Soedin.*, **1970**, *12A*, 1585.
12. Kleinstueck, R.; Leupold, M.; Marquardt, G. *Ger. Offen DE 2834171* **1980**.
13. Pulasaari, J.K.; Weber, W.P., *Polym. Prepr.*, **1998**, *39(2)*, 583.
14. Pulasaari, J.K.; Weber, W.P., *Polym. Prepr.*, **2000**, *41(1)*, 133.
15. Wu, T.C. *US Patent 3,876,677* **1975**.
16. Kishita, H.; Yamaguchi, K.; Yoshida, A. *US Patent 4,898,958* **1990**.
17. Kishita, H.; Takano, K.; Yamaguchi, K.; Takago, T. *US Patent 5,202,453* **1993**.
18. Yu, A.; Yuzhelevskii, E.B.; Kagan, E.G.; Dmokhovskaya, E.B., *Khim. Geterotsikl. Soedin.*, **1967**, 951.
19. Chang, P.-S.; Buese, M.A., *Chemistry of Materials*, **1993**, *5*, 983.
20. Chang, P.-S.; Hughes, T.S.; Zhang, Y.; Webster, Jr., G.R.; Poczynok, D.; Buese, M.A., *J.Polym. Sci., Polym. Chem. ed.*, **1993**, *31*, 891.
21. Chang, P.-S.; Buese, M.A., *J. Am. Chem. Soc.*, **1993**, *115*, 11475.

Chapter 27

Silicone Graft Copolymers with Acrylonitrile, Chloroprene, Styrene, Methylmethacrylate, and an Olefin

Allan H. Fawcett, Andrew B. Foster, Majid Hania, Marcia Hohn, Judge L. Mazebedi, Gerry O. McCaffery, Eddie Mullen, and Declan Toner

School of Chemistry, The Queen's University of Belfast, Belfast BT9 5AG, Northern Ireland, United Kingdom

Abstract. We have reacted methylmethacrylate, styrene, chloroprene and acrylonitrile monomers with siloxane polymers bearing thiol groups using a free radical initiator and have performed a competing thiol-ene reaction with an olefins. One siloxane has a thiol group attached to each silicon atom, the second has thiol groups linked to about 5% of the silicons, and the third, the α, ω-polymer, has a thiol group at each end. At one extreme of the reaction conditions we used – by choosing an olefin, or by using low monomer concentration and high thiol concentration – the olefin or monomers were linked singly to the sulfur by the thiol-ene reaction, and at the other extreme, many became attached to the sulfur atoms when the thiol group behaved as a chain transfer agent, so that graft copolymers formed. The turbidity of cast films and the presence of two glass transitions in the DSC traces are taken as evidence for two phases in the solid states.

Introduction.

The present work developed from an attempt to use the thiol-ene reaction to control the growth of chloroprene popcorn polymers by placing a multifunctional thiol, I, in a liquid within which seeds of the popcorn were growing[1], and caused us to characterize by high field proton and C-13 NMR spectroscopy the products we obtained. We then recognized that the grafted molecules might have interesting structures and properties, and widened the study to other common monomers. As a means of simplifying the spectroscopy and of completing the reaction of thiol groups we introduced olefins. We have used other thiol-bearing silicones, II and III, as a means of changing the thiol group proportion, for kinetic purposes, and as they may offer the opportunity of obtaining graft and block copolymers by free radical reactions. Though the molecular weights of the silicones are below 30,000, a value that is commonly considered to provide good microphase separation in the bulk, they are not so high that the detection of the fates of the thiol groups becomes difficult with the NMR technique. The present report is an account of the NMR spectroscopy of the chains grafted onto thiol (mercaptan) groups by guide to possible syntheses and functionalisation, rather than a description of the successful preparation of useful materials.

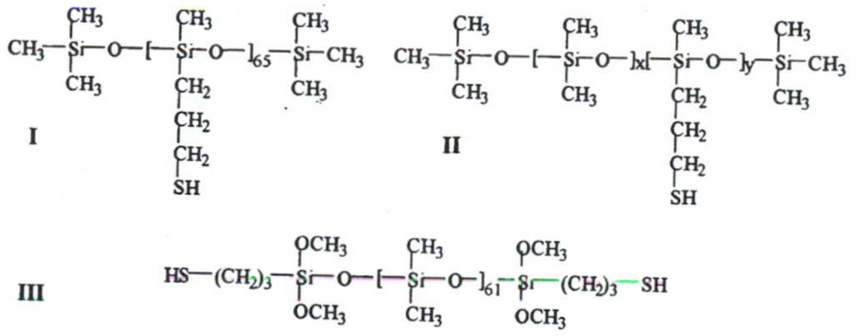

Scheme I. The thiol-bearing silicones of this study.

$$\sim\text{S-H} + \text{I}^{\cdot} = \sim\text{S}^{\cdot} + \text{I-H} \tag{1}$$
$$\sim\text{S}^{\cdot} + \text{CH}_2=\text{CH-R} = \sim\text{S-CH}_2\text{-CHR}^{\cdot} \tag{2}$$
$$\sim\text{S-CH}_2\text{-CHR}^{\cdot} + \text{CH}_2=\text{CHR} = \sim\text{S-CH}_2\text{-CHR-CH}_2\text{-CHR} \tag{3}$$
$$\sim\text{S-CH}_2\text{-CHR}^{\cdot} + \sim\text{S-H} = \sim\text{S-CH}_2\text{-CHR-H} + \sim\text{S}^{\cdot} \tag{4}$$

Scheme II. Some individual free radical reactions.

Experimental

The Siloxanes. A siloxane, **I**, poly(mercaptopropyl methyl siloxane) or PMPMS, bearing a thiol groups on each silicon, was obtained from United Chemical Technologies, Bristol, Pa. The second, **II**, having a thiol on about 4.7% of the Si, was generously provided by Dow-Corning (x + y is unknown), and the last, **III**, was purchased from Genesee Polymers, Flint, Michigan (GP506). From the ^{1}H NMR spectrum we estimated there to be 63.1 silicon atoms per two propane thiol end groups in our sample of **III**.

The Reactions. The free radical processes were performed at 50°C in tubes sealed under vacuum after three freeze-pump-thaw processes, using azobisisobutyronitrile (AIBN) as initiator. In the following example (EM8(5)), 5.0 mls of the siloxane polymer, **III** were placed with 550 mg of AIBN, 5.0 mls of methyl methacrylate and 15 mls of chloroform in a carius tube, which was sealed under vacuum after the oxygen had been removed, and was left in an oven at 50°C. The product was precipitated in an excess of acidified methanol, which was decanted off after two hours and reprecipitated from a solution in chloroform. Solvent was finally removed in a vacuum. A full recovery was difficult when the products were liquids. Details are recorded in Table 1.

Characterization. The polymers were characterized by high field proton and 13-C NMR spectroscopy using a Brucker Avance DRX 500 instrument, using solutions of 5 – 10 mg of material in chloroform (CDCl$_3$). Some molecular weights were obtained with a GPC system callibrated with polystyrene standards. Films were cast from solutions in chloroform, which was first allowed to evaporate off slowly, and was finally removed in a vacuum. DSC measurements were made with a Perkin Elmer DSC 6 at a heating rate of 10°C min^{-1}.

Results and Discussion

A.   NMR Characterization of the Silicone Polymer Structures.

The slicones had only a small number of proton shifts, that shown in Figure 1 from **I** providing an example. The protons had a consistent correspondence, for each of the areas of the methylene shifts was within 1% of the area predicted from the formula given, compared to that of the methyl groups (**a**). Thus we are sure that the observation that the area of the thiol proton, **e**, at 1.4 ppm, was smaller than expected by 14%, a shortfall we attribute to the presence of -S-S- links formed by oxidation. A smaller peak at 2.7 ppm accounts for 4% of these links as one kind of -CH$_2$-S-S-CH$_2$- structure (perhaps a cyclic form), but the other shifts are not distinguished. For **II** the 13-C side bands on each side of the main chain methyl protons at 0.1 ppm were almost as prominent as the main bands at 0.6, 1.7 and 2.5 ppm from the $Si-CH_2-CH_2-CH_2-S$ group, and that at 1.3 ppm from the thiol poton (-S-H). The 7 protons, in the four regions had areas of 1.000, 1.152, 0.908 and 0.503, numbers which showed the reproducibility of the analytical method was of the order of 10%. Taking the structure to be that of Scheme I, and ignoring end groups, we found that $y/(x+y) = 0.047$. These assignments allow shifts to be recognized and measured in the products we obtained by free radical reactions at the thiol groups.

Table 1. Feeds for the Free Radical Reactions of the Mercapto Silicone Fluids with the Monomers within Carius Tubes

Reaction	Silicone	Monomer[a]		Solvent		AIBN	Yield[b]
GM4	I 2.68g	Styrene	2.08 g	CHCl₃	30 ml	0.10g	2.1g
MH1(20)	III 5.0 ml	Styrene	20 ml	-	-	0.50g	14.7g
MH2(5)	III 5.0 ml	Styrene	5 ml	CHCl₃	15 ml	0.50g	6.84g
MH3(1)	III 5.0 ml	Styrene	1 ml	CHCl₃	19 ml	0.50g	5.71g
EM7(20)	II 5.0 ml	MMA	20 ml	-	-	0.55g	13.74g
EM11(1)	II 5.0 ml	MMA	1 ml	CHCl₃	19 ml	0.53g	1.25g
EM6(20)	III 5.0 ml	MMA	20 ml	-	-	0.55g	13.87g
EM8(5)	III 5.0 ml	MMA	5 ml	CHCl₃	15 ml	0.51g	4.52g
EM10(1)	III 5.0 ml	MMA	1 ml	CHCl₃	19 ml	0.54g	1.64g
JLM4-2A	I 2.7 g	Chlor	0.5 g	CHCl₃	20 ml	0.10g	~2g
ABF156	I 2.7 g	Chlor	8.9 g	-	-	0.10g	6.28g
JLM5	II 13.4 g	Chlor	8.9 g	CHCl₃	125 ml	0.50g	12.8g
ABF311	III 5.0 ml	Chlor	20 ml	-	-	0.50g	8.56g
ABF303	III 5.0 ml	Chlor	5 ml	CHCl₃	15 ml	0.50g	7.81g
ABF302	III 5.0 ml	Chlor	1 ml	CHCl₃	19 ml	0.50g	4.70g

a) MMA: methylmethacrylate; Chlor: chloroprene. b) reaction time 24 hrs.

Table 2. Feeds for some Competition Free Radical Reactions of the Mercapto Silicone Fluid I, PMPMS[a].

Reaction	Silicone	M_1	M_2	time	Yield
DT2	I 5.0ml	AN[b] 5.00 g	-	24h	6.91g
DT14	I 5.0ml	AN 5.00 g	Sty 2.77g	2h	3.42g
DT15	I 5.0ml	Sty 2.77 g	H-1[c] 2.23g	2h	1.77g

a) Each feed was in 15 mls of chloroform, and included 0.50 g AIBN;

322

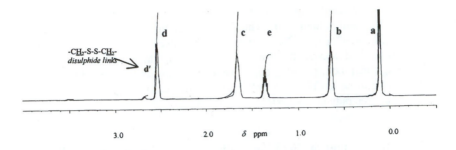

Figure 1. *A proton NMR spectrum of* **I** *in CDCl₃, showing the silicone polymer SiMe₂ signals,* **a,** *beside the TMS signal, and other assignments ordered from the backbone.* **e** *is the mercaptan proton.*

Table 3. Structures of the Siloxane Polymer Derivatives.

Polymer	Silicone	Monomer Grafted	$<x>$ per S^a	$<x>$ per graft	M_n by NMR	M_p GPC
EM7(5)	II	MMA	124	360	-	
EM11(1)	II	MMA	3.6_7	12	-	
EM6(20)	III	MMA	336	-	72,100	
EM8(5)	III	MMA	$24._3$	-	9,720	
EM10(1)	III	MMA	2.5_4	6.4	5,890	
GM4	I	Styrene	0.94	-	13,900	
MH1(20)	III	Styrene	51.4	-	14,450	44,200
MH2(5)	III	Styrene	14.8	-	7,910	4,640
MH3(1)	III	Styrene	6.7	-	6,240	2,400
JLM4-2A	I	Chlor	0.29	1.13	10,400	
ABF156	I	Chlor	3.2_9	6.72	27,800	
JLM5	II	Chlor	9.1_5	23.0	-	
ABF311	III	Chlor	34.9	58.2	10,200	
ABF303	III	Chlor	$24._1$	37.5	9,130	
ABF302	III	Chlor	6.8	11.1	6,060	

a) this ignores the thiol groups that remained when feed [M]/[S-H] was low:

B.   Analysis of Composition and Polymer Structures by NMR.

The Methylmethcrylate Copolymers.

A simple NMR spectrum was provided by a typical product, EM6(20), that obtained by grafting methylmethacrylate monomer onto the silicone polymer, **III**, which has thiol groups only at the ends. The spectum, shown in Figure 2, is essentially a combination of the spectrum of the silicone polymer and that of a free radical polymethylmethacrylate. Thus the signals near 0.0 ppm come from the -O-Si(-Me)$_2$-O- protons, and the remainder of the signals are readily assigned to methacrylate residues C-Me (1.0 ppm), the main chain methylene (1.9 ppm) and side chain -O-Me at 3.6 ppm. The methylene signal shows little evidence of the AB doublet assigned long ago to *meso* dyads[2], an observation confirmed by the relative intensities of the tactic diads shown by the C-Me region, whose peaks are labeled mm, mr+rm, and rr[2], and confirm the expected atactic nature of the PMMA portions. For such methylmethacrylate copolymers, the area of the -O-CH$_3$ protons provided a good measure of that residue, as the signal at 3.6 ppm was well isolated, and the area of the shifts near 0.2 ppm measured the silicone content. Thus values of the number of MMA residues per S were obtained. While the mean value of MMA runs per S atom was 2.5$_4$ in polymer EM10(1) that had been made from **I**, it was clear in the high field ^{1}H NMR spectrum below that the area of the S-CH$_2$-CMeCO$_2$Me- protons, at 2.6-2.7 ppm was just 40% of the area of the –CH$_2$-S- protons at 2.4 ppm, so that the

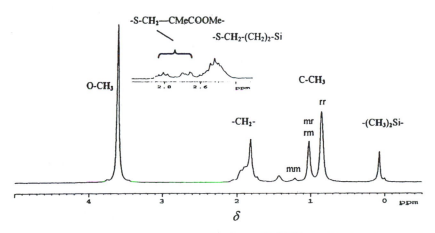

Figure 2. *A proton NMR spectrum of polymer EM6(20) in CDCl$_3$, showing the silicone polymer SiMe$_2$ signals beside the TMS signal, and other assignments from the MMA residues. The three lines come from rr, rm and mm stereochemical dyads in the PMMA blocks, and show syndiotacticity predominates[2]. In the insert we show signals from the protons of polymer EM10, on each side of the pendent S graft sites.*

mean length of the MMA blocks that had been grown from the ends of the α, ω-polymer in reaction (3) of Scheme II is larger, at 6.4 per graft. Chain transfer of an S-H hydrogen atom to the end the growing block (reaction (4)) was delayed by the remoteness of the second thiol group on the same silicone and by the diluting effect of the large volume of solvent (see Table 3).

From the areas of the peaks from grafted monomers and those of the silicone polymer III, whose degree of polymerization was known, we found values of number average molecular weights for the grafted products (Table 3).

The Styrene Copolymers.

The M_n in Table 3 were obtained from the proton NMR spectra, assuming that transfer was the only factor controlling styrene graft lengths. For example in the case of reaction GM4, that of styrene being reacted with an equimolar quantity of siloxane polymer, I, in a large volume of chloroform, the integral areas of the 5 aromatic protons and of the 5 protons on the carbons linked directly to the Si gave a 0.93_6 fraction of styrene residues per silicon. The main chain proton peak of styrene runs that is found at 1.8 ppm in polystyrene was almost negligible and the side chain aromatic peak at 6.6 ppm from polystyrene was absent, so long runs of styrene residues had not formed: the monomers had undergone a thiol-ene reaction in reactions (1), (2) and (4), and styrene radicals had not propagated. Even then the molecular weight of the silicone was substantially increased, as Table 3 records. For the reactions with the α, ω-polymer, again the styrene content was obtained from the area of the aromatic protons. After the area of the $(C\underline{H}_3)_2Si$ had been obtained, and taking the number of such groups per chain as 61, were we able to estimate that the mean number of styrene residues per end was 6.7 for MH3(1), 14.8 for MH2(5), and 51.4 for MH1(20). As the initial [M]/[S-H] increased by a factor of 20, so that (4) became more frequent than (3), $<x>_n$ increased by a factor of just 8. The total molecular weight of the polymers were estimated, and are recorded in Table 3, and are only slightly related to the GPC values, presumably because of termination reactions, molecular shape and calibration issues.

The Chloroprene Copolymers.

In these chloroprene copolymers, orientation and structural isomers may be found, as the proton NMR spectra of Figure 3 demonstrate. The proportion of chloroprene grafts is estimated from the methyl shifts near 1.7 – 1.8 ppm and 2.1 ppm. In the latter region the shifts are singlets, as the adjacent carbon bears the chlorine atom. Futher upfield coupling to the adjacent olefinic proton is evident. Both regions display *cis* and *trans* orientations of the main chain double bond, the latter providing the more prominent features, as in the chloroprene homopolymer. The region between 2.2 and 2.7 ppm contains the main chain methylene shifts and so provides a measure of the chloroprene content, as does the olefinic region futher downfield. An indication of the progressive nature of grafting is provided by the signal at 1.35 ppm from the remnant of the thiol proton: the larger proportion of this, relative to the $-CH_2-CH_2-CH_2-Si$ signal is evident in the lower spectrum. When the spectra of the polymers with ~1 and ~3 grafted chloroprenes per sulfur are compared, it is clear that the shifts of the singly-grafted residues are found slightly downfield of those from end groups to longer side chains, presumably from the influence of the sulfur atom nearby; after considering the S-H areas we found that the mean length of the grafted chains was ~7 residues. These are sufficiently long that the main chain chloroprene residues give rise to shifts near 2.4 ppm which reflect the regioregularity: we have labeled the methylenes as M_1M_1 for the normal placements (taking the chains to grow from the left), when the structure is $-CH_2-CHCl=CH-CH_2-CH_2-CHCl=CH-CH_2-$, and use the labels $M_1'M_1$ and M_1M_1' when respectively the first and the second residues are reversed in orientation[3]. Other types of chloroprene placement are negligible.

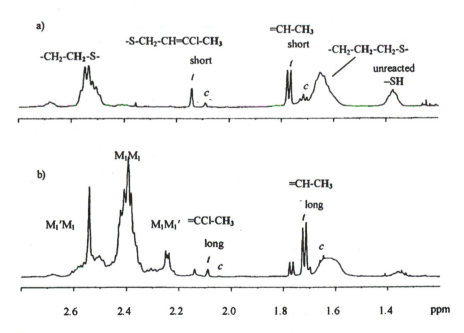

Figure 3. The ^1H NMR spectra of polymers obtained by grafting chloroprene onto siloxane I at 500MHz. Part a) JLM4-2A, <x> = 1.13, and the grafts on the S are predominantly in single residues; part b) ABF156, <x> = 6.72, and longer chloroprene grafts are present. Trans $\sim CH_2-CCl=CH_2-CH_3$ conformations predominate at both type of ends.

The Acrylonitrile and Competition Systems.

The proton NMR spectrum of polymer DT2 showed new shifts at 2.0 and 3.0 ppm, where the methylenes and methane protons of polyacrylonitrile occur, as well as new fine structure near 2.8 ppm. After allowing for four protons in a $-S-CH_2-CH_2-CN$ structure and in the end group of any longer acrylonitrile sequences, we estimated that 0.94 acrylonitriles had grafted per sulfur, but that per graft there were 2.1 residues, for not quite half of the thiol protons had been consumed. In the reaction that formed polymer DT14, styrene and acrylonitrile were in competition with each other. The styrene residues present were estimated from the signal of the five aromatic protons, and the acrylonitrile residues from the integrals of the peaks between 1.5 and 3.5 ppm, after allowing for the shifts of the silicone polymer itself and the methylene and any methine protons of the styrenes and acrylonitriles. Styrenes predominated over acrylonitrile residues by a factor of 7.6 (Table 4), despite the acrylonitrile monomer moles being 4.4 times higher in the feed. Since equimolar feeds of these two monomers provide polymers containing similar proportions of the two residues[4], it is probable that the disparity derives from the greater rate in which styrene monomer undergoes the reaction (2) of Scheme II with the electronegative sulfur radical. The 13-C spectrum contained a number of peaks near 125 and 147 ppm from nitrile and C_1 atoms of the phenyl groups respectively, indicating that the acrylonitrile and styrene residues enjoyed influences from sequence and tacticity effects, despite the grafts being apparently so short that their mean length was just 3.

The polymer, DT15, made from silicone I in a competition between styrene and hex-1-ene, showed the usual aromatic proton shifts from styrene residues and a shift at 0.90 ppm from the methyl group of the $-S(-CH_2)_5-CH_3$ grafts. From these areas the styrene content was thus about 12 times that of the olefin, which reacts only by the thiol-ene reaction. In the part of the 13-C NMR spectrum that we show in Figure 4, there are two small peaks, at 148 and 150 ppm just downfield of the main shift at 146.4 ppm from the phenyl C_1 carbons of single styrene grafts: $-S-CH_2-CH_2$-phenyl. The minor peaks were assigned to styrene duos: $-S-CH_2-CH(phenyl)-CH_2-CH_2$-phenyl, the first, and more heavily sub-

Table 4. Structures of some siloxane Polymer Derivatives.

Polymer	Silicone	Monomer grafted		$<x>$ per S	$<x>$ per graft[a]	M_n by NMR
DT2	I	AN: 0.84		0.94	2.1	9,690
DT14	I	AN: 0.099	Sty: 0.75	0.85	3.0	14,450
DT15	I	Sty: 0.62	H-1: 0.051	0.77	-	7,910

a) this allows for the free thiol groups that remained.

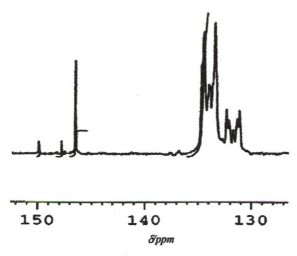

Figure 4. *The 13-C signals from the aromatic carbons of the styrene residues grafted onto polymer DT15. Below 140 ppm are the C₁ phenyl signals, that at 146.5 ppm coming from single styrene grafts, and others from pairs of styrenes.*

stituted C_1 being downfield. Since about 22% of the styrenes are in such pairs (and ignoring the trace at 147.5 ppm) there are about 0.55 S-styrene links to the silicone per sulfur atom, and after allowing for the feed composition, it is thus found that styrene monomer reacts about 11 times as fast as the olefin in reaction (2). Propagation reactions (3) are less frequent than chain transfers (4), by a factor of about 7, chain transfers probably being enhanced over mean field rates by a neighbouring group effect.

C.   Properties of the Grafted Polymers

When cast as films the polymers with styrene and methylmethacrylate grafts of high molecular weight, such as MH1(20), EM6(20) and EM7(5) were turbid in appearance: a second silicone phase was dispersed within the glassy polymer. The same feature was seen in a film of 9 parts of a commercial polystyrene and one part of MH1(20). In Figure 5 we show a DSC trace from polymer MH1(20), where two glass transitions may be seen, at 11°C and 52°C. Silicone-rich and polystyrene-rich phases are detected. Just as the transition in the polystyrene-rich phase is lowered below that of a normal polystyrene, so the transition in the silicone rich phase is raised by the presence within it of some polystyrene.

The chloroprene systems displayed a single glass transition, generally close to that of polychloroprene (-49°C), and much above that of the silicone itself (eg. I at –91°C): thus JLM4-2A(I), ABF156(I) and JLM5(II) had T_g's of -59°, -63° and –74°C respectively. When the second rubber is found in short runs, there is no second phase.

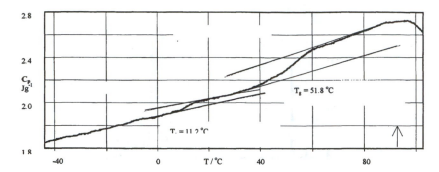

Figure 5. *A DSC plot for Polymer MH1 (20), showing steps from silicone-rich and polystyrene-rich phases. Since the transitions are well away from those of silicone and styrene homopolymers (marked by the arrow), the two blocks of this system are partlymiscible.*

D.   Conclusions

We have shown how monomers such as styrene, acrylonitrile and methylmethacrylate may be free radically grafted onto silicones bearing thiol groups in the same manner as olefins. High field proton and 13-C NMR spectroscopy allows the products to be identified, to observe how the thiol group is consumed and to measure the lengths of grafted chains. Singly grafted residues of chloroprene are readily recognised in both regio-orientations by proton NMR spectroscopy, and competitions between two monomers or with an olefin may be measured. With high values of [M]/[S-H], long enough grafts may be obtained to produce two phase solid systems, even when the silicone blocks are short.

References

1)   Brough, N., Burns, W., Fawcett, A.H., Foster, A., Harrod, S., Poon, L-W., *A C S Polymer Preprints*, **1998**, *39.2*, 374.
2)   Bovey, F.A.; Tiers, J., *J. Polym. Sci.*, **1960**, *44*, 173.
3)   Burns, W., Fawcett, A.H., Foster, A.B., *A.C.S. Polym. Preprints*, **2001**, *42.1*.
4)   Tirrell, D., *Copolymerization*, in Encyclopedia of Polymer Science and Technology, Ed. J. I. Kroschwitz,   John Wiley, **4**, 192 (1986) Wiley, Chichester.

Chapter 28

Synthesis and Properties of Poly(butylene terephthalate)–Poly(dimethylsiloxane) Block Copolymers

David A. Schiraldi

KoSa, P.O. Box 5750, Spartanburg, SC 29304

Abstract

Silanol-terminated poly(dimethyl siloxanes) were demonstrated to copolymerize into PBT under normal polyester polymerization conditions. The resultant polymers exhibited enhanced chemical resistance, and only modest loss of molded mechanical properties up to 15% PDMS content in the copolymer. At higher PDMS levels, the polymer exhibited no cohesive properties in molded parts. Model studies suggest that the actual mode of chemical reactivity is for the silanol end groups to form silanol ethers with the excess diol present in the reaction system, or more likely, with hydroxyl end groups of growing polyester chains. These silanol-diol ethers can serve effectively as chain extenders, and reduce the overall polymerization times.

Starting around 1990, several reports of polyester-polysiloxane copolymers began to appear in the literature. These efforts appear to have been focused on the production of materials possessing improved antistick/antiblock and mold release properties (1-5), improved thermal properties (6), and improved fiber properties (7). In each of these initial studies, the authors' intent was to combine the useful mechanical properties of polyesters with the extremely low surface energies of polysiloxanes, for extruded and/or injection molded applications. In order to effectively incorporate the polysiloxane groups into the polyester backbone, the most common approach was to endcap the polysiloxane with terminal hydroxyalkyl groups, thereby creating Si-O-C linkages in the final polymer products. These bonds were found to be somewhat susceptible to hydrolysis, and further modification of the polymer with phenolic monomers was employed to reduce problematic hydrolysis (8). Alternatively, end capping of the polysiloxane with an epoxy group was employed to once again generate a Si-O-C linkage in the synthesis of polyester-polysiloxane copolymers. The objective of current work was to directly produce polyester-polysiloxane block copolymers from silanol-terminated polysiloxanes. This effort was undertaken with the knowledge that silanols would not be expected to be highly reactive. If successful in producing the desired materials, evaluation of mechanical properties and solvent resistance of these polymers would be undertaken, and comparisons to the analogous copolyesterether elastomers, Figure 1, could be made.

Experimental

Materials
All reagents were used without further purification.

Laboratory Polymerizations
In order to test the viability of these polyester-polysiloxane copolymerizations, laboratory batches were produced using 4.0 moles dimethyl terephthalate (DMT), 9.0 moles 1,4-butanediol, silanol-terminated poly(dimethyl siloxane) (PDMSDS), and tetrabutyl titanate catalyst. Laboratory polymerizations proceeded in a typical manner, except that polycondensation times to the desired

melt viscosities decreased with increasing PDMSDS levels. Given that no problems were encountered, several of these polymers were scaled up in a pilot plant batch autoclave.

Copolyesterether elastomer

Copolyestersiloxane copolymer

Figure 1. Copolyesterester elastomer and corresponding copolyestersiloxane copolymer

Pilot Scale poly(butylene terephthalate)-poly(dimthylsiloxane) Polymerizations

Given in Table 1 are formulations for PBT-PDMS copolymers produced in a 10 gallon, 316-ss pilot plant autoclave. For each batch, Dimethyl terephthalate (DMT, KoSa), 1,4-butanediol (ISP), PDMSDS (Huels), and tetrabutyltitanate (DuPont, 70.0 gm) were charged to the reactor under nitrogen. The reaction mixture was heated to approximately 170°C, at which time methanol evolution commenced. The reaction temperature was allowed to slowly increase with time; the theoretical methanol was removed over approximately 90 min, with a typical batch temperature of 210°C at that point. The batch temperature was then increased to 250°C, and vacuum was applied. An ultimate reactor pressure of ca. 1 Torr was obtained for each batch. The progress of the polymerization reaction was followed by monitoring electrical current required to maintain an agitation rate of 10 rpm (current being proportional to viscosity) – when the target viscosity was reached, the polymer was extruded under nitrogen pressure into a water trough, and cut into cylindrical pellets for further processing.

Injection Molding and Testing of Polymers

The pilot plant produced polymers were injection molded using a Boy 22-S molding machine, operating with an average melt temperature of 250°C. A "family" mold containing one ASTM type IV tensile bar and an ASTM flexural bar was used in the molding machine – the mold temperature was maintained at approximately 80°C using a recirculating oil supply. After a 48 hour period during which tensile and flexural bars were allowed to equilibrate under laboratory temperature and humidity, the bars were tested in tension, and 3-point flexural bending using a Tineous-Olsen universal tester, and were manually tested for Rockwell D hardness using a needle-penetration type tester. For solvent resistance studies, tensile and flexural bars were immersed in the appropriate fluids in sealed glass jars held at room temperature. Sample bars were removed at prescribed times, blotted with a paper towel, allowed to dry in a laboratory hood for 1 hour at room temperature, then were tested on the universal tester.

Results and Discussion

Polymer Synthesis

The formulations tested on a pilot scale are given in Table 1, and the synthetic scheme used is given in Figure 2. Brilliant white polymers were obtained in all cases, in polymerizations that proceeded in a manner typical for PBT (except for

Table 1. Pilot Plant Scale Syntheses of PBT-PDMS Copolymers

SAMPLE/ Wt % PBT	DMT (lb)	1,4-BDO (lb)	PDMS (lb)	MW PDMS	Poly Time (min)
(1) 85	13.3	9.2	2.4	1750	40
(2) 85	15.7	10.9	3.0	4200	55
(3) 95	17.0	11.8	0.94	1750	55
(4) 95	16.9	11.8	1.0	4200	100
(5) 62	13.3	8.8	8.0	1750	15
(6) 62	12.8	8.8	8.4	4200	15

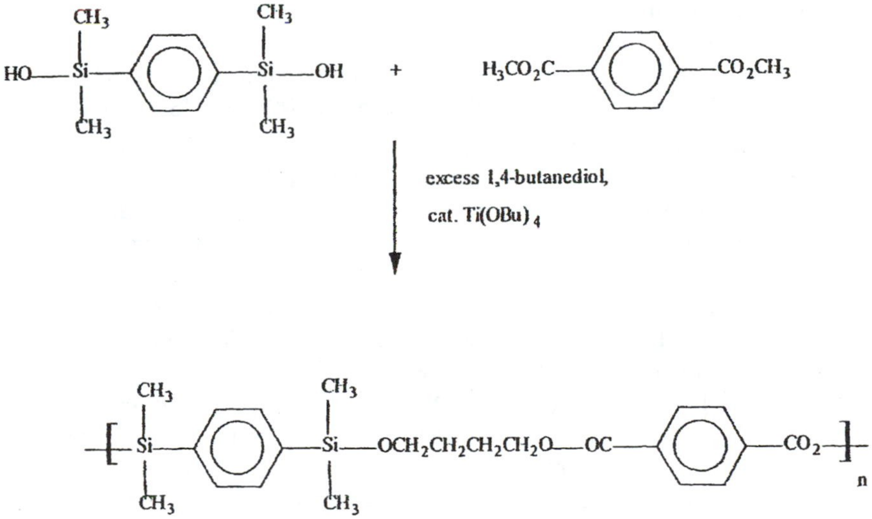

Figure 2. PBT-PDMS Synthetic scheme

polymerization times). Polymerization times necessary to achieve the desired melt viscosity decreased monotonically with increasing PDMSDS content. Unlike systems which contain monomers with functionalities greater than two, these polyester-polysiloxane polymers flowed well and appeared to be tough and ductile. These polymers were found to be completely insoluble in all solvents tested, so it was not possible to obtain nmr spectra of the PBT-polydimethyl siloxane copolymers. Given that chemical resistance of polymers is generally a desirable attribute, we took the insolubility of these polymers in nmr solvents such as trifluoroacetic acid to be positive, if annoying properties. Analysis of reactor overheads indicated only trace amounts of polysiloxane had been removed from the reactor. The T_m values, given in Table 2, are consistent with the expected melting point depressions that would result from full incorporation of the polysiloxane. Soxhlet extraction of ground polymer samples with refluxing chloroform resulted in negligible reduction in weight, so we conclude that the PDMSDS was in fact incorporated into the PBT copolymer. This differs from a literature report that suggested only a fraction of PDMS, terminated with hydroxyalkyl groups, are incorporated into polyesters (9).

Injection Molding and Polymer Properties

Samples 1-6 and a PBT control, 7, were all injection molded into ASTM flexural and tensile test bars. The PBT control and PBT-PDMS copolymers containing 85-95% PBT all produced molded bars that exhibited toughness, and appeared suitable for mechanical testing. The molded parts produced from the 62% PBT copolymers fibrillated upon bending, and could be separated into strands. It is our belief that PBT and polysiloxane phases are incompatible with one another, and this phase separation led to loss of mechanical integrity. Because these 62% PBT samples lacked cohesiveness, they were not further tested. Tensile, flexural, and hardness properties of samples 1-4 and 7 are given in Table 2. Noteworthy among these mechanical tests is that (i) the molecular weights of PDMSDS used (1750 and 4200) generally had little effect on polymer properties, and (ii) tensile strength, flexural modulus, and hardness all declined monotonically with polysiloxane content. A comparison of tensile strengths for these reactor products with those produced by another group using reactive extrusion with epoxy-terminated PDMS and PBT is given in Figure 3 (5). As can be seen in this comparison, the reactive extrusion products appear to have higher tensile strengths, measured using a similar test method. It is not readily obvious what differences in structures were obtained using the reactive extrusion and melt polymerization processes, though it is possible that the former produced extremely high molecular weights via chain extension, thereby leading to increased tensile strength with added bisepoxy additive levels.

Table 2. Polymer/Molded Part Properties for PBT-PDMS Copolymers

SAMPLE/Wt % PBT	MW-PDMS	T_m °C	Tensile Strength (KPsi)	Flexural Modulus (KPsi)	Elongation Break (%)	Hardness Rockwell D
(1) 85	1750	221.2	8.3	202	70	70
(2) 85	4200	220.3	8.7	202	50	71
(3) 95	1750	221.4	10.6	232	40	77
(4) 95	4200	221.3	10.6	224	260	76
(7) 100	----	223.4	12.4	252	80	80

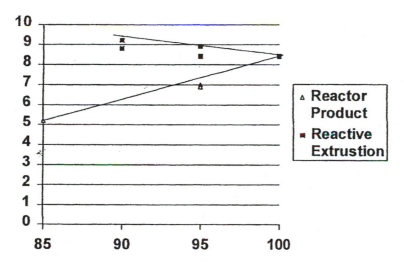

Figure 3. Comparison of Reactor and Reactive Extrusion Product Tensile Strengths (Kpsi) as a Function of % PBT in PBT-PDMS Copolymers

Solvent Resistance of PBT-PDMS Copolymers

A major objective of this work was to determine whether relatively low levels of PDMS copolymerized into PBT would significantly improve its solvent resistance. Molded bars of PBT and of the 95% PBT/5% PDMS 1750 copolymer were immersed in a wide spectrum of solvents, with samples withdrawn for tensile and flexural testing after 7 and 14 days immersion. These results, given in Tables 3 & 4, show that neither polymer is significantly affected by immersion in common alcohols or hydrocarbons, and that the PBT-PDMS copolymer is less susceptible to attack by ketones and chloroform. This small improvement in solvent resistance of PBT which results from copolymerization of PDMS blocks, is consistent with the increased resistance to nmr solvents noted earlier. In general, semicrystalline polyesters are resistant to most solvents, with ketones and halogenated solvents posing the most significant challenge to these polymers. The incorporation of PDMS blocks did improve resistance in this one area of weakness. An interesting question not answered in this study is whether incorporation of PDMS blocks would significantly improve solvent resistance in amorphous and low crystallinity polyesters; amorphous polyesters are extremely susceptible to attack by many families solvents, with low crystallinity polyesters being somewhat more resistant (10)

Silanol Reactivity

Analysis of the PBT-PDMS copolymers was severely hampered by our inability to find a suitable nmr solvent. A model monomer was copolymerized into PBT, in hopes of obtaining a structure which was soluble in nmr solvents, and would demonstrate the presence of silanol esters in the backbone of the PBT copolymers. 1,4-bis(dimethyl silanol)benzene (BDSB) was copolymerized into PBT, following the same procedures used for laboratory synthesis of PBT-PDMSDS copolymers. When equimolar amounts of BDSB and DMT were polymerized in the presence of 2.5 molar excess 1,4-butanediol, the resultant polymer, which was soluble in CDCl3, showed a 1:2:1 ratio of terephthalate:butanediol:BDSB, with the butanediol resonances shifted from their normal locations in PBT. We believe that this ratio and nmr chemical shift difference suggests formation of a butanediol bisether of BDSB, which then acts as a diol to condense with DMT, as is illustrated in Figure 4. The new resonances were observed at 4.25, 3.76, 1.88 and 1.65 ppm (*vs.* TMS) and coincide with the patterns/chemical shifts predicted by simulation software.

Table 3. Effect of Solvents on PBT and PBT-PDMS Copolymers - Tensile

Solvent	Flexural	Strength	(Psi)	
	PBT	PBT	PBT-PDMS	PBT-PDMS
	7 days	14 days	7 days	14 days
None	8,300		6,700	
IPrOH	8,400	8,400	6,600	6,800
MeOH	8,300	8,000	7,100	6,900
EtOH	8,400	8,400	7,200	6,900
Acetone	7,300	7,200	6,000	5,800
MEK	8,000	7,500	6,500	6,200
EtOAc	7,900	7,700	6,700	6,500
Hexane	8,400	8,300	7,200	7,200
Toluene	8,100	8,100	6,900	6,800
Motor Oil	8,300	8,500	7,200	7,200
Cloroform	5,400	5,600	7,200	7,300

Table 4. Effect of Solvents on PBT and PBT-PDMS Copolymers - Flexural

Solvent	Flexural	Strength	(Psi)	
	PBT	PBT	PBT-PDMS	PBT-PDMS
	7 days	14 days	7 days	14 days
None	12,400		10,600	
IPrOH	12,900	12,900	11,000	11,000
MeOH	10,800	10,300	9,500	9,300
EtOH	11,100	11,200	9,600	9,500
Acetone	11,100	11,000	8,500	8,700
MEK	11,700	10,800	9,600	8,900
EtOAc	10,500	11,300	8,800	8,500
Hexane	12,500	12,300	10,600	10,800
Toluene	11,000	10,800	9,300	8,900
Motor Oil	11,300	11,300	9,700	9,600
Cloroform	5,800	4,700	4,300	3,300

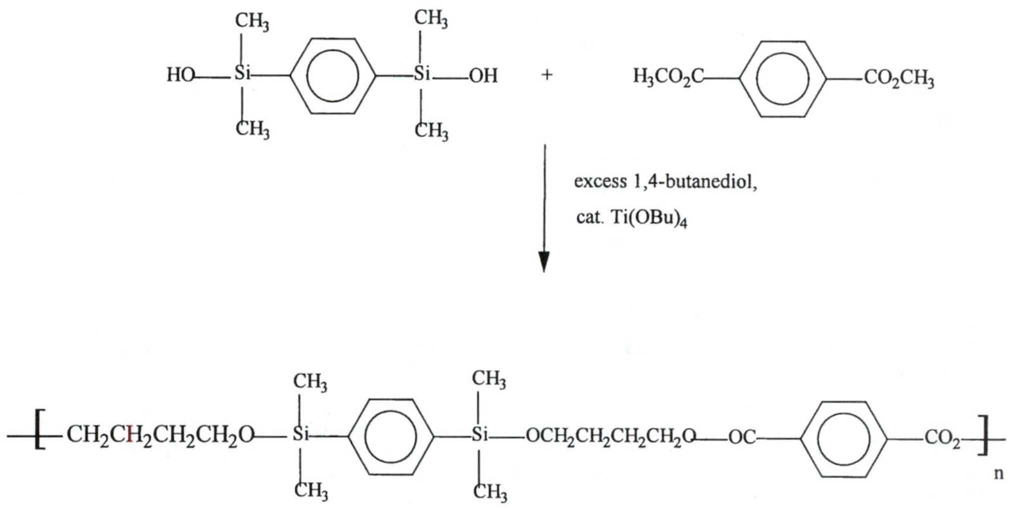

Figure 4. Model Silanol Polymerized in the presence of Terephthalate

Heating neat BDSB with excess 1,4-butanediol in the presence of tetrabutyl titanate produced a polymeric substance, which we have tentatively identified as the linear polyether of these two reactants, Figure 5. Hence, we believe that the mode of reactivity of silanol-terminated poly(dimethyl siloxane) in polyester synthesis is to form ether links with the diol solvent; these hydroxy-terminated polysiloxane oligomers then copolymerize into the polyester backbone. Assuming that the ether formation reaction is comparable to that of polycondensation reactions under polymerization conditions, the disilanols can perhaps be thought of as chain extenders, and this may be the root of the reduced polymerization times observed when PDMSDS is added to PBT polymerization reactions.

Conclusions

Silanol-terminated poly(dimethyl siloxanes) were demonstrated to copolymerize into PBT under normal polyester polymerization conditions. The resultant polymers exhibited enhanced chemical resistance, and only modest loss of molded mechanical properties up to 15% PDMS content in the copolymer. At higher PDMS levels, the polymer exhibited no cohesive properties in molded parts. Model studies suggest that the actual mode of chemical reactivity is for the silanol end groups to form silanol ethers with the excess diol present in the reaction system, or more likely, with hydroxyl end groups of growing polyester chains. These silanol-diol ethers can serve effectively as chain extenders, and reduce the overall polymerization times.

References

(1) Haken, J. K.; Harahap, N.; Burford, R. P. *J. Chromatography*, 1990, *500*, 367.
(2) Ginnings, P. R. U.S. Patent 4,496,704, **1995**.
(3) Fleury, E.; Michaud, P.; Tabus, L.; Vovelle World Patent 9318086, **1993**.
(4) Young, D. J.; Murphy, G. J.; Deyoung, J. J. U.S. Patent 5,132,392 **1992**.
(5) Nakane, T., Hijikata, K.; Kagayama, Y. Takahashi, K. U.S. Patent 4,927,895, **1990**.

1,4-bis(dimethyl silanol) benzene

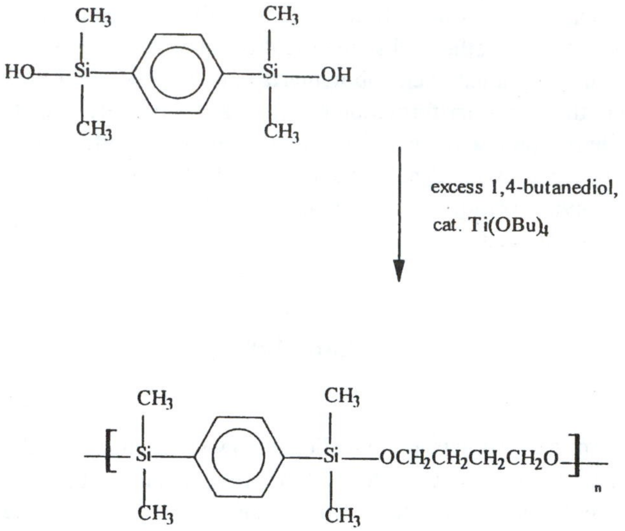

excess 1,4-butanediol,

cat. Ti(OBu)$_4$

Figure 5. Model Silanol Polymerized in the Absence of Terephthalate

(6) Matsukawa, K.; Inoue, H. *J. Polym. Sci., Polym. Lett.*, **1990**, *28*, 13.

(7) Ostrozynski, R. L.; Greene, G. H.; Merrifield, J. H. U.S. Patent 4,766,181, **1988**.

(8) Yamamoto, N.; Mori, H.; Nakata, A.; Suehiro, M. U.S. Patent 4,894,427, **1990**.

(9)   Mikami, R.; Yoshitake, M.; Okawa, T. U.S. Patent 5,082,916, **1992**.

(10) Schiraldi, D. A., *J. Ind. Eng. Chem., submitted for publication.*

Chapter 29

The Synthesis of Vinyloxyacetate–Siloxane Copolymer and Its Absorption onto E Glass Fibers

Rosalind P. Ma, Christopher Le-Huy, Leanne G. Britcher, and Janis G. Matisons

Ian Wark Research Institute, University of South Austraila, Mawson Lakes Campus, SA 5095, Australia

Polyvinyl acetate is used as a film former in sizing solutions used to coat glass fibers and is often chosen for its availability, price, and the ability to modify its properties. The 1-[(2-vinyloxy)ethoxy]ethylacetate monomer was synthesised by Zhang, which could be polymerized at room temperature in the presence of the Lewis acid, zinc chloride, to form a highly branched polymer with remnant vinyl groups. The polymer was then hydrosilated with triethoxysilane and a polysiloxane. The triethoxysilane/vinyloxyactate hyperbranched polymer was applied to the E glass fiber and SEM, DRIFT and XPS was used to analyze the adhesion.

Fiber sizes play a critical role in the manufacture of glass fiber reinforced composites. The size is applied to the glass fiber filaments online after they have been formed and cooled to protect them during subsequent processing as well as meeting various requirements set by both the glass fiber and the composite manufacturers. A size usually consists of a coupling agent, film former(s), lubricant, surfactant and an antistatic agent (see Table I). Further additives may be required depending on the characteristics of the final composite (1). Among the components of the size, coupling agents and film formers are considered to be critical to the success of, not only the manufacture of the glass fiber, but also of the composite. The film former must bind the glass fiber filaments into strands, impart the required handling characteristics, and protect both the filaments and the strands from damage during the manufacturing process.

Table I. Silane Agent Ingredients

Ingredients	Amount in Water
Polymer Film Former(s) incl. plasticizer	3.5 – 15.0 %
Lubricants	0.1 – 0.3 %
Antistatic Agents	0 – 0.3 %
Silane Coupling Agent	0.3 – 0.6 %

Polyvinyl acetate (PVAc) is commonly used as the film former in size formulations (1). It is chosen for its cost, availability and ease of modification of its properties to give the required final effect. PVAc is formed by free radical polymerization of emulsified vinyl acetate in water and the molecular weight or particle size can be controlled. Generally the higher the molecular weight the greater the cohesive strength and the higher the softening point of the film former. However, at the same time the particle size can not be too large as this will make it difficult for the particles to penetrate the fiber bundles. Additives in the emulsion as well as in the size formulation can also affect the characteristics of the film former. For example, plasticizers, which are generally added to improve its flexibility and thermoplasticity, can also affect the softening point of the film former. Development in film former technology over the years has resulted in a vast range of film formers, including copolymers with PVAc, to meet the needs of customers.

When glass fibers are incorporated into polymer resins the film former chosen should also be soluble in it. During impregnation of the resin the film former dissolves into the resin, leaving behind the silane coupling agent on the fiber. The coupling agent, on the other hand, plays a very different role in the glass fiber sizing process.

Glass fiber reinforced composites in the early days suffered from delamination, due to diffusion and reaction of water at the glass-matrix interface (2-4). Coupling agents were introduced to eliminate this problem by generating a water-resistant interface between the inorganic reinforcing fiber and the organic matrix. If designed properly, coupling agents can also react with groups of the polymeric resin in the composite, and become strongly bonded during the curing process (5–9).

For more than a decade, we have investigated the adhesions of silanes and siloxanes onto E glass fibers (10-13). Silanes may interact with the glass fiber surface initially through hydrogen bonding after which, condensation reactions at the glass surface or with each other, generates a siloxane structure. Functionalised siloxanes, like silanes, are also capable of adhering to the glass surface (11). Britcher, et. al. successfully hydrosilated vinyltrimethoxy silane or vinyltris(2-methoxyethoxy)silane onto siloxanes containing SiH groups (12). The adhesion of these siloxane coupling agent analogues to glass fibers was compared with its silane counterpart as well as the OH terminated polydimethylsiloxane (PDMS). They proved that the interactions of the siloxanes were similar to that of the silane, and both showed greater interaction than that of the PDMS. Additionally, Provatas (1998) showed that siloxanes, functionalized with amino groups, also helped improve adsorption of siloxanes to glass surfaces, despite having no hydrolyzable groups *(13)*.

Conventional silane coupling agents form bonds to the oxide surface which are hydrolyzable, and the rate of hydrolysis increases dramatically with both increasing temperature and pH *(12)*. A long flexible polymer backbone, such as a siloxane however, may help to prevent water ingress, and the concomitant rupture of hydrolyzable silane oxide bonds. Siloxanes, unlike silanes, can be multifunctional, with groups along the polymer chain, which are, not only used to attach the polymer to the oxide or glass surface, but also to improve the hydrophobicity and polarity of the treated surface and thereby improve the strength and durability of the coating. Additionally, the long flexible siloxane backbone allows for a continuous coverage of the surface, as the siloxane polymer is flexible enough to adjust to the flaws of the oxide surface.

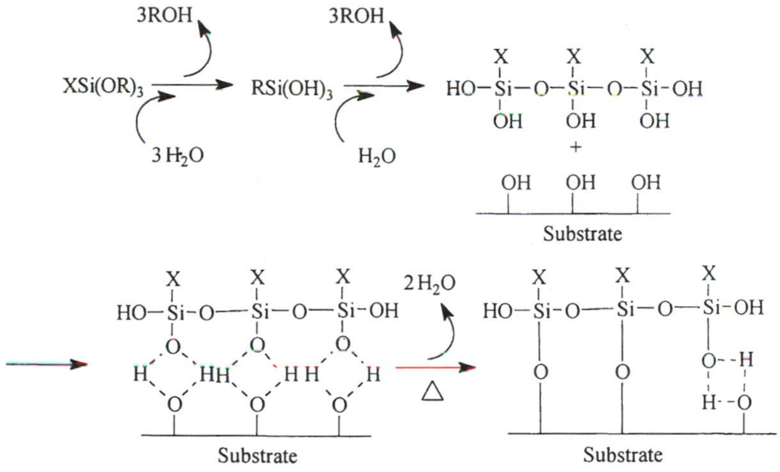

Figure 1. Hydrolysis of silanes on glass surfaces

Silane hydrolysis (see Figure 1) is thought to proceed in two steps. The first, the displacement of alkoxy groups from Si-OR bonds (or alternatively halogen groups from Si-X, where X = Cl, Br, or I) by water to form Si-OH bonds. Condensation of Si-OH bonds, forming Si-O-Si (siloxane) bonds, then occurs in a second slower step.

Nevertheless, as most silane coupling agents are applied from aqueous solution, some condensed 'siloxane type species' are thought to be present. Given that perspective, we thought to examine whether the functions of both film former and coupling agent could be combined by using a suitable multifunctional siloxane polymer, containing some functional groups that can attached to the glass surface, and other functional groups that can associate with the composite matrix.

Since earlier work has shown that siloxanes can adhere to glass surfaces if appropriately functionalized, the focus of this paper is to look at attaching the film former, PVAc to the coupling agent. Two products were synthesized, PVAc polymer was attached to triethoxysilane and similarly the same PVAc polymer was also attached to polydimethylsiloxane. The triethoxysilane product was applied onto the glass fiber and the coating was then characterized by Scanning Electron Microscopy (SEM), Diffuse Reflectance Fourier Transform Infrared (DRIFT) spectroscopy, and X-ray Photoelectron Spectroscopy (XPS).

Experimental

Materials

Ethylene glycol divinyl ether (Aldrich), triethoxysilane (Aldrich) and zinc chloride in 1.0M solution (Aldrich) was used as received. Acetic acid (Aldrich) was dried azeotropically with toluene, and toluene (Ace Chemicals) was dried over sodium and benzophenone. Hexachloroplatinate hexahydrate (IV) (Sigma, 40wt.% Pt) was made to a 1% solution in diried THF (distilled from sodium/benzophenone).

Instrumentation

Nuclear Magnetic Resonance (NMR) Spectroscopy.

^1H nuclear magnetic resonance (NMR) and ^{13}C NMR spectra of all products was obtained with a Varian Gemini Fourier Transform NMR Spectrometer (200MHz). Samples were prepared in dilute deuterated chloroform (Cambridge Isotope Laboratories) with an internal standard at 7.25ppm for ^1H NMR and 77ppm for ^{13}C NMR.

Gel Permeation Chromatography (GPC).

A Waters 2690 Separation Model equiped with a differential refractometer, Model RI 2410, detector was used to determine molecular weight (M_w) of the polymer. The column used was a Waters Styragel HR 5E, 7.8 × 300mm column, with THF as the solvent and polystyrene was used as linear standards.

Diffuse Reflectance Fourier Transform Infrared (DRIFT) Spectroscopy.

DRIFT spectra were measured at room temperature analyzed with a single beam Nicolet Magna Model 750 Spectrometer in the wavenumber region 4000-650cm^{-1}, with the use of a MCT-A liquid nitrogen cooled detector. The interferogram was apodized using the boxcar method installed in the FTIR software (OMNIC FTIR software version 2.0). Signal to noise ratio is generally better than 100 – 1, using 256 scans.

The fibers were mounted parallel to each other in a specially constructed sample holder, which was placed in a Spectra Tech diffuse reflectance apparatus [10]. Each glass fiber sample was scanned 256 times, with the fibers at an angle of 90°C, with respect to the direction of the infrared beam, for maximum signal to noise ratio. The background spectrum was taken from high purity (IR) grade KBr powder (Merck), placed in a sample cup and leveled to the top of the cup using a spatula. No pressure was applied to the KBr powder in the cup, when packing or leveling.

Scanning Electron Microscopy (SEM).

A Cambridge stereoscan 100 SEM was used for scanning electron micrographs. The treated fibers were coated with a thin (~ 200 Å) evaporated carbon layer to reduce the effects of charging.

X-ray Photoelectron Spectroscopy (XPS).

The XPS data were analyzed using a Perkin-Elmer PHI 5100 XPS system with a concentric hemispherical analyzer and a MgKα X-ray source functioning at 300W, 15kV, and 20mA. High vacuum pressure achieved during analysis varied from 10^{-8} to 10^{-9} Torr. The angle between the X-ray source and the analyzer was fixed at 54.6°. Surface charging was corrected to the adventitious carbon 1s peak (284.6eV). The glass fibers were carefully cut and placed on a metallic sample holder with a molybdenum cover plate securing the fibers. Care was taken such that the X-ray beam was only on the fibers and not on the molybdenum cover plate.

Method

Synthesis of 1-[(2-vinyloxy)ethoxy]ethyl acetate

Acetic acid (99%, 4.20g, 70mmol) was added dropwise to a stirred solution of ethylene glycol divinylether (9.14g, 80mmol) at 60°C under nitrogen. After addition was completed, the mixture was refluxed for 5hrs. The required 1-[(2-vinyloxy)ethoxy]ethyl acetate was isolated by vacuum distillation (85-90°C; 10 Torr). Yield 75%

Synthesis of poly 1-[(2-vinyloxy)ethoxy]ethyl acetate

$ZnCl_2$ (2cm^3 of 1.0M solution in diethyl ether) was added dropwise (30min) to a stirred solution of 1-[(2-vinyloxy)ethoxy]ethyl acetate from above (5g, 29mmol) and toluene (15cm^3) and stirring was maintained for 24hours under nitrogen. The reaction was quenched with methanol containing 25vol. % ammonia (5cm^3), then washed with 10% aqueous sodium sulfate, followed by water, and then brine, before being dried over $MgSO_4$. Finally, the solvent removed in vacuo to obtain poly 1-[(2-vinyloxy)ethoxy]ethyl acetate in 95 % yield. GPC data: $M_w = 945$g/mol.

Synthesis of poly 1-[(2-vinyloxy)ethoxy]ethyl acetate triethoxysilane

A mixture of poly 1-[(2-vinyloxy)ethoxy]ethyl acetate (5g, 29mmol), triethoxysilane (4.7g, 890mmol), toluene (10cm^3), and the hexachloroplatinate hexahydrate catalyst (0.1cm^3) was refluxed under nitrogen at 80°C for 24 hours. The resultant mixture was placed under high vacuum to remove excess reagents and the solvent. Poly 1-[(2-vinyloxy)ethoxy]ethyl acetate triethoxysilane was isolated in 43% yield as a clear colourless liquid.

Synthesis of poly 1-[(2-vinyloxy)ethoxy]ethyl acetate polydimethylsiloxane.

A mixture of the poly 1-[(2-vinyloxy)ethoxy]ethyl acetate (1,06g) Si-H terminated polydimethyl siloxane (5g, Mw=1688g/mol) , toluene (100cm^3) and the hexachloroplatinate hexahydrate catalyst (0.1cm^3) was refluxed under nitrogen for 32 hours and then allowed to cool. The solvent was removed in vacuo leaving the product, in 41% yield, as a clear colourless oil.

Application of poly 1-[(2-vinyloxy)ethoxy]ethyl acetate triethoxysilane to E glass fibers

The poly 1-[(2-vinyloxy)ethoxy]ethyl acetate triethoxysilane (8.12g), synthesized above, was mixed with toluene (262g) to form a 3wt.% silane solution. Glass fibers (15g) were soaked in this solution for 16 hours and then allowed to dry at room temperature. The glass fibers were then washed with a sequence of solvents (40 – 50cm^3) as follows: - dichloromethane; toluene; hexane; wet acetone; and then dichloromethane again. The washed fibers were then allowed to dry at room temperature again

Results and Discussion

Synthesis of polyvinylacetate hyperbranched polymer.

The polyvinylacetate hyperbranched polymer were first was synthesised by Zhang (1997) [14], through the self condensation of the monomer, 1-[(2-vinyloxy)ethoxy]ethyl acetate, see Scheme 1. This monomer was initially synthesised by the addition of equimolar amounts of dry acetic acid and divinylethylene glycol. Our own resultant mixture was vacuum distilled to remove any excess reagents and any solvent from the reaction, as the presence of any excess divinylethylene glycol will lead to gelation during the polymerisation stage. A 75% product yield was recovered based on ^{1}H NMR data. The ^{1}H NMR (Figure 4) revealed, a doublet at 1.4ppm due to the CH$_3$ next to the CH, a singlet at 2.05ppm from the CH$_3$ next to the carbonyl carbon, a multiplet at 3.6 – 4.0ppm from the CH$_2$-CH$_2$ sequence of the ethylene glycol, a doublet of doublets at 4.1ppm due to the CH$_2$ of the vinyl group, a quartet at 5.9ppm from the CH attached to the CH$_3$ group, and a doublet of doublets from the CH of the vinyl groups at 6.4ppm.

Polymerization was initiated by the addition of a Lewis acid, zinc chloride, to a dilute solution of the monomer. The living cationic polymerisation reaction proceeds at room temperature and is readily terminated by quenching with methanol. The mechanism of the reaction is thought to involve the 'RCOOH/ZnCl$_2$' system [15], where the ZnCl$_2$ complexes with the ester group thereby creating a stabilized carbocation (see Scheme 2.).

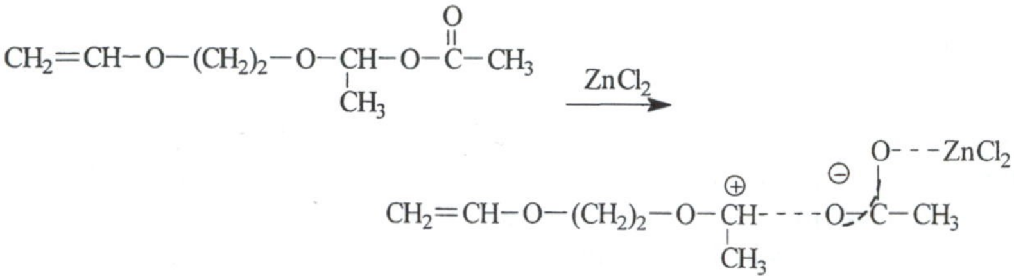

$$CH_2=CH-O-(CH_2)_2-O-\underset{\underset{CH_3}{|}}{CH}-O-\overset{\overset{O}{\|}}{C}-CH_3 \quad \xrightarrow{ZnCl_2}$$

$$CH_2=CH-O-(CH_2)_2-O-\underset{\underset{CH_3}{|}}{\overset{\oplus}{CH}}----O\overset{\ominus}{\underset{}{}}\overset{\overset{O---ZnCl_2}{|}}{C}-CH_3$$

Scheme 1. Stabilising Effects of the Lewis Acid

The nucleophilicity of the RCOO- group depends on the electron withdrawing power of the R group. Generally, the more acidic the R group, such as R = CF3 or CCl3, the faster the polymerisation reaction and the narrower the molecular weight range of the product (15). Once the carbocation is created, it reacts with another monomer at the vinyl end in a 'living manner' until the reaction is terminated eg. by adding methanol. Every addition of a monomer to the chain, brings in an additional initiating site that can be activated by more ZnCl2 resulting in multiple branches, characteristic of hyperbranched molecule (19). It is also important to note that there always remains one vinyl group on every hyperbranched molecule formed. The molecular weight of our synthesized polymer was determined by GPC calibrated against polystyrene linear standards to be 945g/mol., which equates to a hyperbranched tetramer. The ^{1}H NMR (Figure 3.) shows changes, which can be seen, compared to that of the monomer. Firstly, the CH3 attached to the CH has divided into two different peaks, differentiating the internal CH3 and the peripheral CH3 (labelled A and B in Scheme 1. and Figure 3.) on the polymer branch. Secondly, the peaks from the vinyl groups at 4.1ppm and 6.4ppm decrease significantly due to the fact that the majority of the vinyl groups are consumed in the reaction.

Synthesis of Polyvinyloxyacetate/Triethoxysilane, Hyperbranched Polymer

Both the Si-H terminated polydimethylsiloxane (PDMS) and triethoxysilane (TEOS) were separately hydrosilated onto the vinyl terminated polyvinyloxyacetate in the presence of a platinum catalyst (H2PtCl6) (see Scheme 3). Hydrosilation involves the addition of a Si-H group to an unsaturated bond. Two possible adducts can be formed according to the 'anti-Markonikov's rule'(see Scheme 4.), however, the proportion of the major α-

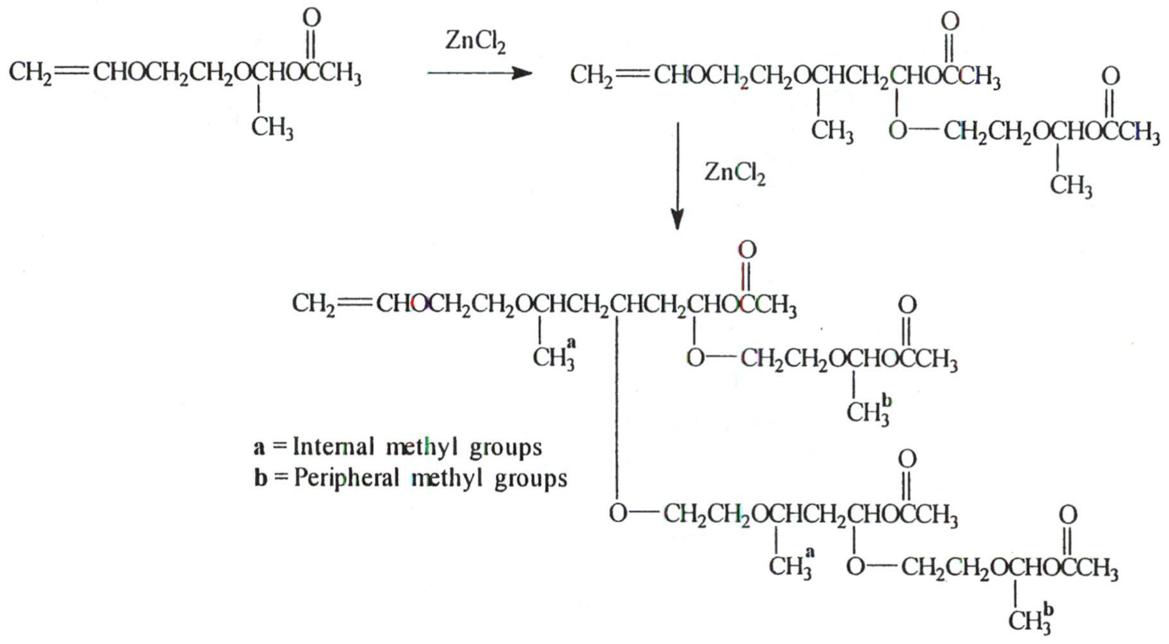

Scheme 2. Synthesis of the vinyloxyacetate hyperbranched polymer

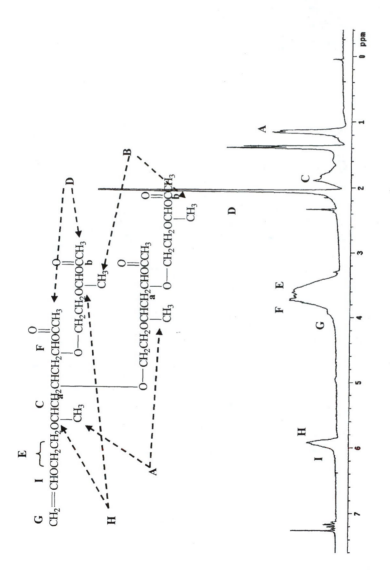

Figure 2. Proton NMR of Polyvinyloxyacetate

adduct formed is dependent on the reaction conditions, the catalyst, the solvent, and the steric as well as electronic influences of the R group attached to the central Si atom of the Si-H group.

Hydrosilation of polyvinyloxyacetate using TEOS successfully isolated only the α-adduct by ^{1}H NMR. The ^{1}H NMR of the product (which was very similar to polyvinyloxyacetate) is shown in Figure 4. The peak at 0.6ppm is from the Si-CH$_2$ protons formed in the hydrosilation, and peak 1.2ppm was due to the CH$_3$ in the triethoxy group. The peak seen at 6.5ppm in the unreacted polymer due to the vinyl group has now totally disappeared indicating that the hydrosilation was completed successfully.

The absorption of the silane onto the glass surface is evident in the infrared spectra, shown in Figure 5. The ability of DRIFT to analyze the surface of E glass fibers below 1600cm^{-1} is made difficult by the strong absorbances of the glass. The peaks between 2960 and 2850 cm^{-1} indicate the presence of CH$_2$ and CH$_3$ groups. Unhydrolyzed TEOS also appears to have adsorbed on the fiber surface at 1470cm^{-1} *(12)*. The adsorption band at 2671cm^{-1} is from the first overtone of the B-O stretching vibration present on the glass surface *(17)*.

SEM results (Figure 6.) showed the coating adhered smoothly and uniformly, in most places, onto the glass surface, however there are some regions of thicker coating where ridges of the coating can be seen. This is rather unusual since the corresponding size on commercially coated fibers generally appear as spots on the fiber surface rather than a smooth continuous coating. The spots of coating can then build bridges with adjacent glass fiber filaments thus linking them together (see Figure 7.). The coating remains quite thin as the diameter of the fibers remains at 11 - 13μm.

Adsorption of Poly 1-[(2-vinyloxy)ethoxy]ethyl acetate triethoxysilane onto E Glass Fibers

The poly 1-[(2-vinyloxy)ethoxy]ethyl acetate triethoxysilane was applied to the glass fiber by soaking in a 3%wt solution of the silane in toluene, the fiber was then washed with a series of solvents (dichloromethane; toluene; hexane; wet acetone; and dichloromethane again) to remove any excess physisorbed silane and to test the adhesion of the chemisorbed silane to the glass surface. In general, all the solvents chosen for the washing process are all known to dissolve siloxanes very well, and wet acetone is known to be the most effective solvent in removing physisorbed silane coupling agent *(6)*.

To gain an understanding of the bonding mechanisms and coating integrity on the glass surface the fibers were analysed by surface analysis techniques. DRIFT was used as an initial check to determine qualitatively how much adsorption of the silane to the glass surface had occurred.

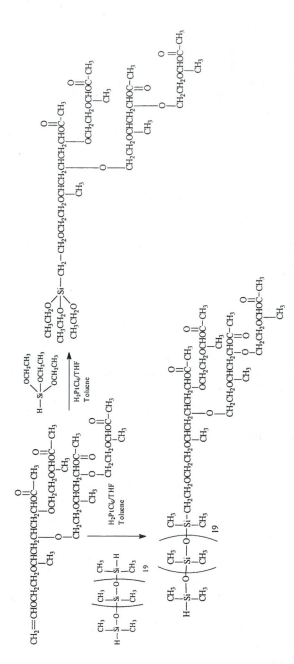

Scheme 3. Hydosilation of polyvinyloxyacetate with tetramethyldisiloxane of polydimethylsiloxane

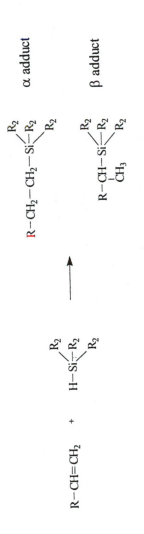

Scheme 4. Hydrosilation Reaction

354

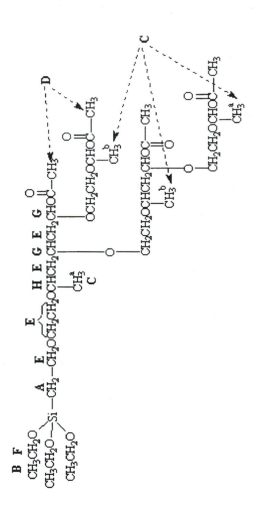

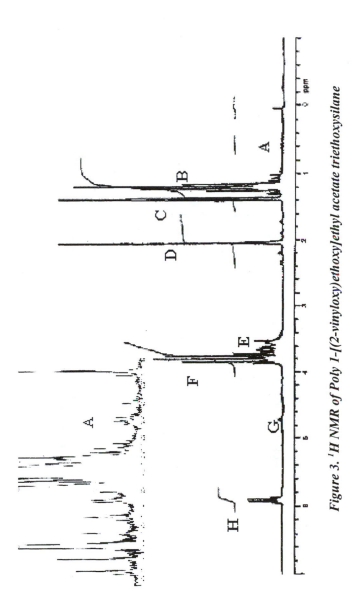

Figure 3. ^1H *NMR of Poly 1-[(2-vinyloxy)ethoxy]ethyl acetate triethoxysilane*

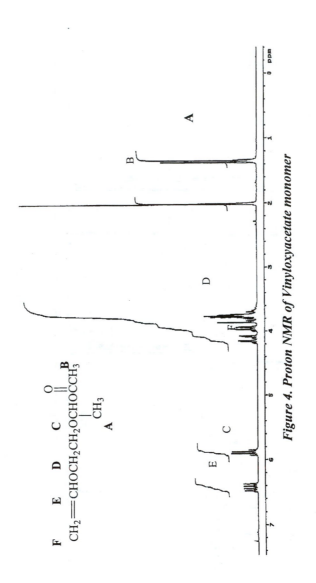

Figure 4. Proton NMR of Vinyloxyacetate monomer

357

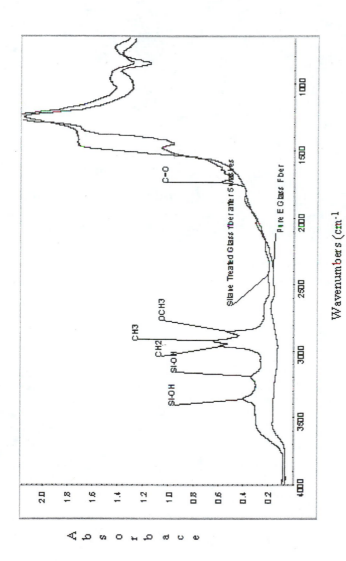

Figure 5. DRIFT of Vinyloxyacetate/Triethoxysilane treated glass fiber

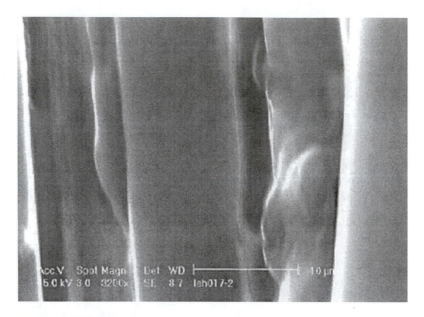

Figure 6. SEM micrograph of poly 1-[(2-vinyloxy)ethoxy]ethyl acetate triethoxysilane coated E glass fibers after the complete washing process.

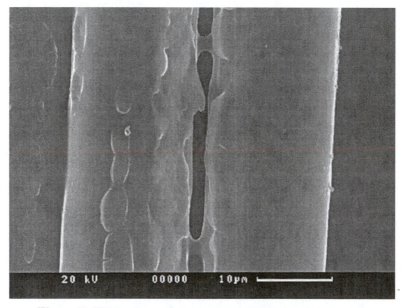

Figure 7. SEM micrograph of commercially coated glass fibers

XPS analysis was also used to examine the detailed surface composition as this technique is capable of giving the surface chemistry of the top few nanometers *(18)* (DRIFT looks at depths of several microns). The glass fiber was analyzed before and after a 5min etch with argon ions. The etching process is known to remove a depth of 5nm using a calibrated sputter rate on standard thin Ta_2O_5 films *(16)*. The value of an etching process is that it provides depth profiling information helpful in differentiating surface contamination from real surface composition effects.

The carbon content of a clean glass fiber is adjusted to 284.6eV to compensate for surface charging effects inherent to the technique. The silicon binding energy is slightly lowered at 102.2eV (pure quartz is at 103.7eV) and this is due to the presence of the sodium, calcium and aluminium oxide components in E glass fibers. Additionally, the presence of surface moisture also lowers the Si bonding energy due to the formation of thin $Si(OH)_x$ surface films.

The calcium level provides the most information on the adsorption of siloxanes or silanes to glass fibers. The combined low binding energy (347eV) and high sensitivity factor (1.58) of calcium, means that the calcium 2p photoelectrons are not as severely attenuated by siloxane or silane coatings, as the other glass elements are. The XPS results shown in Table 2 reveals the Ca concentration decreases significantly on the coated fiber, both before or even after etching (as compared to the untreated clean glass fiber). Such results indicate good coating coverage

Synthesis of Polyvinyloxy acetate/Polydimethylsiloxane.

Following the success of the hydrosilation with TEOS to the hyperbranched polymer, a second hydrosilation reaction was attempted with a $Si(Me)_2H$ terminated polydimethylsiloxane. Figure 8. reveals the ^1H NMR of the product, poly 1-[(2-vinyloxy)ethoxy]ethyl acetate/polydimethysiloxane. The $SiCH_2$ peak formed from the hydrosilation appears at 0.85ppm, and once again there are no vinyl groups remaining. The peak at 4.8ppm is from residual SiH groups that did not hydrosilate on the PDMS.

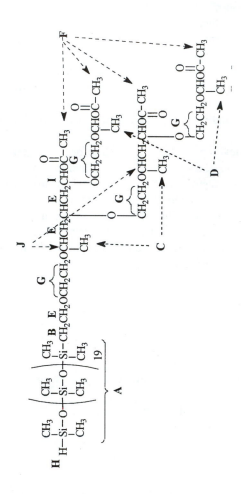

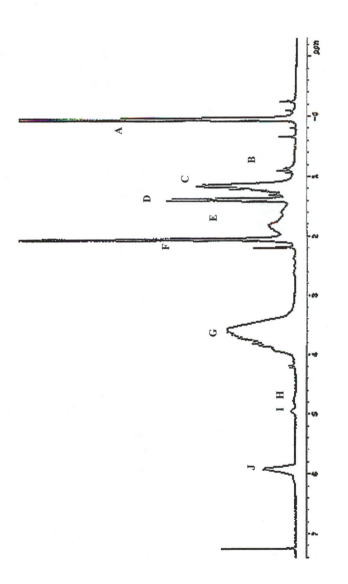

Figure 8. ^1H *NMR of Polyvinyloxy acetate/Polydimethylsiloxane.*

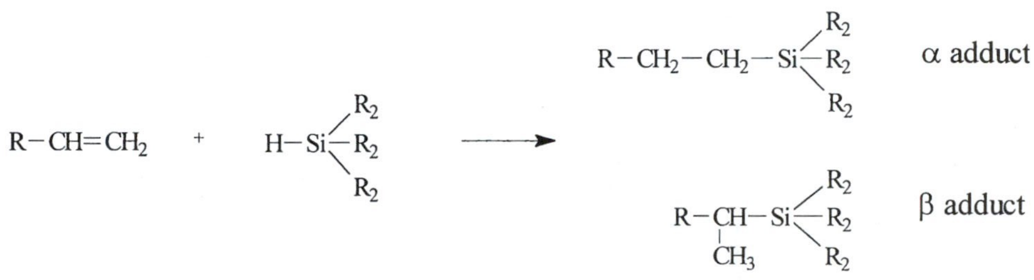

$$R-CH=CH_2 \quad + \quad H-Si \begin{matrix} R_2 \\ R_2 \\ R_2 \end{matrix} \quad \longrightarrow \quad R-CH_2-CH_2-Si \begin{matrix} R_2 \\ R_2 \\ R_2 \end{matrix} \quad \alpha \text{ adduct}$$

$$R-CH-Si \begin{matrix} R_2 \\ R_2 \\ R_2 \end{matrix} \quad \beta \text{ adduct}$$
$$\overset{|}{CH_3}$$

Scheme 4. Hydrosililation Reaction

Table II. Atomic Concentration of Elements of E Glass Fiber Samples

Element	Binding Energy (eV)	Untreated Clean Glass Fiber	Silane Treated - No Sputter	Silane Treated - Sputtered
C1s	284.60	29.00	56.90	38.20
O1s	531.50	50.00	25.90	39.10
Al2p	74.00	3.50	1.30	3.10
Si2p	102.20	12.00	15.60	17.70
Ca2p	347.70	2.80	0.40	0.80
Na1s	1071.90	0.54	< 0.05	< 0.05

Conclusion

A film former, poly 1-[(2-vinyloxy)ethoxy]ethylacetate was successfully hydrosilated to triethoxysilane and a Si-H terminated polydimethylsiloxane.

Poly 1-[(2-vinyloxy)ethoxy]ethylacetate triethoxysilane was applied to E glass fibers. Surface analysis techniques, such as SEM, DRIFT and XPS, revealed that the poly 1-[(2-vinyloxy)ethoxy]ethylacetate triethoxysilane effectively adhered to the glass surface and gave a consistent film coverage on the fibers.

References

1. Loewenstein, K.L. The Manufacturing Technology of Continuous Glass Fibers; Elsevier: Amsterdam, 1993.
2. Atkins, A.G.; J. Mater. Sci. 1975, 10, 819.
3. Bader, M.G.; Bailey J.E.; Bell, I.; In Materials Science Research; Kriegel, W.W., Palmour, H., Eds., Vol. 5, Plenum Press: New York 1972.
4. Outwater, J.O.; J. Adhes. 1970, 2, 242.
5. De, S.K.; White, J.R.,; Short Fiber-Polymer Composites, Woodhead Publishing Limited: Cambridge, England, 1996.
6. Plueddemann, E.P.; Silane Coupling Agents; Plenum Press: New York, 1982.
7. Angst, D.L.; Simmons, G.W.; Langmui,r 1991, 7, 2236.
8. Park, J.M.; Subramanian, R.V.; J. Adhes. Sci. Technol., 1991, 5, 459.
9. Rosen, M.R.; Goddard, E.D.; Proceedings of the 34th Annual Technical Conference, SPI Reinforced Plastics/Composites Int., Sect. 19-E, 1979.
10. L.G. Britcher, D. Kehoe, and J.G. Matisons 2000, 'Direct spectroscopic measurements of adsorption of siloxane polymers onto glass fiber surfaces', in Silane and other coupling agents, ed. K.L. Mittal , VSP Publishing, The Netherlands, pp. 99-114.
11. Britcher, L.G.; Kehoe, D.C.; Matisons, J.G.; Smart, R. St.C.; and Swincer, A.G.; Langmuir 1991, 9(7), 1609.
12. Britcher, L.G.; Kehoe, D.C.; Matisons, J.G.; and Swincer, A.G., Macromol. 1995, 28(9), 3110.
13. Provatas, A.; Matisons, J.G.; Smart, R. St. C., Langmuir 1998, 14, 1656.
14. Zhang, H.; Ruckenstein, E., Polymer Bulletin 1997, 39, pp399 – 406.
15. Kamigaito, M.; Sawamoto, M.; and Higashimura, T., Macromolecules, 1991, 24(4), 3988.

16. Bruckner, R.; Chun, H. U.; Goretzki, H.; and Sammet, M.J., Non Cryst. Solids, 1980, 42, 49.

17. Haaland, D.M., Appl. Spectrosc. 1986, 40, 1152.

18. Arora, P.S. Matisons, J.G., Provatas, A., and Smart, R. St.C., Langmuir, 1995, 11(6), 2009.

19. Frechet, J.M.J., and Hawker, C.J., 'Synthesis and properties of dendrimers and hyperbranched polymers', in Comprehensive Polymer Science, 2nd Suppl., Ch3, 71.

Reinforcing Fillers

Chapter 30

Silicification and Biosilicification

Part 2. Silicification at pH 7 in the Presence of a Cationically Charged Polymer in Solution and Immobilized on Substrates

Siddharth V. Patwardhan[1], Michael F. Durstock[2], and Stephen J. Clarson[1,*]

[1]Department of Materials Science and Engineering, University of Cincinnati, Cincinnati, OH 45221–0012
[2]Air Force Research Laboratory, Wright-Patterson Air Force Base, Dayton, OH 45433

Biosilicification is facilitated by proteins and occurs under modest conditions in an aqueous medium (pH 7 and ambient temperature). Silicification at neutral pH *in vitro* has been shown to occur in the presence of various cationically charged synthetic macromolecules in solution. Here, the synthesis of silica from an aqueous silica precursor in the presence of poly(allylamine hydrochloride) (PAH) and polyacrylic acid (PAA) both in solution and immobilized on substrates is investigated. The results show that the formation of ordered silica structures under these modest conditions was favored for the PAH and the PAH-PAA in solution but neither for PAA in solution nor when the polymers were immobilized as PAH-PAA bilayers on flat substrates. It is possible that the immobilization of the PAH by the electrostatically self-assembly (ESA) technique may allow it to retain its catalytic function, while not allowing it to fulfill its role as a template / structure directing agent. The silica structures were characterized using scanning electron microscope (SEM) and energy dispersive spectroscopy (EDS). For the PAH system in solution silica spheres were seen and for PAH-PAA in solution hexagonal silica structures were observed in co-existence with silica spheres. The results presented herein may be helpful in elucidating biosilicification mechanism(s) and should lead to a better understanding of silicification.

Introduction

Investigations of silicification at neutral pH and under ambient conditions are important due to their close relationship with biosilicification, which also occurs under such mild conditions. The silicified structures that are formed in biological systems are highly sophisticated and are species specific. Examples of organisms, which form ornate silica structures, include diatoms, sponges, grasses and higher plants [1].

Studies of biosilicification have led to the isolation of silaffin proteins from the diatom *Cylindrotheca fusiformis* that catalyze silica formation and the amino acid primary sequence was determined. Silicatein proteins from sponges have been investigated [2, 3] and related studies on the key amino acids from silica forming proteins in grasses have also been described [4].

To verify the specificity and significance of the aforesaid proteins responsible for silicification in biological systems, a variety of synthetic polymers including polypeptides and diblock copolypeptides have been investigated for their role in silicification [5, 6, 7].

A synthetic polymer poly(allylamine hydrochloride), PAH, that is cationically charged under the conditions for silicification at neutral pH, has been studied in detail [8, 9]. It was demonstrated that PAH can facilitate the formation of nanometer and micrometer sized spherical silica particles under mild conditions from an aqueous solution of a silica precursor. It was shown by Energy Dispersive Spectroscopy (EDS) and Fourier Transform Infra Red Spectroscopy (FTIR) that the PAH was incorporated into the final silica structures. In the absence of PAH the reaction mixture gelled in one day. These results indicate that PAH acts as a catalyst as well as a template or structure directing agent for silicification [8]. In this context, Tacke has described how macromolecules facilitate silica formation via scaffolding [9]. In further investigations, various parameters that govern the silica synthesis and morphology of the silicified products were studied [10]. Among the important parameters were the reaction time, the precursor concentration and the precursor pre-hydrolysis time. Small Angle Light Scattering (SALS) experiments on the polymeric solutions in the reaction medium (a buffer) revealed the existence of polymer domains that have a periodic correlation distance [11]. When another cationically charged polymer was added to the reaction mixture disc-like silica particles were observed rather than spherical silica particles seen with just PAH. It is postulated that the two polymers may affect the chain conformations of each other, thus altering the product morphology [5]. Experiments *in vitro* using a mixture of polyamines and silaffins extracted from diatoms resulted in similar disc-like structures [12]. A shear force was externally applied to the reaction mixture and this resulted in the formation of fiber-like silica structures for the PAH system. This might be due the orientation of the polymer in the solution under externally applied shear [13]. Apart from fiber-like structures, various other silica structures were also synthesized and they were found to be amorphous [5]. Further investigations of these structures and the conditions under which they formed are in progress.

With this background in mind, some questions still remain unanswered such as: can silicification under such modest conditions be carried out heterogeneously? What silica morphologies may result upon the use of a mixture of a positively and a negatively charged polymer?

Here we have made use of PAH and polyacrylic acid (PAA) in an attempt to answer these questions. In one case they were coated by electrostatically self-assembled (ESA) onto a flat substrate, which was then dipped into a silicic acid solution. In another case, a mixture of PAH and PAA in solution was exposed to a silica precursor solution. The results were compared to the PAH system at neutral pH and under ambient conditions.

Experimental

1. Chemical Reagents:

Tetramethoxysilane (TMOS), 99+%, was used as the silica precursor. Hydrochloric acid (HCl) was used for the TMOS pre-hydrolysis. The polymers used for silicification in solution were poly(allylamine hydrochloride), PAH, (Molecular Weight = 15,000 g mole^{-1}) and polyacrylic acid, PAA, (Molecular Weight = 50,000 g mole^{-1}, 25 % solution in water). The reaction medium used was a potassium phosphate buffer (pH 7.0). Deionized ultra filtered (DIUF) water was used for washing the samples. All reagents were used as received without any further purification. The details of the reagents used for the ESA experiments are described in the respective section.

2. Silica Synthesis in Solution using PAH or PAA only:

Tetramethoxysilane (TMOS) was chosen as the silica precursor and potassium phosphate buffer as a solvent. A stock solution of 1mM HCl in DIUF water was prepared and was used for all the reactions. The TMOS solution in 1 mM HCl and the polymer solution in buffer were always freshly prepared for each experiment, as the TMOS solution was found to gel within 24 hours. A typical reaction mixture contained 80 μL of the buffer, 20 μL polymer solution (50 mg/ml in the buffer for PAH and as received for PAA) and 10 μL TMOS solution, which was pre-hydrolyzed in HCl solution. All the reactants were measured and added to micro sample polypropylene test tubes (1.5 mL). The tubes were then closed and shaken well to thoroughly mix the reactants in each case. All the reactions were carried out at 20^0 C, atmospheric pressure and neutral pH.

After the desired reaction time, the samples were centrifuged at 14,000 x G force for 5 minutes. It was observed that a white solid precipitated in the tubes in the case of PAH only. The liquid was removed and DIUF water was then added to the tubes. The precipitate was then re-dispersed in the DIUF water. This washing of samples was repeated three times to remove any free polymer, which ensures that the reaction has been terminated. This dispersion was diluted further and 2-4 drops of this solution were placed on an aluminum SEM sample holder in each case. The solution was then left to dry under ambient conditions overnight.

3. Silica Synthesis in Solution using PAH + PAA:

The reaction mixture contained 80 µL of the buffer, 10 µL PAH solution in the buffer (50 mg/ml), 10 µL of PAA (used as received) and 10 µL TMOS solution, which was pre-hydrolyzed in HCl solution. All the reactions were carried out at 20^0 C, atmospheric pressure and neutral pH. Rest of the procedure was as described in the previous section.

4. Preparation of PAH-PAA Bilayers on Substrates:

A set of experiments was carried out to study silicification in the presence of PAH when it was immobilized on flat silicon and silica glass substrates. This was achieved by sequentially electrostatically self adsorbing alternate layers of polyallylamine hydrochloride (PAH) and poly(acrylic acid) (PAA) onto either a glass or silicon wafer substrate, as described briefly below and in full detail elsewhere [14, 15].

For the ESA bilayers the PAH used had a molecular weight of 60,000 g/mole and was obtained from Polysciences. The PAA had a molecular weight of 240,000 g/mole and was obtained from Scientific Polymer Products. Solutions were made of each polymer with deionized water with a resistivity of at least 18 Mohm•cm obtained from a Milli-Q filtration setup. The pH of both solutions was adjusted to 3.5 by the addition of a small amount of HCl.

First, a substrate was dipped into a 10^{-2} M aqueous solution of PAH for 15 minutes. The substrate was then rinsed in three separate water baths for 1-2 minutes each with the result being that the surface was now positively charged due to the adsorbed layer of PAH. It was then dipped into a 10^{-2} M aqueous solution of PAA for 15 minutes and again rinsed in the three separate water rinse baths. The adsorption of a thin layer of the PAA renders the surface negatively charged. This process of alternating between polycation (PAH) and polyanion (PAA) solutions, with a rinsing step in between, was then consecutively repeated to build up a film of the desired thickness. The films used comprised of 10.5 or 11.5 bilayers and had PAH as the outermost layer.

5. Silica Synthesis on PAH-PAA Coated Substrates:

The PAH-PAA coated substrates were dipped into either a 1 M or 0.1 M pre-hydrolyzed TMOS solution, removed after 5 minutes, washed with DIUF water and then dried. The samples were then examined by SEM for silica formation on the substrates.

6. Product Characterization:

The products obtained in each case were characterized by Scanning Electron Microscopy (SEM) and Energy Dispersive Spectroscopy (EDS). A palladium-gold alloy was vacuum evaporated onto the dried samples. They were then investigated using a Hitachi S-4000 Field Emission Scanning Electron Microscope at the Advanced Materials Characterization Center (AMCC), Department of Materials Science and Engineering (MSE), University of Cincinnati (UC). EDS analysis was performed using an OXFORD ISIS system attached to the SEM.

Results and Discussion

1. Silica Synthesis in Solution using PAH:

PAH is able to facilitate the formation of ordered silica structures from a silica precursor in aqueous solution at neutral pH and the observed structures include spheres when the system is unperturbed and fibers when the system is under flow / shear [6, 8, 11, 13].

Representative silica structures from an unperturbed reaction solution at pH 7 and under ambient conditions are shown in Figure 1. The silica spheres were in the size range of 800 nm – 2 µm.

2. Silica Synthesis in Solution using PAA:

Further, to investigate the activity of PAA for facilitating the silica formation, we carried out similar experiments with PAA as described above for PAH. The PAA did not facilitate the precipitation of ordered silica structures at neutral pH and formed gel in a day [6].

3. Synthesis in Solution using PAH+PAA:

When the PAH and PAA solutions were mixed they formed a turbid white solution, which did not precipitate. Upon addition of the silica precursor solution to this mixture silica was seen to precipitate after 5 minutes reaction time. A representative SEM micrograph of the resulting silica is shown in Figure 2.

Two kinds of morphologies were observed to coexist: spheres (~ 100-200 nm diameter) and hexagons (~ 350-700 nm length of sides). Energy dispersive spectra (EDS) revealed that both types of structures were composed of silica. Similar co-existence of different silica structures was reported in an investigation of the poly-L-lysine system [5]. Detailed studies on the nature of the hexagons and the mechanism of their formation are in progress.

4. Synthesis on Substrates (ESA bilayers of PAH-PAA):

There was no evidence for the controlled formation of silica on the PAH-PAA coated substrates except for a few irregular shaped silica regions. This silica found on the substrates is most likely gel that had simply adhered to the surface and not formed due to the PAH immobilized on the substrate. If this hypothesis is wrong, then the silica should have been found uniformly covering the substrates as the PAH was coated uniformly. Furthermore, it should have been well structured as seen when PAH was used in solution [8-10].

5. Homogeneous versus Heterogeneous Silicification:

It was observed that PAH was incorporated into the silica structures during silicification in solution and hence the cationically charged polymer not only acts as a catalyst for the silica formation but also has the role of template / structure directing agent. In Figure 3 it is shown how silica may grow over the charged polymer (PAH) in the silica precursor solution / reaction medium thus illustrating its scaffolding role.

Figure 1. A representative scanning electron micrograph of silica structures formed upon the use of PAH in solution at neutral pH. Bar = 1 μm.

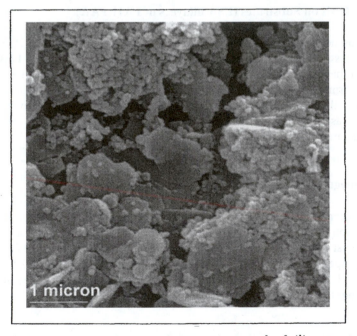

Figure 2. Representative scanning electron micrograph of silica structures formed upon the use of PAH and PAA in solution at neutral pH.

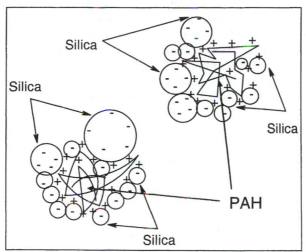

Figure 3. The scaffolding role of PAH in solution in facilitating silicification at neutral pH.

The failure of the formation of silica on PAH coated substrates may be due to the non-availability of the charges on the protonated amines of the PAH as the PAA may be tying up the PAH by ionic and hydrogen bonds [13, 15]. It is possible that the immobilization of the PAH by the ESA technique may allow it to retain its catalytic function, while not allowing it to fulfill its role as a template / structure directing agent. It might also suggest that the silica formation is favorable in homogeneous systems and not in the heterogeneous system investigated here.

It was hypothesized that the PAH + PAA reaction mixture may not have facilitated the formation of ordered structures silica but would gel due to the non-availability of charged PAH which might be tied by PAA as discussed for the PAH+PAA bilayers. This experiment indicates the importance of the availability of charge(s) present on macromolecules in the silicification and the templating / structure directing role of the polymer in solution.

Conclusions

Silicification at neutral pH in the presence of a cationically charged polymer in solution and immobilized on substrates is described here. It is proposed that the silica formation may not be favorable in the heterogeneous systems considered above due to immobilization of the cationic polymer and the elimination of its role as a structure directing agent. The results discussed herein indicate the importance of availability of active sites (charges in this case) present on macromolecules in silicification. A novel hexagonal morphology was seen when a mixture of PAH and PAA was used in solution while ordered silica spheres were seen in the case of unperturbed PAH solutions. These results may help elucidate the mechanism(s) of biosilicification and lead to a better understanding of silicification, in general. The use of synthetic polymers in synthesizing novel morphologies has many potential applications in the field of bioinspired materials science.

Acknowledgements

We thank DAGSI for providing the financial support for this research. We appreciate the help and expertise in the SEM analysis provided by Niloy Mukherjee. We also thank Dr. Jeff Baur for several helpful discussions.

References

1. Simpson, T. L.; Volcani, B. E. Eds., *Silicon and Siliceous Structures in Biological Systems*, New York: Springer-Verlag, 1981.
2. Kroger, N.; Deutzmann, R.; Sumper, M. *Science* **1999**, *286*, 1129.
3. Shimizu, K.; Morse, D. E. In: *Biomineralization: From Biology to Biotechnology and Medical Application*, E. J. Baeuerlein, ed., Wiley-VCH: New York, 2000, pp 207.
4. Harrison (formerly Perry), C. C., *Phytochemistry* **1996**, *41(1)*, 37.
5. Patwardhan, S. V.; Mukherjee, N; Clarson, S. J. *J. Inorg. Organomet. Polym.* **2001**, *11(3)*, 193.
6. Patwardhan, S. V.; Clarson, S. J. *Silicon Chemistry* **2002**, in press.
7. Cha, J. N.; Stucky, G. D.; Morse, D. E.; Deming, T. J. *Nature* **2000**, *403*, 289.
8. Patwardhan, S.V.; Clarson, S.J. *Polym. Bull.* **2002**, *48(4-5)*, 367.
9. Tacke, R. *Angew. Chem. Int. Ed.* **1999**, *38(20)*, 3015.
10. Patwardhan, S. V.; Mukherjee, N; Clarson, S. J. *Silicon Chemistry* **2002**, *1(1)*, 47.
11. Patwardhan, S. V., Masters Thesis, Department of Materials Science and Engineering, University of Cincinnati, OH, USA, **2002**.
12. Kroger, N.; Deutzmann, R.; Bergsdorf, C.; Sumper, M. *PNAS* **2000**, *97(26)*, 14133.
13. Patwardhan, S. V.; Mukherjee, N; Clarson, S. J. *J. Inorg. Organomet. Polym.* **2001**, *11(2)*, 117.
14. Shiratori, S. S.; Rubner, M. F. *Macromolecules* **2000**, *33*, 4213.
15. Decher, G. *Science* **1997**, *277*, 1232.
16. Durstock, M. F.; Rubner, M. F. *Proceedings of SPIE* **1997**, *3148*, 126.

Chapter 31

Diffusion-Controlled Titanate-Catalyzed Condensation of Alkoxysilanes in Nonpolar Solvents

Xiaobing Zhou, Sanlin Hu, Nick E. Shephard, and Dongchan Ahn

Dow Corning Corporation, 2200 West Salzburg Road,
Midland, MI 48686–0994

The diffusion-controlled titanate-catalyzed condensation of alkoxysilanes has been studied in non-polar solvents using ^{17}O and ^{18}O labeling techniques. Incorporation of the oxygen isotopes into reaction intermediates and products *via* labeled water vapor provided a tag for direct detection of TiOSi, SiOSi and TiOTi species by ^{17}O NMR and FT-IR. By monitoring reactions between alkoxysilanes and titanium *t*-butoxide (TtBT), the fates of the reactants were determined. The key intermediate linkage that facilitates formation of siloxanes was found to be the TiOSi bond. Studies with model compounds have shown that TiOSi groups can be converted to SiOSi species through either transesterification or redistribution routes, suggesting that the reaction of TiOSi species with silanols may be insignificant in the formation of siloxanes. Based upon these results, a catalytic cycle has been proposed for diffusion-controlled titanate-catalyzed condensation of alkoxysilanes.

Introduction

Titanate-catalyzed condensation reactions of alkoxysilanes to form siloxane materials have been used in many industrial processes that utilize sol-gel chemistry. Sol-gel processing is used to prepare primers, protective coatings, photocatalytic zeolites, fillers and numerous other intermediates for ceramic manufacture. Motivated by such applications, most fundamental studies of sol-gel process chemistry have been carried out in wet polar solvents (1). Alternatively, our research focuses on such reactions conducted in dry nonpolar solvents exposed to moisture. These conditions are frequently encountered in the cross linking of hydroxyl functional polymers. Commercially important examples utilizing such chemistry include silicone networks for protective coatings, primers, adhesives and sealants (2). The catalytic mechanism, however, still remains somewhat unclear due to the lack of convincing experimental data (3). In this study, we utilize the latest labeling techniques and a novel sample arrangement to investigate the reaction mechanism under diffusion-limited conditions.

Mechanistic studies of the hydrolysis and condensation chemistry of organic titanates and other systems including alkoxysilanes have recently been made easier by the development of both solution and solid state ^{17}O NMR techniques (4-9). ^{17}O NMR analysis often requires enrichment of the ^{17}O isotope in hydrolyzed or condensed products such as metal oxo complexes or siloxanes due to the low sensitivity (relative sensitivity = 2.91 x 10^{-2}) and low natural abundance (0.037%) of the nucleus. In some cases, isotopic enrichment can be achieved *in situ* simply by using ^{17}O enriched water as a starting material. Instead of the direct addition of liquid water, in this study water is allowed to diffuse into reaction mixtures of organic titanates and alkoxysilanes so that condensation reactions can be monitored over a long period of time. This approach allows improved control of the rates of hydrolysis and condensation of hydrolytically unstable titanates. Direct addition of water results in rapid formation of insoluble Ti-O-Ti species that are difficult to characterize due to their low solubilities and polymeric nature. This combination of diffusion-controlled experiments with labeled water has enabled a direct mechanistic study of the titanate catalyzed alkoxysilane condensation reaction.

Experimental

Materials. Titanium *t*-butoxide (TtBT), methytrimethoxysilane (MTMS), and octamethylcyclotetrasiloxane (D$_4$) were purchased from Aldrich. Octamethyltrisiloxane (OS-20) is a Dow Corning product. MTMS, D$_4$ and OS-20 were dried over 4 Å molecular sieve before use. Titanium trimethylsiloxide

(Ti(OSiMe$_3$)$_4$) and silanol terminated polydimethylsiloxane (MW = 400 – 700) were purchased from Gelest and used as received. 20% ^{17}O enriched water and 95% ^{18}O enriched water were purchased from Isotec and used without further purification. Pentane was dried and distilled from sodium/benzophenone. The 0.1 N NaOH solution was purchased from Aldrich.

Fourier Transform Infrared Spectroscopy (FT-IR). Infrared spectra were recorded at a resolution of 4 cm^{-1} on a Nicolet 5SXB FT-IR spectrometer with purge of dry nitrogen. KBr disks were prepared with solid samples. Relative peak areas for semi-quantitative analysis of TiOSi redistribution kinetics were obtained by normalizing the integrated area of the 923 cm^{-1} band by the nearby Si-Me band at 1270 cm^{-1}.

Nuclear magnetic resonance spectroscopy (NMR). The solution NMR data were collected on a Varian INOVA 400 spectrometer operating at 54.203 MHz for ^{17}O, at 79.402 MHz for ^{29}Si, and at 100.507 MHz for ^{13}C. A Nalorac 10 mm dual broadband probe was used with H$_2$O-^{17}O as an external reference for ^{17}O NMR. A special capacitor stick insert was used to make the probe tunable to ^{17}O frequency. To optimize spectral resolution, the probe was shimmed with D$_2$O before running test samples. All subsequent runs were conducted neat with samples unlocked. The experimental parameters were optimized to enable accumulation of FIDs within two minutes for a spectrum with sufficient signal-to-noise ratio for kinetic studies of fast reactions. A standard single pulse sequence with Waltz-16 composite pulse decoupling was employed with a 10 μs 90° pulse and an acquisition delay of 200 ms. In most cases, the spectra were recorded after 500 scans. Solid state ^{13}C (spin rate = 5,000 Hz) and ^{29}Si (spin rate = 4000 Hz) NMR data were collected using Varian's CP/MAS probe in a 7-mm ZrO$_2$ rotor.

Diffusion controlled hydrolysis and condensation reactions. Reactions were run in 10 mm screw thread NMR tubes. In a typical experiment at ambient temperature, TtBT (0.50 ml, 1.3 mmol), MTMS (0.56 ml, 3.9 mmol) and D$_4$ (1.5 ml) were pre-mixed in a tube. A small ball of cotton (0.042 g) was lodged in the headspace of the tube about 15 cm above the meniscus of the solution (Figure 1). The tube was then capped with a Teflon/Silicone septum and ^{17}O or ^{18}O labeled water (0.15 ml, 8.5 mmol) was injected into the cotton. The reactions were complete after two weeks at 25 °C. The reactions were monitored with NMR without spinning the NMR tube.

Titanium siloxide resin. Silanol terminated polydimethylsiloxane (20 g, containing approximately 60 mmol OH groups) in 100 ml anhydrous pentane was added dropwise to TtBT (5.62 ml, 15 mmol) in 50 ml anhydrous pentane. After stirring for 10 minutes, volatile species were removed *in vacuo* (25°C, 10^{-2} mmHg) to leave a viscous liquid that thickened further with time.

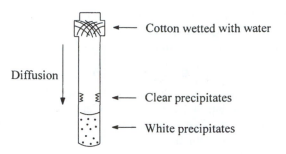

Figure 1. Experimental setup for the diffusion controlled condensation of alkyltrimethoxysilanes

Condensed MTMS Gel. The base-catalyzed condensed MTMS gel was prepared by quickly mixing MTMS and excess 0.1 N NaOH solution. A clear gel was immediately formed and dried *in vacuo* (25°C, 10^{-2} mmHg) for 2 hr.

Results and Discussion

Reactions Observed with NMR

Direct addition of water. Attempts to add ^{17}O enriched water directly into the reaction mixture of TtBT, MTMS and OS-20 led to immediate formation of a white precipitate. In the liquid phase, a strong ^{17}O NMR signal (0.0 ppm) for bulk water and a smaller peak (-12 ppm) for water in solution were detected. Since most organic titanates are much more reactive with water than alkoxysilanes the white precipitate is attributed to hydrolyzed and condensed titanium oxo alkoxides, $Ti(O)_x(O^tBu)_y$, derived from TtBT. Thus, water added to the reaction mixture as a separate liquid phase leads to a very rapid interfacial reaction with titanates yielding insoluble titanium oxo alkoxides that are no longer good catalysts for the condensation of alkoxysilanes.

Vapor phase water addition. Condensation appeared to begin shortly after the labeled water was added to initiate the diffusion of moisture, as a clear gel was formed within two hours. In the first ^{17}O NMR spectrum recorded at ambient temperature (25°C), three peaks appear at 300.4, 55.2 and -15.6 ppm and are assigned to TtBT, D_4 and MTMS respectively (Figure 2). Sharp new signals at around 240 ppm in subsequent spectra can be assigned to titanium siloxides (TiOSi), of which the exact structures are not known at this time (*7, 10*). The narrow linewidths of both new signals suggest the initial titanium siloxides may still be small molecules. Both peaks gradually evolved into broad featureless bands within 2 days. Simultaneously, another group of new signals at 60 ppm also grew with time. These can be assigned to various siloxane (SiOSi) structures (*10*). Over the next 3 days, the siloxanes became dominant in the reaction mixture while the titanium siloxides only grew moderately, and a third group of new species, titanium oxo (TiOTi) compounds, were formed. The titanium oxo compounds show unique ^{17}O NMR bands consistent with prior assignments of μ^2 (OTi$_2$, 760 ppm), μ^3 (OTi$_3$, 530 ppm) and μ^4 (OTi$_4$, 380 ppm) structures (*4*). ^{17}O spectra collected from 100 h to 140 h at 60°C clearly reflect these changes (Figure 3). These titanium oxo compounds are evidently complex macromolecular structures, as indicated by the broad multimodal peaks. In the later stages of the experiment at 25°C, the NMR tube began to fill with a white precipitate, and the solution ^{17}O NMR signals broadened and decayed relative to the D$_4$ signal. This signal decay is attributed to the precipitation of ^{17}O enriched TiOTi, TiOSi and SiOSi solids that have peaks too broad for detection by the solution NMR setup.

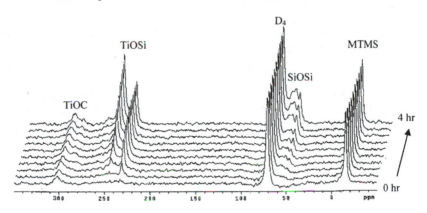

Figure 2. ^{17}O NMR data – diffusion controlled condensation of MTMS at 25°C

The diffusion-controlled condensation reactions were also conducted at elevated temperatures to obtain information on the relative rates of reactions. At 90°C, for example, ^{17}O NMR data show that fewer titanium siloxides and more siloxanes were formed within the first 2 hours than at 25°C (Figure 4). These trends continued until the reactions were complete. This phenomenon implies that higher temperatures favor the conversion of titanium siloxides to siloxanes. Titanium siloxides can be converted to siloxanes *via* either transesterification with methoxysilanes or redistribution into TiOTi and SiOSi species (*6*). Both routes have been further investigated and relative reaction rates have been measured. It appears that transesterification is more favored at elevated termperature since TiOTi species were not detected along with the siloxanes in the early stage at 90°C. Further, at 90°C the reaction mixture eventually gelled after the reaction was stopped, forming a small amount of clear precipitate above the reaction mixture.

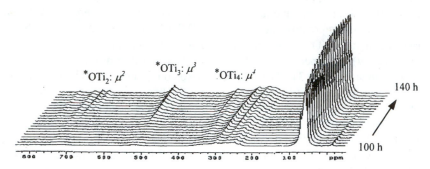

Figure 3. ^{17}O NMR data – diffusion controlled condensation of MTMS at 60°C

Siloxane formation. As observed in Figures 2, 3 and 4, the siloxane products have a complex ^{17}O NMR pattern. Closer examination of the SiOSi region reveals that MTMS is condensed stepwise, although the reaction rate for each step is comparable such that a mixture of siloxane products is generated (Figure 5). At 90°C, the ^{17}O NMR bands of the siloxane products were initially dominated by a strong signal at 55 ppm in the first 4 hours. Over time, the dominant peak position shifts to 60 ppm (from 6 to 16 h), and then to 65 ppm after 18 hours. Similar transitions have been observed when the reactions were conducted at 25°C and 60°C. Thus it appears that MTMS first condenses to the T^1 unit, which was then further condenses to the T^2 and T^3 structures. The sequential formation of T^1, T^2 and T^3 structures has been verified with ^{29}Si NMR, a method commonly used to differentiate siloxane structures (11).

$$MeSi(OMe)_3 \xrightarrow[H_2{}^*O]{[Ti]} Si^*O-\underset{\underset{OMe}{|}}{\overset{\overset{Me}{|}}{Si}}-OMe \xrightarrow[H_2{}^*O]{[Ti]} Si^*O-\underset{\underset{{}^*OSi}{|}}{\overset{\overset{Me}{|}}{Si}}-OMe \xrightarrow[H_2{}^*O]{[Ti]} Si^*O-\underset{\underset{{}^*OSi}{|}}{\overset{\overset{Me}{|}}{Si}}-{}^*OSi$$

$$\qquad\qquad\qquad\qquad T^1 \qquad\qquad\qquad\qquad T^2 \qquad\qquad\qquad\qquad T^3$$

Further, a semi-quantitative analysis of the ^{17}O NMR data over the first 7 h reveals that the rates of siloxane formation from TiOSi intermediates increased dramatically with temperature (Figure 6. A). However, the temperature dependence of TiOSi bond formation appears more complex (Figure 6. B). In these diffusion-controlled titanate-catalyzed reactions, TiOSi peaks were observed from the first spectrum after water vapor introduction and continuously thereafter. These are believed to be the key intermediates in the formation of the siloxanes (*3*). There are several competing reactions that determine the transient concentration of TiOSi species, including 1) formation from the reactants TiOC (*e.g.*, TtBT) and SiOC (*e.g.*, MTMS), 2) transesterification with SiOC to SiOSi and TiOC and 3) redistribution to SiOSi and TiOTi. The relative effects of these competing reactions at any given time and temperature may be related to the non-monotonic temperature dependence of the TiOSi concentration shown in Figure 6 B).

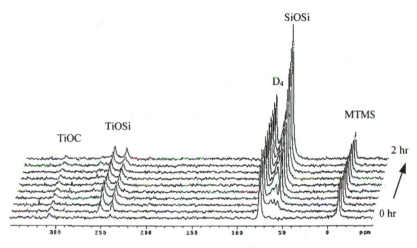

Figure 4. ^{17}O NMR data –diffusion controlled condensation of MTMS at 90°C

The diffusion controlled condensation of MTMS was complete after two weeks at 25°C when the TiOTi, TiOSi, TiOC and MTMS signals disappeared in the solution ^{17}O NMR spectra. The resulting clear gel (roughly 0.05 g) and white solid precipitate (roughly 0.7 g) were separated and characterized using a variety of techniques including solid state NMR and FT-IR (spectra not shown). Both the clear gel and the white precipitate were found to be amorphous by XRD. The clear gel is composed of almost pure T^2 and T^3 siloxanes, containing only a trace amount of titanium species as indicated by a Si:Ti molar ratio of 20:1. The white solid is a composite of titanium oxo alkoxides and siloxanes, and has a Si:Ti ratio of 3:1. The clear gel contains almost no TiOMe species, while the white precipitate has a rather large amount of TiOMe species. Both products contain residual SiOMe and TiO^{t}Bu species, suggesting incomplete hydrolysis and condensation.

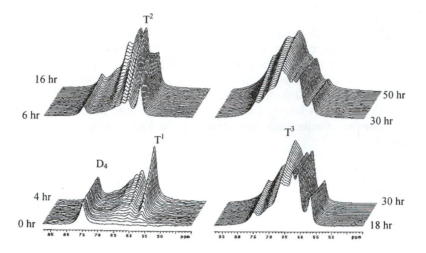

Figure 5. ^{17}O NMR data – stepwise formation of siloxanes at 90°C

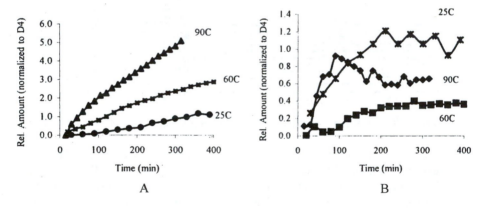

Figure 6. SiOSi (A) and TiOSi (B) formation as a function of time at different temperatures

Reactions observed with FT-IR

In the FT-IR spectra of the clear gel and white precipitate, SiMe (1270 and 2965 cm^{-1}), SiOMe (2835 cm^{-1}), TiO^{t}Bu (1370 - 1790 cm^{-1}) and SiOSi (1000 and 1050 cm^{-1}) were readily identified. The broad bands at 512 and 590 cm^{-1} in the white precipitate can be assigned to the TiOMe group which is absent in the clear gel. The IR absorptions at 778 and 844 cm^{-1} may be caused by the different vibrational modes of the SiMe group, although a definite assignment of the two bands is difficult at this time.

The IR spectra of both solid phase products show a band at 913 cm^{-1} (Figure 7: C and E). However, when 95 atom % ^{18}O water was used in the same reaction, the 913 cm^{-1} band was not observed in the IR spectra of the clear and white precipitates. Instead, a new peak appears at 882 cm^{-1} (Figure 7: D and F). The 31 cm^{-1} red shift is typical of an isotopic effect due to ^{18}O labeling. The bands at both 913 cm^{-1} and 882 cm^{-1} were thus assigned to the same TiOSi bond. This assignment was further confirmed by comparison against IR spectra of a condensed MTMS gel formed in basic conditions and of neat Ti(OSiMe$_3$)$_4$ (Figure 7: A and B, respectively). The condensed MTMS gel was prepared by mixing MTMS and excess 0.1N NaOH solution. It has no TiOSi bond and thus does not absorb infrared at 913 cm^{-1}. On the contrary, Ti(OSiMe$_3$)$_4$ has a strong IR band at 913 cm^{-1}. The similar isotopic effect has been observed for the siloxane (SiOSi) IR bands near 1041 cm^{-1}. Detection of the TiOSi bond by IR in both the clear and the white precipitate indicates that TiOSi species were formed in solution during the early stages of the reactions (in keeping with the solution ^{17}O NMR data), incorporated into the network, and eventually precipitated along with other species.

The TiOSi bond has previously been reported to have low thermal stability in a sol-gel derived SiO$_2$/TiO$_2$ glass by solution ^{17}O NMR (6). It was believed that titanium siloxides in the network decompose to give siloxanes (SiOSi) and titania (TiOTi). In order to better understand the chemistry of the reaction intermediates and products in a diffusion-controlled environment, a model titanium siloxide network was prepared from TtBT and a low MW silanol terminated polydimethylsiloxane containing equimolar levels of TiO^{t}Bu and SiOH groups. The freshly prepared titanium siloxide material was initially a clear viscous liquid in which both the residual unreacted silanol (3296 cm^{-1}) and TiOSi bond (923 cm^{-1}) were detected by IR (Figure 8. (3 h)). Disappearance of the silanol signal within one day at 25 °C was accompanied by the formation of a clear, high viscosity viscoelastic gel. After gelation, the network was continuously monitored with IR while stored in N$_2$ at 25°C. It was found that the TiOSi IR band shrank with time, and completely disappeared within 6 months (Figure 8. 224 d). No significant changes were observed for other IR bands.

A semi-quantitative analysis of the IR data provides insight to the kinetics of the redistribution process. The 923 cm^{-1} bands in these spectra were integrated against the nearby SiMe bands at 1270 cm^{-1}, and the resulting ratios were plotted as a function of time (Figure 9). From this curve, the half-life $t_{1/2}$ of the titanium

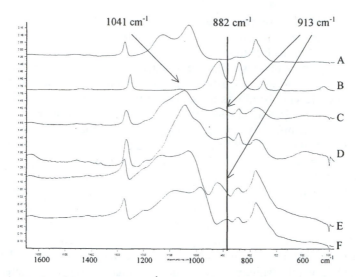

Figure 7. IR data (500 – 1600 cm⁻¹) - detection of the TiOSi bond observed in the solid products of titanate-catalyzed MTMS condensation A. Condensed MTMS gel using NaOH catalyst, B. Ti(OSiMe₃)₄, C. White precipitate, D. ^{18}O labeled white precipitate, E. clear gel, and F. ^{18}O labeled clear gel

siloxides is estimated to be 19 hr at 25°C. At elevated temperatures, the redistribution of the TiOSi bond occured faster.

The TiOSi bond in the network was quite moisture stable. When the sample was exposed to air (RH ~ 40 %), the TiOSi IR band still decreased at the same rate as in the inert environment. While dissolution of the titanium siloxide network in most common solvents was very slow, it could be dissolved in excess MTMS or the silanol precursor within one hour. Once mixed with MTMS, the TiOSi bond was reacted and the siloxane derivatives of MTMS were formed. Titanium methoxides were also detected in the reaction products. Dissolution of the titanium siloxide network in the silanol precursor was followed by the slow formation of a highly viscous liquid and the release of water. An IR analysis shows that the intensity of TiOSi band was almost unaffected over a long period of time after dissolution, suggesting that hydroxy transfer from silicon to titanium is kinetically insignificant, although siloxy exchange between titanium siloxides and silanols may occur. The titanium siloxides may also catalyze the condensation of the silanols to form higher MW linear PDMS and water.

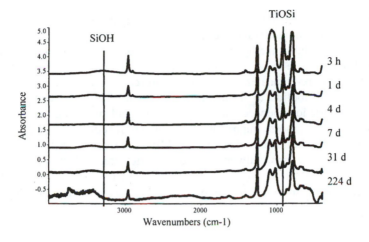

Figure 8. IR data – redistribution of the titanium siloxide network at 25°C

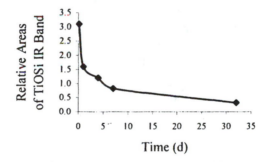

Figure 9. Semi-quantitative kinetic analysis - redistribution of the titanium siloxide network at 25°C

Proposed reaction mechanism

Based upon the preceding results, we propose the catalytic cycle shown in Scheme I for diffusion-limited condensation of alkoxysilanes. Organic titanates, $Ti(OR)_4$, are well known to be very sensitive to hydrolysis. It is reasonable to expect TtBT to be hydrolyzed first when moisture diffuses into the reaction mixture. Although not detected by ^{17}O NMR, the first intermediate should therefore be hydroxy titanates, TiOH, as proposed in earlier mechanistic studies (*3*). These rapidly attack methoxysilanes, SiOMe, to form the prevalent titanium siloxides, TiOSi, and release methanol. This condensation reaction actually activates methoxysilanes since titanium siloxides are more reactive. Titanium siloxides can then undergo transesterification with methoxysilanes to form either soluble or insoluble methyl titanates, TiOMe, and siloxanes, SiOSi. Soluble methyl titanates are more sensitive to moisture than other alkyl titanates and can revert to hydroxy titanates, TiOH, if hydrolyzed. Titanium siloxides can also undergo redistribution to form the siloxanes and TiOTi species. The facile metathesis reactions between titanium alkoxides and silicon alkoxides should be considered the most important crosslinking reactions. Meanwhile, the redistribution of titanium siloxides should also be considered as a competing crosslinking route, especially at the initial reaction stage when titanium siloxides may exist at a higher concentration. The titanate catalysts can be deactivated in side reactions by forming titanium oxo species or insoluble titanium alkoxides.

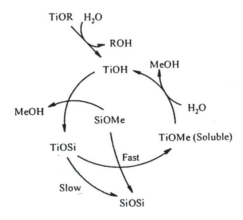

Scheme I. Proposed mechanism for the titanate-catalyzed condensation of alkoxysilanes

Summary

We have utilized a simple experimental scheme for vapor phase delivery of ^{17}O and ^{18}O labeled water into titanate catalyzed alkoxysilane reaction mixtures dissolved in non-polar solvents. This technique has enabled direct observation of reaction intermediates by NMR and FT-IR. We have verified that the condensation of alkoxysilanes in diffusion-controlled reactions is catalyzed by organic titanates. Titanium siloxides, as co-condensation products, have been directly detected as the key intermediates leading to formation of siloxanes through either transesterification with alkoxysilanes or reversion to Ti-O-Ti through redistribution. A catalytic cycle has been proposed based upon these learnings.

References

(1) Brinker, C.J.; Scherer, G.W. *Sol-gel Science*, Academic Press: New York, **1990**

(2) Feld, R; Cowe, P. *The Organic Chemistry of Titanium,* Washington Butterworths: London, **1965**, pp 171

(3) Brook, M.A. "Silicon in Organic, Organometallic and Polymer Chemistry". John Wiley & Sons, New York, **2000**, pp 285-286

(4) Day, V.W.; Eberspacher, T.A.; Klemperer, W.G.; Park, C.W.; Rosenberg, F.S. *J. Am. Chem. Soc.* **1991**, *113*, 8190

(5) Bastow, T.J.; Doran, G.; Whitfield, H.J. *Chem. Mater.* **2000**, *12*, 436

(6) Delattre, L.; Babonneau, F. *Chem. Mater.* **1997**, *9*, 2385

(7) Hoebbel, D.; Nacken, M.; Schmidt, H.; Huch, V.; Veith, M. *J. Mater. Chem.* **1998**, *8*, 171

(8) Farkas, I.; Bányai, I.; Szabó, Z.; Wahlgren, U.; Grenthe, I. *Inorg. Chem.* **2000**, *39*, 799

(9) Alam, T.M.; Celina, M.; Assink, R.A.; Clough, R.L.; Gillen, K.T.; Wheeler, D.R. *Macromolecules* **2000**, *33*, 1181

(10) Hoebbel, D.; Reinert, T.; Schmidt, H. *J. of Sol-Gel Sci. & Tech.* **1996**, *6*, 139

(11) Hoebbel, D.; Nacken, M.; Schmidt, H. *J. of Sol-Gel Sci. & Tech.* **1998**, *13*, 37

Author Index

Subject Index

T